Advances in twistor theory

Research Notes in Mathematics
Sub-series in Mathematical Physics

L P Hughston & R S Ward (Editors)

University of Oxford

Advances in twistor theory

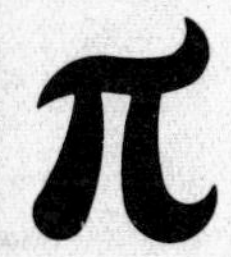

Pitman Advanced Publishing Program

SAN FRANCISCO · LONDON · MELBOURNE

PITMAN PUBLISHING LIMITED
39 Parker Street, London WC2B 5PB

North American Editorial Office
1020 Plain Street, Marshfield, Massachusetts 02050

North American Sales Office
FEARON PITMAN PUBLISHERS INC.
6 Davis Drive, Belmont, California 94002

Associated Companies
Copp Clark Pitman, Toronto
Pitman Publishing New Zealand Ltd, Wellington
Pitman Publishing Pty Ltd, Melbourne

AMS Subject Classifications: (main) 83-XX, 32-XX
(subsidiary 14-XX, 81-XX

Library of Congress Cataloging in Publication Data

(Research notes in mathematics ; 37)
Bibliography: p.
Includes index.
1. Twistor theory. I. Hughston, L. P., 1951- II. Ward, Richard Samuel, 1951- III. Series.
QC173.75.T85A38 530.1'42 79-17800
ISBN 0-8224-8448-X

Manufactured in Great Britain

US ISBN 0-8224-8448-X
UK ISBN 0-273 08448 8

Preface

Twistor theory provides a reformulation of the ideas of relativistic physics. It offers striking insights into the nature of gravitation, and suggests a new approach to the mechanisms that are responsible for the interactions of elementary particles. Twistor theory demonstrates the fundamental role played in mathematical physics by algebraic geometry, sheaf cohomology, and the theory of deformations of complex analytic structures, and it has given rise to powerful computational tools which should be of use in many areas of applied mathematics.

In 1976 a group of us at the Mathematical Institute, Oxford began to issue a series of informal publications that went by the name of *Twistor Newsletters*, outlining developments in the subject, and intended — originally — to have a rather limited distribution. However, demands increased steadily, and the point was reached eventually where it seemed appropriate that we should prepare a portion of the *Twistor Newsletter* material for the needs of a wider readership of mathematicians and physicists, with the hope of stimulating further research into this area.

The articles included here represent the work of eighteen authors, and span — for the most part — a period of three years. It will thus be found that this volume presents a kaleidoscopic view of twistor theory, embracing many aspects of the subject, as seen from many different angles and at various stages of its development. With the requirements of those who may be previously unacquainted with twistor theory particularly in mind, we have supplemented the volume with a selection of introductory material, and have included at its end a comprehensive bibliography. §§2.2, 2.6, 3.7 and a portion of §3.9 appear also in *Complex Manifold Techniques in Theoretical Physics*, edited by D.E. Lerner & P.D. Sommers (Pitman, 1979). §5.14 (*Combinatorial Quantum Theory and Quantized Directions*) was written in 1961 and appears as an appendix in Penrose's 1966 Cambridge University Adams Prize essay. We hope that the reader will be impressed with the essential unity of most of the topics discussed in what follows and will be able to capture a useful glimpse of this fascinating subject, with which so many of us, mathematicians and physicists alike, both here and elsewhere, are becoming increasingly familiar.

Thanks are due to many of our friends and colleagues at Oxford, particularly R. Penrose, for helpful assistance and advice given throughout the preparation of this volume. We are indebted to T.R. Hurd for preparing a large number of the figures for us. Gratitude is expressed to the publishers for their courtesy and efficiency, and we would also like to offer our thanks to Ms. T. Moss, who typed the manuscript.

L.P.H., R.S.W.

Oxford, July 1979

List of Contributors.

M.G. Eastwood, Mathematical Institute, Oxford.

M.L. Ginsberg, Mathematical Institute, Oxford

A.P. Hodges, Mathematical Institute, Oxford.

S.A. Huggett, Mathematical Institute, Oxford.

L.P. Hughston, Mathematical Institute, Oxford.

T.R. Hurd, Mathematical Institute, Oxford.

R.O. Jozsa, Mathematical Institute, Oxford.

D.E. Lerner, Department of Mathematics, University of Kansas, Lawrence, Kansas.

J.P. Moussouris, IBM Thomas J. Watson Research Center, Yorktown Heights, New York.

E.T. Newman, Department of Physics, University of Pittsburgh, Pittsburgh.

R. Penrose, Mathematical Institute, Oxford.

Z. Perjés, Central European Research Institute for Physics, Budapest.

A. Popovich, Mathematical Institute, Oxford.

M. Sheppard, Mathematical Institute, Oxford.

G.A.J. Sparling, Department of Physics, University of Pittsburgh, Pittsburgh.

Tsou S.T. Wadham College, Oxford.

R.S. Ward, Merton College, Oxford.

N.M.J. Woodhouse, Wadham College, Oxford.

Contents

1 An introduction to twistor theory

One of the great mysteries of basic physics concerns understanding the precise nature of the role played by the continuum concept. Space-time, for example, is modelled on a four-dimensional real continuum : but it appears quite unavoidable that such a model must break down at distances of the order of 10^{-33} cm, the so-called *Planck length*. In fact, there are good reasons for believing that such a breakdown might occur even at elementary particle lengthscales, i.e. $\sim 10^{-13}$ cm (cf. Penrose 1979). It might be argued that space-time should ultimately be understood in terms of purely *combinatorial* mathematics. Some evidence for this view is provided by the theory of *spin-networks* (see Penrose 1971a, 1971b, 1972; also §§5.14 - 5.17 in this volume), which builds up non-relativistic geometry in terms of combinatorial concepts. This theory also involves in an intimate way another continuum, namely the *complex* continuum which one encounters in quantum mechanics and which is essential to the notion of linear superposition of states.

Twistor theory may be viewed – at least to a certain extent – as an attempt to formulate and implement a fully relativistic analogue of spin-network theory. At the outset this involves a reformulation of the geometry of special relativity. It also involves the essential use of the complex continuum and, in particular, the concept of holomorphic (i.e. complex-analytic) functions. The primary objects in the theory are *twistors* – the points of space-time appear only as secondary, derived objects. This suggests an approach to the hitherto largely intractible problem of *quantizing gravity*, namely one in which space-time points become "fuzzy", uncertain objects (cf.

Penrose 1972, Penrose 1975a pp.268-278). Similarly, twistor theory hints at a way in which we may eventually be able to come to terms with the problems of the *structure and classification of space-time singularities* — for we know from the singularity theorems that under circumstances which we can expect to be satisfied in our own universe the structure of space-time *must* break down (e.g. at the big bang): but what is space-time to be "replaced" with, when it goes singular? Space-time points, as we mentioned before, are *secondary* objects in the geometry of twistor space, and thus we can envisage the possibility of space-time points simply *disappearing* in the regions of twistor space corresponding to singularities in space-time. Thus, although in the neighbourhood of a singularity space-time ceases to be a valid notion, it should nevertheless be possible to discuss physical properties of the "singular region" by referring instead to various aspects of the associated regions in twistor space.

The geometry of twistor space is rather intricate — even in the absence of gravitation — and since many of the relevant ideas and associated notational devices are used freely and without a great deal of explanation in the chapters which follow, we shall take the time here to outline some of the notions which can be regarded as particularly basic. Much of the material immediately following will be familiar already to relativists, but not necessarily familiar to mathematicians and physicists in general.

Space-time, Spinors, and Conformal Geometry

Minkowski space-time M^I is the real manifold R^4, equipped with the Lorentzian line element

$$ds^2 = \eta_{ab}\, dx^a\, dx^b := (dx^0)^2-(dx^1)^2-(dx^2)^2-(dx^3)^2. \qquad (1)$$

Here R denotes the field of real numbers; the four numbers $x^a = (x^0,x^1,x^2,x^3)$

are the standard coordinates on R^4; and the Einstein convention of summation over repeated indices is employed (as it will be throughout this volume). The *Poincaré group* P is the 10-dimensional Lie group of diffeomorphisms of M which preserve the metric (1). P is a semi-direct product of the group of translations of R^4 and the *Lorentz group* O(1,3), which has four components. The identity-containing component $O_+^\uparrow(1,3)$ of O(1,3) is doubly connected and has as its universal covering space the group SL(2,C) of 2 × 2 unimodular complex matrices [C denotes the field of complex numbers]:

$$SL(2,C) \xrightarrow{2-1} O_+^\uparrow(1,3). \tag{2}$$

Spinors are the objects which are acted on by this covering group SL(2,C). The letters A,B,... (ranging over 0,1) and A',B',... (ranging over 0',1') are used for spinor indices. Spinor indices may be raised and lowered by using the Levi-Civita spinors ε^{AB}, ε_{AB}, $\varepsilon^{A'B'}$ and $\varepsilon_{A'B'}$ (defined by their skew-symmetry, together with the scaling $\varepsilon^{01} = \varepsilon_{01} = \varepsilon^{0'1'} = \varepsilon_{0'1'} = 1$):

$$\begin{aligned} \xi^A &= \varepsilon^{AB}\,\xi_B\,, \quad \xi_A = \xi^B\,\varepsilon_{BA}, \\ \xi^{A'} &= \varepsilon^{A'B'}\,\xi_{B'}, \; \xi_{A'} = \xi^{B'}\,\varepsilon_{B'A'}\,. \end{aligned} \tag{3}$$

Every pair of spinor indices AA' (one unprimed and one primed) corresponds to a 4-vector index a. A 4-vector v^a corresponds to the spinor $v^{AA'}$ defined by

$$\begin{bmatrix} v^{00'} & v^{01'} \\ v^{10'} & v^{11'} \end{bmatrix} = \frac{1}{\sqrt{2}} \begin{bmatrix} v^0+v^1 & v^2+iv^3 \\ v^2-iv^3 & v^0-v^1 \end{bmatrix} \tag{4}$$

(this is not the only possible correspondence, but it is the one adopted in this book). We write $v^a = v^{AA'}$, $F_{ab} = F_{AA'BB'}$ etc. Here clearly, the indices a, A,... cannot be thought of as assuming *numerical* values: they must be regarded as *abstract indices* (see Penrose 1968) so that, for example, v^a is a vector rather than a collection of four components. The distinction between real and abstract indices is not very crucial in flat space-time, but becomes more important in curved space-time.

The operation of complex conjugation on spinors interchanges primed and unprimed indices. Given a spinor ξ^A, we may take its product with its complex conjugate $\bar{\xi}^{A'}$, and obtain the vector

$$v^a = \xi^A \bar{\xi}^{A'}. \tag{5}$$

As may easily be checked from (4), this vector is a future-pointing null vector; conversely, *every* future-pointing null vector may be obtained in this way. Another way of seeing that v^a is null is to observe that

$$v^a v_a = (\xi^A \xi_A)(\bar{\xi}^{A'} \bar{\xi}_{A'});$$

but $\xi^A \xi_A = \varepsilon^{AB} \xi_B \xi_A = 0$, since $\varepsilon^{AB} = -\varepsilon^{BA}$. Here we have used the fact that raising and lowering spinor indices with the ε's is consistent with raising and lowering vector indices with the metric (1) and its inverse. We may write:

$$\eta_{ab} = \varepsilon_{AB} \varepsilon_{A'B'} \quad \eta^{ab} = \varepsilon^{AB} \varepsilon^{A'B'}. \tag{6}$$

Let $F_{ab} = F_{[ab]}$ be a 2-form (square brackets denote skew-symmetrization over the indices they enclose; round brackets denote symmetrization). Then F_{ab} has the spinor decomposition

$$F_{ab} = \phi_{AB} \varepsilon_{A'B'} + \bar{\phi}_{A'B'} \varepsilon_{AB} \tag{7}$$

with $\phi_{AB} = \phi_{(AB)}$. So, for example, Maxwell's equations $\nabla_{[a}F_{bc]} = 0$ and

$\nabla^a F_{ab} = 0$ become, in spinor language,

$$\nabla^{AA'} \phi_{AB} = 0 \tag{8}$$

(cf. Penrose 1965a). Here $\nabla_{AA'}$ is the spinor translation of the derivative operator ∇_a $(= \partial/\partial x^a)$.

For further information about the 2-component spinor formalism — and, in particular, the details of its formulation in *curved* space-time — see, for example, Pirani 1965, Penrose 1968, Penrose and MacCallum 1972, Penrose 1975.

A map of Minkowski space-time M^I to itself, which preserves its *conformal* structure (i.e. which sends its metric η_{ab} to $\Omega^2\eta_{ab}$, for some nowhere-zero function Ω), is called a *conformal* map. The conformal diffeomorphisms of M^I are generated by the Poincaré transformations, the dilatations

$$x^a \longmapsto kx^a, \qquad k > 0 \text{ constant}, \tag{9}$$

and the inversions

$$x^a \longmapsto (x^b x_b)^{-1} x^a. \tag{10}$$

Actually this is not quite true, since the inversions (10) are singular on the null cone of the origin, where $x^b x_b = 0$. To remedy this slight difficulty append to M^I a *null cone* $\mathscr{I}$ *at infinity*. [See Penrose 1963, 1965a, & 1965b. A somewhat more streamlined account of I is outlined in Penrose 1968. Also see the treatment given in Hawking & Ellis 1973.] This gives us a *compact* space $M = M^I \cup \mathscr{I}$ (with topology $S^1 \times S^3$). M does not possess a metric — however, it does possess a flat *conformal* metric. The point at the vertex of $\mathscr{I}$ is called I; the inversion (10) interchanges $\mathscr{I}$ (the null cone of I) and the null cone of the origin $x^a = 0$. The *conformal group* C(1,3) is the group of all conformal diffeomorphisms of M to itself. C(1,3) is a 15-dimensional

Lie group generated by the Poincaré group, (9) and (10). Note that under C(1,3), every point of M (including I) is on an equal footing with every point of M. A crucial fact, as far as twistor theory is concerned, is the existence of the pair of two-to-one covering maps

$$SU(2,2) \xrightarrow{2-1} SO(2,4) \xrightarrow{2-1} C_{+}^{\uparrow}(1,3), \tag{11}$$

where $C_{+}^{\uparrow}(1,3)$ is the identity-containing component of C(1,3). One definition of twistors is that they are elements of the natural representation space C^4 for SU(2,2).

Null Geodesics

Consider the space PN of (unscaled) null geodesics in compactified Minkowski space-time M (the concept of a null geodesic is conformally invariant, so the conformal structure of M is sufficient to define null geodesics). Every null geodesic in ordinary Minkowski space-time M^I is also a null geodesic in M; in addition to these, M also contains the 2-sphere's worth of null geodesics which are the generators of $\mathscr{I}$. We may write $PN = PN^I \cup I$, where $PN \cong S^3 \times S^2$, $PN^I \cong R^3 \times S^2$ is the space of null geodesics in M^I, and $I \cong S^2$ is the sphere's worth of generators of $\mathscr{I}$.

Let us introduce coordinates on PN^I as follows. Let $(\omega^A, \pi_{A'})$ be a pair of spinors, with $\pi_{A'} \neq 0$. Let γ be the locus of points x^a in M^I such that

$$\omega^A = ix^{AA'}\pi_{A'}. \tag{12}$$

It is easily seen that if we impose the condition

$$\omega^A \bar{\pi}_A + \bar{\omega}^{A'}\pi_{A'} = 0, \tag{13}$$

then γ is a null geodesic, i.e. a point of PN^I; furthermore *every* point of PN^I arises in this way (see Penrose 1967, p.349). The freedom in $(\omega^A, \pi_{A'})$

is clearly

$$(\omega^A, \pi_{A'}) \mapsto (\lambda\omega^A, \lambda\pi_{A'}), \tag{14}$$

with $0 \neq \lambda \in C$ [since equation (12) is homogeneous in $(\omega^A, \pi_{A'})$]. So as coordinates on PN^I we may use the pair $(\omega^A, \pi_{A'})$, with $\pi_{A'} \neq 0$, satisfying (13), and modulo (14).

One may consistently regard the null geodesics at infinity as corresponding to the pairs $(\omega^A, 0)$, with $\omega^A \neq 0$, modulo (14) (Penrose 1967, p.350). Thus the points of PN are in one-to-one correspondence with the set

$$\{Z^\alpha | Z^\alpha \neq 0,\ Z^\alpha \bar{Z}_\alpha = 0\},$$
$$\text{modulo } Z^\alpha \mapsto \lambda Z^\alpha \qquad (0 \neq \lambda \in C), \tag{15}$$

where we have written $Z^\alpha = (\omega^A, \pi_{A'})$.

The index α ranges over the values 0,1,2,3. For components we have

$$Z^0 = \omega^0 \quad Z^1 = \omega^1 \quad Z^2 = \pi_{0'} \quad Z^3 = \pi_{1'}$$
$$\bar{Z}_0 = \overline{Z^2} = \bar{\pi}_0 \qquad \bar{Z}_1 = \overline{Z^3} = \bar{\pi}_1$$
$$\bar{Z}_2 = \overline{Z^0} = \bar{\omega}^{0'} \qquad \bar{Z}_3 = \overline{Z^1} = \bar{\omega}^{1'} \quad .$$

The condition $Z^\alpha \bar{Z}_\alpha = 0$ is easily seen to be the same as (13).

The space

$$PT = \{Z^\alpha \neq 0\}, \text{ modulo } Z^\alpha \to \lambda Z^\alpha \tag{16}$$

is just complex projective 3-space CP^3 (see the Appendix to this chapter). PT is called *projective twistor space;* its points are projective twistors. The points of PN are called projective *null* twistors; PN is given by the equation $Z^\alpha \bar{Z}_\alpha = 0$. See Figure A. This hypersurface PN divides PT into two

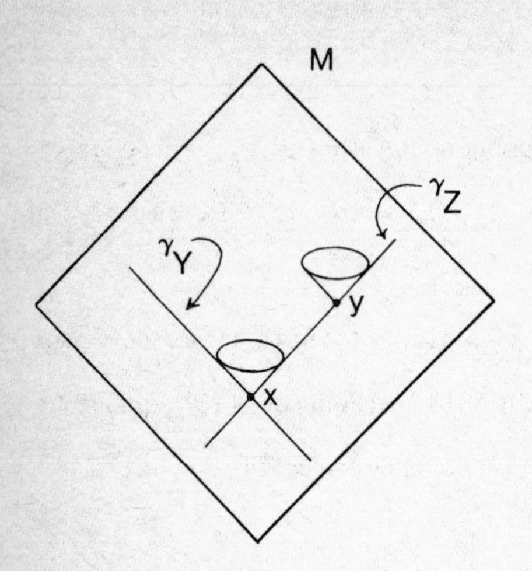

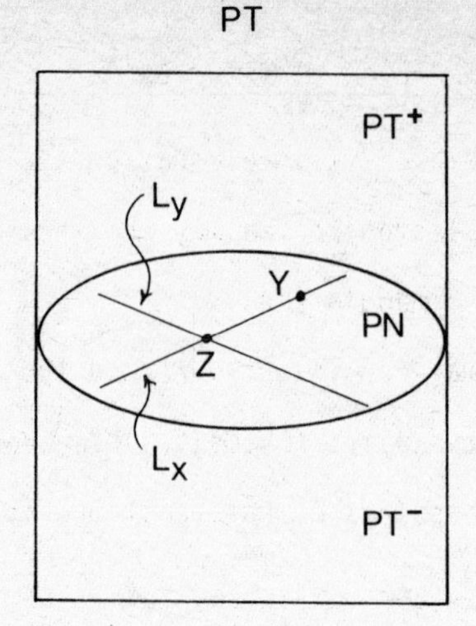

Figure A

halves, namely PT^+ (where $Z^\alpha \bar{Z}_\alpha > 0$) and PT^- (where $Z^\alpha \bar{Z}_\alpha < 0$). The points Z of PN correspond to null geodesics γ_Z in M; conversely, a point x in M may be represented by the sphere's worth of null geodesics through it, i.e. by a real 2-sphere L_x in PN (see Figure A). In fact, we see from (12) that this sphere L_x is a *complex projective line* in PT; L_x is intrinsically a Riemann sphere. The projective lines in PT which lie entirely in PN correspond precisely to the points of M. Thus

$$\text{point } x \in M \leftrightarrow \text{line } L_x \text{ in PN},$$

$$\text{null geodesic } \gamma_Z \in M \leftrightarrow \text{point } Z \text{ in PN}.$$

Clearly, x and y are null-separated points in M if and only if the corresponding lines L_x and L_y in PN *intersect*. (Cf. Figure A). Also, if the null geodesics γ_Z and γ_Y in M intersect (at, say, the point x), then the line L_x joining the points Z and Y in PN lies entirely in PN; and conversely.

Robinson Congruences

But what about *non*-null twistors, i.e. those not on PN? One can also interpret these in the space-time M, namely as certain twisting congruences of null geodesics called *Robinson congruences* (cf. Penrose 1967 §IV; see Penrose 1975 p. 291 for a "picture" of a Robinson congruence). A Robinson congruence is a special example of a *geodesic shear-free* congruence.

A spinor field $\xi^A(x)$ on M defines a congruence of null curves (namely the integral curves of the vector field $v^a = \xi^A \bar{\xi}^{A'}$), and this congruence is said to be *shear-free* if ξ^A satisfies the equation

$$\xi^A \xi^B \nabla_{AA'} \xi_B = 0. \tag{17}$$

A congruence of null geodesics clearly corresponds to a 3-surface Γ in PN; how does one recognize, from looking at Γ, whether or not the congruence is shear-free?

Theorem (Kerr). The 3-surface Γ in PN corresponds to an analytic shear-free congruence in space-time if and only if Γ is the intersection with PN of a *holomorphic* complex 2-surface in PT.

This rather important theorem provides an illustration of the role played by holomorphic functions in twistor theory. (For further details see Penrose 1967 §VIII; cf. also Hughston 1979 pp. 126-129.) The geodesic shear-free condition is of significance in connection with the construction of null solutions to Maxwell's equations (Robinson 1961). It is also of considerable significance, when generalized to curved space-time, in connection with solving Einstein's equations.

An especially intriguing point about the Kerr theorem is that it gives us an example of the way in which complex analytic methods can be used with great effectiveness to obtain general solutions to various *non-linear* problems in physics. Indeed, "curved twistor space" techniques should be of wide interest to mathematicians — even apart from the immediate applications of these techniques to physical questions. These matters are discussed in Chapter 3, and in references cited therein.

The Infinity Twistors

The space $T = \{Z^\alpha \neq 0\}$ (*not* factored out by the relation of proportionality) is called *non*-projective twistor space. Considering $T \cong C^4 - \{0\}$ as a complex vector space with the origin deleted, we may construct its dual T^*, and the general tensor product $T \otimes \ldots \otimes T \otimes T^* \otimes \ldots \otimes T^*$, with p factors of T and q factors of T^*. An element of this tensor product is called a twistor of valence [p,q]. As an example we mention two important twistors, the *infinity twistors*, of valence [2,0] and [0,2] respectively; they are given by

$$I^{\alpha\beta} = \begin{bmatrix} \varepsilon^{AB} & 0 \\ 0 & 0 \end{bmatrix},$$

$$I_{\alpha\beta} = \begin{bmatrix} 0 & 0 \\ 0 & \varepsilon^{A'B'} \end{bmatrix}.$$

In other words, $I^{01} = -I^{10} = I_{23} = -I_{32} = 1$, with all other components zero. These infinity twistors are the objects which break conformal invariance: the conformal group SU(2,2) acts on T [$Z^\alpha \bar{Z}_\alpha$ being the Hermitian form with signature (++--) preserved under this action], and the subgroup of SU(2,2) which preserves $I^{\alpha\beta}$ is precisely the Poincaré group.

From the matrix expressions given above, it should be evident that the infinity twistors satisfy the following relations:

$$I^{\alpha\beta} = \overline{I_{\alpha\beta}} \quad , \qquad I^{\alpha\beta} I_{\beta\gamma} = 0. \tag{18}$$

Moreover, we also have

$$I^{\alpha\beta} = \tfrac{1}{2}\,\varepsilon^{\alpha\beta\gamma\delta}\,I_{\gamma\delta}\ ,\quad I_{\alpha\beta} = \tfrac{1}{2}\,\varepsilon_{\alpha\beta\gamma\delta}\,I^{\gamma\delta}, \tag{19}$$

where $\varepsilon^{\alpha\beta\gamma\delta}$ is the totally skew-symmetric twistor, with $\varepsilon^{0123} = 1$.

The Klein Representation

We now want to introduce another interpretation of a twistor, namely as an object in *complexified Minkowski space-time* CM^I. CM^I is the four-dimensional complex manifold C^4, equipped with the complex holomorphic metric $ds^2 = \eta_{ab}dx^a dx^b$, $x^a = (x^0,x^1,x^2,x^3)$ being the four complex coordinates on C^4. In other words, one analytically extends Minkowski space-time by simply allowing its coordinates to become complex.

Suppose that $(\omega^A,\pi_{A'})$ is a pair of spinors with $\pi_{A'} \neq 0$ and let Z be the locus of points x^a in CM^I such that

$$\omega^A = ix^{AA'}\,\pi_{A'} \tag{20}$$

(cf. equation 12). It is obvious that Z is a complex 2-plane in CM^I; and it is also clear that Z is *totally null:* every vector v^a tangent to Z has the form $v^{AA'} = \lambda^A\,\pi^{A'}$ for some λ^A, and is therefore null (cf. equation 5). We call such a 2-plane Z an *α-plane.* Another way of thinking about Z is that it consists of points $x \in CM^I$ where the spinor field

$$\mu^A(x) = \omega^A - ix^{AA'}\,\pi_{A'} \tag{21}$$

vanishes. The field $\mu^A(x)$ satisfies

$$\nabla^{A'(A}\mu^{B)} = 0 \tag{22}$$

and *every* solution of (22) has the form (21); cf. Penrose 1967 §V. Equation (22) is called the *twistor equation* (of valence [1,0]). There are analogues of (22) for other valences. For example, a dual twistor W_α has spinor components $W_\alpha = (\pi_A, \omega^{A'})$; from them one constructs a spinor field

$$\mu^{A'}(x) = \omega^{A'} + ix^{AA'}\pi_A, \tag{23}$$

which satisfies

$$\nabla^{A(A'}\mu^{B')} = 0. \tag{24}$$

The vanishing of $\mu^{A'}(x)$ defines a totally null complex 2-plane W in CM^I which is called a β-*plane*. A β-plane has tangent vectors of the form $v^a = \pi^A \lambda^{A'}$, with π^A fixed; whereas the vectors tangent to an α-plane have the form $v^a = \lambda^A \pi^{A'}$, with $\pi^{A'}$ fixed.

Twistors $(\omega^A, 0)$ with $\omega^A \neq 0$ may be interpreted as α-planes at infinity, i.e. lying in the complexification $C\mathscr{I}$ of $\mathscr{I}$. By attaching $C\mathscr{I}$ to CM^I we obtain a compact 4-dimensional complex manifold CM, the complexified, compactified Minkowski space-time. The points of projective twistor space PT correspond precisely to the α-planes in CM (see Figure B). And the points x of CM

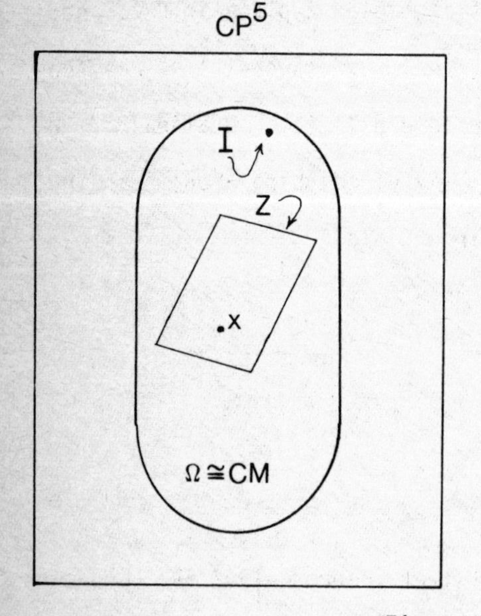

Figure B

correspond to the (complex projective) lines L_x in PT (not necessarily in PN). The space of lines in CP^3 may be thought of as a quadric hypersurface Ω in CP^5: this is known as the *Klein representation* of the lines in CP^3. The quadric Ω is precisely the space CM.

The Klein representation has been over the years an object of intense investigation by algebraic geometers — see, for example, Semple & Roth 1949. Much of this material is now directly applicable to physics. Let us examine the structure of Ω a little bit more closely. The CP^5 in which Ω sits as a quadric is realized explicitly as the projective space of skew-symmetric twistors of valence [2,0]. A skew-symmetric twistor $X^{\alpha\beta} = -X^{\beta\alpha}$ has 6 independent components — and thus up to proportionality it has 5 independent components, corresponding to the 5 dimensions of CP^5. The quadric Ω is defined by

$$\varepsilon_{\alpha\beta\gamma\delta} X^{\alpha\beta} X^{\gamma\delta} = 0, \qquad (25)$$

and it is an elementary exercise to show that equation (25) is a necessary and sufficient condition that

$$X^{\alpha\beta} = Y^{[\alpha}Z^{\beta]} \qquad (26)$$

for some choice of Y^α and Z^β. A skew-symmetric twistor of the form (26) is called *simple*, and represents the line in CP^3 that joins Y^α and Z^β. If Z^β is fixed in (26) and we allow Y^α to vary, the resulting values of $X^{\alpha\beta}$ define the α-*plane* in Ω corresponding to the point Z^α in CP^3. The α-plane corresponds to the *star of lines* in CP^3 through Z^α.

If $P^{\alpha\beta}$ and $Q^{\alpha\beta}$ represent lines in CP^3 (i.e. lie in Ω), then the condition that these lines intersect is

$$\varepsilon_{\alpha\beta\gamma\delta}\, P^{\alpha\beta}\, Q^{\gamma\delta} = 0.$$

This means that the two points in space-time represented by $P^{\alpha\beta}$ and $Q^{\alpha\beta}$ are complex null-separated.

It follows from equations (18) and (19) that $I^{\alpha\beta}$ is simple. The point in CM which it represents is in fact the *vertex of* $C\mathcal{J}$. This line lies in PN, as is expressed by the reality condition $I^{\alpha\beta} = \overline{I_{\alpha\beta}}$. Lines in CP^3 which meet $I^{\alpha\beta}$ correspond to points on $C\mathcal{J}$. These are the "points at infinity" in complexified compactified Minkowski space-time. Thus, using (19), we see that $X^{\alpha\beta}$ represents a point on $C\mathcal{J}$ if it satisfies

$$X^{\alpha\beta}\, I_{\alpha\beta} = 0.$$

If this condition is *not* satisfied, then $X^{\alpha\beta}$ represents a point in CM^I, i.e. an ordinary "finite" point in complex space-time. In that case we can scale $X^{\alpha\beta}$ so as to put $X^{\alpha\beta}\, I_{\alpha\beta} = 2$, and we can derive the following standard matrix representation for $X^{\alpha\beta}$:

$$X^{\alpha\beta} = \begin{bmatrix} -\tfrac{1}{2}x_d x^d\, \varepsilon^{AB} & ix^A{}_{B'} \\ -ix_{A'}{}^B & \varepsilon_{A'B'} \end{bmatrix},$$

where $x^{AA'}$ is the space-time point corresponding to $X^{\alpha\beta}$.

Of special interest in connection with physical applications are points lying in the so-called *forward tube* of CM. These are complex points $x^a = v^a - iw^a$ with v^a, w^a real and w^a timelike future-pointing. Such points correspond to lines in PT that lie entirely in PT^+ (as does the line L_x in Figure B, for example). This sort of analysis can be pursued, of course, at much greater length. See, in particular, Penrose 1967 §§III, VI, IX, & X; also see Penrose 1975 pp. 278-303, Hughston 1979 chapter 2, and Penrose & Ward 1979 §2.1.

Appendix

This is a brief summary of some facts about complex manifolds, line bundles and vector bundles. For definitions, proofs and further details, the reader may wish to consult some of the *mathematical background reading* listed at the end of this chapter.

Complex projective n-space CP^n is a compact n-dimensional complex manifold defined to be the quotient $(C^{n+1} - \{0\})/\sim$, where the equivalence relation $\sim$ is that of proportionality:

$$z^j \sim y^j \iff z^j = \lambda y^j \text{ for some } 0 \neq \lambda \in C.$$

Here $z^j = (z^0,\ldots,z^n)$ are the standard coordinates on C^{n+1}. The simplest example, namely CP^1, is just the Riemann sphere.

A holomorphic vector bundle of rank k over a complex manifold X is a holomorphic allocation of a complex linear space $F_x \cong C^k$ to each point $x \in X$. F_x is called the *fibre* over x. A *section* of the bundle is a holomorphic map which takes each point x to a point of the fibre F_x over x.

A vector bundle of rank 1 is called a *line bundle*. The line bundles over a complex projective n-space CP^n are classified by the integers: we may denote them L(p), where p is an integer. Sections of L(p) over regions of CP^n correspond to functions $f = f(z^j)$ of the homogeneous coordinates on CP^n which are homogeneous of degree p, i.e.

$$f(\lambda z^j) = \lambda^p f(z^j) \text{ for all } 0 \neq \lambda \in C.$$

Mathematical Background Reading

Atiyah, M. & MacDonald, I.G. 1969 *Introduction to Commutative Algebra.* Addison-Wesley.

Chern, S.S. 1967 *Complex Manifolds without Potential Theory.* Van Nostrand.

Godement, R. 1973 *Théorie des Faisceaux.* Hermann, Paris.

Grauert, H. & Fritzsche, K. 1976 *Several Complex Variables.* Springer-Verlag.

Griffiths, P., & Adams, J. 1974 *Topics in Algebraic and Analytic Geometry.* Princeton University Press.

Griffiths, P., & Harris, J. 1978 *Principles of Algebraic Geometry.* Wiley & Sons.

Grothendieck, A., & Dieudonne, J. 1971 *Eléments de Géométrie Algébrique* Springer-Verlag.

Gunning, R.C. 1966 *Lectures on Riemann Surfaces.* Princeton University Press.

Gunning, R.C. 1967 *Lectures on Vector Bundles over Riemann Surfaces.* Princeton University Press.

Gunning, R.C. & Rossi, H. 1965 *Analytic Functions of Several Complex Variables.* Prentice-Hall.

Hartshorne, R. 1977 *Algebraic Geometry.* Springer-Verlag.

Hirzebruch, F. 1966 *Topological Methods in Algebraic Geometry.* Springer-Verlag.

Morrow, J. & Kodaira, K. 1971 *Complex Manifolds.* Holt, Rienhard, and Winston.

Mumford, D. 1976 *Algebraic Geometry I : Complex Projective Varieties.* Springer-Verlag.

Semple, J.G. & Kneebone, G.T. 1952 *Algebraic Projective Geometry.* Clarendon Press, Oxford.

Semple, J.G. & Roth, L. 1949 *Algebraic Geometry*. Clarendon Press, Oxford.

Shafarevich, I.R. 1977 *Basic Algebraic Geometry*. Springer-Verlag.

Swan, R.G. 1964 *Theory of Sheaves*. University of Chicago Press.

Walker, R.J. 1978 *Algebraic Curves*. Springer-Verlag.

Wells, R.O. 1973 *Differential Analysis on Complex Manifolds*. Prentice-Hall.

REFERENCES

Hawking, S.W. & Ellis, G.F.R. 1973 *The Large Scale Structure of Space-Time*. Cambridge University Press.

Hughston, L.P. 1979 *Twistors and Particles*. Springer Lecture Notes on Physics, Volume 97. Springer-Verlag.

Penrose, R. 1963 Phys. Rev. Letts. 10, 66.

Penrose, R. 1965a Proc. Roy. Soc. A284, 159.

Penrose, R. 1965b In: *Relativity, Groups, and Topology*, eds. B. DeWitt & C.M. DeWitt. Gordon and Breach.

Penrose, R. 1967 J. Math. Phys., Vol. 8, No. 2, 345.

Penrose, R. 1968 *The Structure of Spacetime*. In: *Battelle Rencontres*, eds. C.M. DeWitt & J.A. Wheeler.

Penrose, R. 1971a In: *Combinatorial Mathematics and its Applications*, ed. D.J.A. Welsh. Academic Press

Penrose, R. 1971b In: *Quantum Theory and Beyond*, ed. T. Bastin. Cambridge University Press.

Penrose, R. 1972 In: *Magic Without Magic*, ed. J. Klauder. Benjamin.

Penrose, R. 1975 In: *Quantum Gravity: An Oxford Symposium*, eds. C.J. Isham, R. Penrose, & D.W. Sciama. Clarendon Press, Oxford.

Penrose, R. 1979 *Singularities and Time-Asymmetry*. In: *General Relativity: An Einstein Centenary Survey*, eds. S.W. Hawking & W. Israel. Cambridge University Press.

Penrose, R. & MacCallum, M.A.H. 1972 Phys. Reports, Vol. 6C, No.4.

Penrose, R. & Ward, R.S. 1979 *Twistors for Flat and Curved Space-Time.* To appear in an Einstein centennial volume, eds. P. Bergmann, J. Goldberg & A. Held.

Pirani, F.A.E. 1965 In: *Lectures on General Relativity.* A. Trautman, F.A.E. Pirani, & H. Bondi, 1964 Bradeis Summer Institute on Theoretical Physics. Prentice-Hall.

Robinson, I. 1961 J. Math. Phys. 2, 290.

Semple, J.G. & Roth, L. 1949 *Algebraic Geometry.* Clarendon Press, Oxford.

2 Massless fields and sheaf cohomology

§2.1 INTRODUCTION *by R.S. Ward*

Let $\phi_{AB\ldots D}(x)$ be a totally symmetric spinor field on Minkowski space-time M. We say that $\phi_{AB\ldots D}$ is a *massless free field* if it satisfies the massless free field equations

$$\nabla^{AA'} \phi_{AB\ldots D} = 0 \tag{1}$$

(see Penrose 1965). The equations (1) include Weyl's neutrino equation $\nabla^{AA'} \phi_A = 0$, Maxwell's equations $\nabla^{AA'} \phi_{AB} = 0$, and the linearized Einstein equations $\nabla^{AA'} \phi_{ABCD} = 0$. One can also have massless fields $\phi_{A'B'\ldots D'}$ with primed indices:

$$\nabla^{AA'} \phi_{A'B'\ldots D'} = 0; \tag{2}$$

and massless scalar fields ϕ, satisfying

$$\Box \phi = 0, \tag{3}$$

where $\Box = \nabla_a \nabla^a$ is the d'Alembertian on M. Equations (1), (2) & (3) are all conformally invariant (Penrose 1965; Penrose & MacCallum 1972).

One may generate solutions of the massless field equations by taking an arbitrary holomorphic function of three complex variables and performing a certain contour integral (Penrose 1968; 1969). In the language of twistor theory, the procedure is as follows.

Let $f(Z^\alpha)$ be a holomorphic function of a twistor Z^α, homegeneous of degree $-n-2$. Let ρ_x denote the operator which "puts the twistor Z^α through the space-time point x^a"; in other words,

$$\rho_x \colon Z^\alpha \longmapsto (ix^{AA'}\pi_{A'}, \pi_{A'})$$

and $$\rho_x \colon f(Z^\alpha) \longmapsto f(ix^{AA'}\pi_{A'}, \pi_{A'}).$$

Now put

$$\phi_{A'\ldots D'}(x) = (2\pi i)^{-1} \oint_\Gamma \rho_x \, \pi_{A'} \ldots \pi_{D'} \, f(Z^\sigma)\Delta\pi, \tag{4}$$

if $n \geq 0$, with n factors of $\pi_{A'}$ in the integrand; or

$$\phi_{A\ldots D}(x) = (2\pi i)^{-1} \oint_\Gamma \rho_x \, \hat{\pi}_A \ldots \hat{\pi}_D \, f(Z^\sigma)\Delta\pi, \tag{5}$$

if $n < 0$, with $-n$ $\hat{\pi}$'s in the integrand. Here $\Delta\pi$ denotes the differential form $\pi_{A'} \, d\pi^{A'}$, and $\hat{\pi}_A$ denotes the differential operator $-\partial/\partial\omega^A$. The contour Γ for the integration is defined as follows. In the projective twistor space PT, the map ρ_x restricts the twistor Z^α to the projective line L_x corresponding to the point x^a; L_x is intrinsically a Riemann sphere (see Chapter 1). The integrands of (4) and (5), being homogeneous of degree zero in $\pi_{A'}$, are defined on this Riemann sphere (except, that is, where the function $\rho_x f(Z^\alpha)$ is singular). The contour Γ in (4) or (5) is taken to be *any* closed curve on L_x which avoids the singularities of $\rho_x \, f(Z^\alpha)$, and which varies continuously with x.

It is easily verified that the fields defined by (4) and (5) are massless free fields. The converse result, that *every* analytic massless free field arises in this way, is somewhat more difficult to establish, but is nevertheless true (see Penrose 1968, 1975). The map which takes one from a massless free field to a twistor function f which generates it, is called the *inverse twistor function*, and is discussed further in §§2.2, 2.7 & 2.8.

There are analogues of the formulae (4) and (5) which generate *massive* fields (such as Klein-Gordon or Dirac fields). However, these are somewhat more intricate than the massless cases. See §4.1 for more details.

One particular class of massless fields are of special interest, namely those which are of *positive frequency*. The positive frequency condition may be stated as follows (see, e.g., Streater & Wightman 1964): a field $\phi_{...}$ on Minkowski space-time is of positive frequency if it can be extended analytically to the *forward tube* CM^+. Here CM^+ consists of those points of complexified Minkowski space-time which have position vectors of the form $x^a - iy^a$, where x^a and y^a are real and y^a is timelike future-pointing. A positive frequency field $\phi_{A'...D'}$ with n primed indices is said to have helicity $\frac{1}{2}n$, whereas a positive frequency field $\phi_{A...D}$ with n unprimed indices has helicity $-\frac{1}{2}n$. [In many of the sections of this Chapter, however, the terms helicity and spin are used even when referring to fields that are not positive frequency. In such cases, "helicity" means half the number of indices, counting primed indices positively and unprimed indices negatively.] The points of CM^+ correspond to lines which lie entirely in the 'top half' PT^+ or projective twistor space (Penrose & MacCallum 1972). Thus a twistor function $f(Z^\alpha)$ will produce a positive frequency massless free field if its singularity set S in PT^+ consists of two components between which a contour Γ may be threaded (see Fig. A). If the line L_x ventures out of PT^+, then Γ is in danger of being 'pinched' between the singularities of $f(Z^\alpha)$, whereupon the space-time field $\phi_{...}(x)$ becomes singular.

For example, S may consist of the union of two planes P and Q in PT, where the line $P \cap Q$ lies in PT^-; for then $P \cap PT^+$ and $Q \cap PT^+$ are disjoint. A suitable twistor function would be

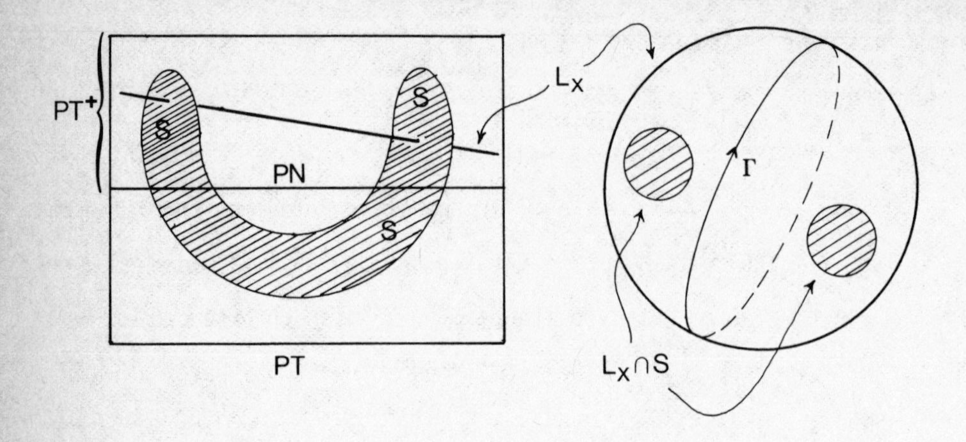

Figure A

$$f(Z^{\alpha}) = \frac{(V_{\alpha}Z^{\alpha})^{j}\ (W_{\beta}Z^{\beta})^{k}}{(P_{\gamma}Z^{\gamma})^{m+1}(Q_{\delta}Z^{\delta})^{\ell+1}}\ , \tag{6}$$

with $m + \ell - j - k = n$ (m,ℓ,j,k non-negative integers). The corresponding space-time field $\phi\ldots$ is called an *elementary state*. As an illustration, consider the case $n = m = \ell = j = k = 0$. Let $p^{A'}(x) = P^{A'} + i\ x^{AA'}P_{A}$ and $q^{A'}(x) = Q^{A'} + i\ x^{AA'}Q_{A}$ be the solutions of the twistor equation $\nabla^{A(A'}{}_{\mu}{}^{B')}=0$ corresponding, respectively, to the twistors P_{α} and Q_{α} (cf. Chapter 1). The contour integral (4) is easily evaluated in this case and yields the answer

$$\begin{aligned}\phi(x) &= \frac{1}{p^{A'}(x)\ q_{A'}(x)} \\ &= \frac{\kappa}{(z^{a}-x^{a})(z_{a}-x_{a})}\ ,\end{aligned} \tag{7}$$

where z^a is the space-time point corresponding to the line $L_z = P \cap Q$, and κ is the constant $2(P_A Q^A)^{-1}$. Elementary states are discussed further in §2.15, and a generalization of them in §2.16.

The twistor-theoretic description of massless free fields is best understood using the language and techniques of *sheaf cohomology*. Most of the articles in this Chapter deal with this cohomological description. Cf. also Hughston (1979) for a self-contained account. The essential result is that the twistor function $f(Z^\alpha)$ should be regarded as an element of a sheaf cohomology group $H^1(X,\mathcal{O}(-n-2))$, where X is an appropriate region of twistor space and $\mathcal{O}(-n-2)$ is the sheaf of germs of holomorphic functions on X, homogeneous of degree -n-2. The basic ideas of sheaf cohomology are sketched in §2.2; for more details, the reader may refer, for example, to Godement (1964); Gunning (1966); Gunning & Rossi (1965); or Morrow & Kodaira (1971).

REFERENCES

Godement, R. 1964 *Topologie Algébrique et Théorie des Faisceaux*. Hermann: Paris.

Gunning, R.C. 1966 *Lectures on Riemann Surfaces*. Princeton: University Press.

Gunning, R.C. & Rossi, H. 1965 *Analytic Functions of Several Complex Variables*. Prentice-Hall: Englewood Cliffs.

Hughston, L.P. 1979 *Twistors and Particles*. Springer Lecture Notes in Physics, Vol. 97. Berlin: Springer-Verlag.

Morrow, J. & Kodaira, K. 1971 *Complex Manifolds*. Holt, Rinehart & Winston: New York.

Penrose, R. 1965 Proc. Roy. Soc. A284, 159-203.

Penrose, R. 1968 Int. J. Theor. Phys. 1, 61-99.

Penrose, R. 1969 J. Math. Phys. 10, 38-39.

Penrose, R. 1975 *Twistor Theory, its Aims and Achievements*. In: *Quantum Gravity*, eds. C.J. Isham, R. Penrose & D.W. Sciama, pp.268-407.

Penrose, R. & MacCallum, M.A.H. 1972 Phys. Reports 6C, 241-315.

Streater, R.F. & Wightman, A.S. 1964 *PCT, Spin and Statistics, and All That*. Benjamin: New York.

§2.2 TWISTOR FUNCTIONS AND SHEAF COHOMOLOGY *by R. Penrose*

Recall (Penrose 1975a,b) the following properties of a twistor function $f(Z^\alpha)$, to be used for generating a zero rest-mass field $\phi_{A'\ldots L'}$, or $\phi_{A\ldots K}$, by means of a contour integral,

$$\phi_{A'\ldots L'}(x) = (2\pi i)^{-2} \oint \rho_x \, \pi_{A'} \cdots \pi_{L'} \, f(Z^\alpha) \, d^2\pi,$$

$$\phi_{A\ldots K}(x) = (2\pi i)^{-2} \oint \rho_x \, \hat{\pi}_A \cdots \hat{\pi}_K \, f(Z^\alpha) \, d^2\pi:$$

(i) The function f is holomorphic on some domain $\mathcal{D}$ of twistor space *not* invariant under SU(2,2) (nor under the Poincaré group, nor the Lorentz group).

(ii) There is a "gauge" freedom G whereby

$$f \mapsto f + h^- + h^+,$$

where $h^\pm$ is holomorphic on some extended domain $\mathcal{D}^\pm (\supset \mathcal{D})$ of twistor space in which the contour γ can be deformed to a point (to the "left" in $\mathcal{D}^-$ and to the "right" in $\mathcal{D}^+$: see Fig. A).

(iii) G depends on the location of γ (and γ is not invariant under SU(2,2)--nor under the Poincaré group nor the Lorentz group--so G is not invariant either).

(iv) By invoking G, then moving γ, then invoking a new G, moving γ again, we can obtain a whole family of equivalent twistor functions, all giving the same field, the entire family being invariant under SU(2,2).

(v) This seems a little nebulous, and, for example, how do we add two such families (to give $\phi_{A'\ldots L'} + \psi_{A'\ldots L'}$), etc. etc?

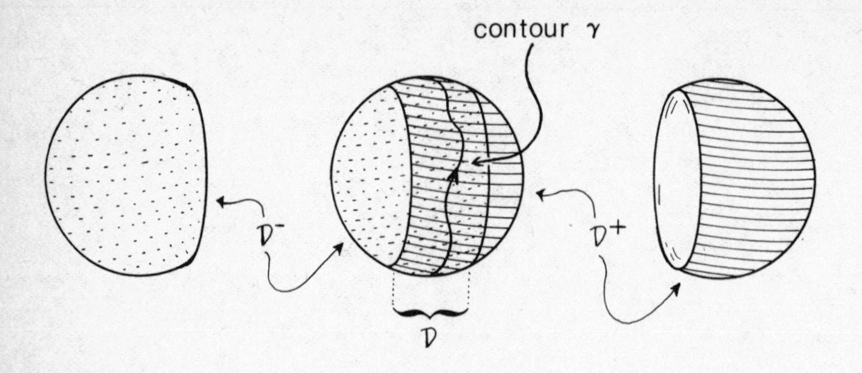

Figure A

A new viewpoint concerning twistor functions has been gradually emerging which makes good mathematical sense of all this: a twistor function is really to be viewed as a representative function (or cocycle) defining an element of a sheaf cohomology group. Now, the twistor theorist, when attacked by a purist for shoddiness in the domains, can counter-attack armed with his sheaf!

Thumbnail sketch of (relevant) sheaf cohomology theory

First let us recall how ordinary (Čech) cohomology works. Let X be a space (a Hausdorff paracompact topological space, say). Cover X with a locally finite system of open sets $\mathcal{U}_i$. We define a *cochain* (with respect to this covering) with coefficients in an additive abelian group G (say, the integers Z, the reals R, or the complex field C) in terms of a collection of elements $f_i, f_{ij}, f_{ijk}, \ldots \in G$, assigned to the various $\mathcal{U}_i$ and their nonempty intersections: f_i assigned to $\mathcal{U}_i$; f_{ij} assigned to $\mathcal{U}_i \cap \mathcal{U}_j$; f_{ijk} assigned to $\mathcal{U}_i \cap \mathcal{U}_j \cap \mathcal{U}_k$; ... and $f_{ij} = -f_{ji}$, $f_{ijk} = -f_{jik} = f_{jki} = \ldots, \ldots$, i.e.

$f_{i\ldots\ell} = f_{[i\ldots\ell]}$. Then

$$\text{0-cochain } \alpha = (f_1, f_2, f_3, \ldots)$$

$$\text{1-cochain } \beta = (f_{12}, f_{23}, f_{13}, \ldots)$$

$$\text{2-cochain } \gamma = (f_{123}, f_{124}, \ldots)$$

(where $\mathcal{U}_1 \cap \mathcal{U}_2$, $\mathcal{U}_2 \cap \mathcal{U}_3$, $\mathcal{U}_1 \cap \mathcal{U}_3, \ldots$, $\mathcal{U}_1 \cap \mathcal{U}_2 \cap \mathcal{U}_3$, $\mathcal{U}_1 \cap \mathcal{U}_2 \cap \mathcal{U}_4, \ldots$ are the non-empty intersections of $\mathcal{U}_i$'s).

Define *coboundary operator* δ as follows:

$$\delta\alpha = (\underset{\text{"}f_{12}\text{"}}{f_2 - f_1}, \ \underset{\text{"}f_{23}\text{"}}{f_3 - f_2}, \ \underset{\text{"}f_{13}\text{"}}{f_3 - f_1}, \ldots)$$

$$\delta\beta = (\underset{\text{"}f_{123}\text{"}}{f_{12} - f_{13} + f_{23}}, \ \underset{\text{"}f_{124}\text{"}}{f_{12} - f_{14} + f_{24}}, \ldots)$$

etc.

(where, again, $\mathcal{U}_1 \cap \mathcal{U}_2, \ldots$ are the non-empty intersections). Then we have $\delta^2 = 0$. We call γ a *cocycle* if $\delta\gamma = 0$; we call γ a *coboundary* if $\gamma = \delta\beta$ for some β. Define the pth *cohomology group* by

$$H^p_{\{\mathcal{U}_i\}}(X,G) = \left(\begin{matrix}\text{additive group}\\ \text{of p-cocycles}\end{matrix}\right) / \left(\begin{matrix}\text{additive group}\\ \text{of p-coboundaries}\end{matrix}\right).$$

<u>Note</u>: $H^p_{\{\mathcal{U}_i\}}(X,G)$, as defined, depends on the covering $\{\mathcal{U}_i\}$. What we should do, to define $H^p(X,G)$, is to take the appropriate "limit" of all these $H^p_{\{\mathcal{U}_i\}}(X,G)$ for finer and finer coverings $\{\mathcal{U}_i\}$ of X. However (for X suitably non-pathological) we can always settle on a particular "sufficiently fine" covering $\{\mathcal{U}_i\}$ where, in effect, there is no "relevant topology" left on each $\mathcal{U}_i$ or intersections thereof (i.e. all the $H^p(\mathcal{U}_i \cap \ldots \cap \mathcal{U}_k, G)$ vanish for all $p > 0$--although, as it stands, this is somewhat unhelpful because

a direct limit is then already involved in the definition of "sufficiently fine"). Then *this* $H^p_{\{U_i\}}(X,G) = H^p(X,G)$. I shall henceforth assume that such a "sufficiently" fine covering has been taken, and that it is countable and locally finite.

Now what does this definition have to do with the familiar "dual" relation to ordinary homology $H_p(X,G)$? How does γ assign values (elements of G) in a linear fashion to p-cycles in X, where γ is some element of $H^p(X,G)$? Intuitively, one adds together the f_{ij} for the shaded regions entered by the 1-cycle κ: see Fig. B. If γ is defined by $(f_{12}, f_{23}, \ldots)$, then $\gamma(\kappa) = f_{42} + f_{23} + f_{31} + \ldots$; the right hand side is a sum of (correctly signed) f's on regions entered by κ. The definition for higher dimensional p-cycles is similar (a complete rigorous discussion is easy enough).

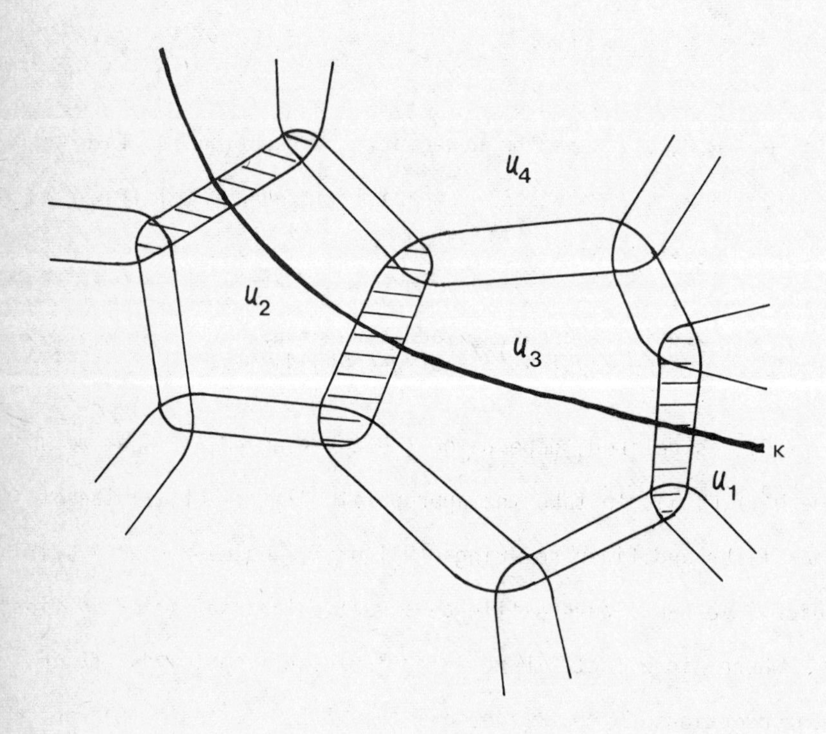

Figure B

What about *sheaf* cohomology though? It's really a rather natural generalization of the above. But first we must rephrase the concept of a cochain slightly. Rather than thinking of f_i as simply an element of G "assigned" to $\mathcal{U}_i$, and f_{ij} "assigned" to $\mathcal{U}_{ij}$, etc., we think of f_i as a *function* defined on $\mathcal{U}_i$ which happens, in the above, to take this *constant* value "f_i" $\in$ G, and we think of f_{ij} as a constant function on $\mathcal{U}_i \cap \mathcal{U}_j$ with values in G, etc. (This has the incidental advantage that the requirement $\mathcal{U}_i \cap \mathcal{U}_j \neq \emptyset$ is now unnecessary.) This is still ordinary (Čech) cohomology. But the generalization to sheaf cohomology is now easily made: the functions $f_i, f_{ij}, f_{ijk}, \ldots$ are now *not* required to be constant. (In fact, we could even allow that the additive group G, in which the values of the f's reside, may vary from point to point in X, or, indeed, the f's need not really be "functions" in the ordinary sense at all--but such situations will not be considered here.) We may require that the f's be restricted in some way--in particular, for the purposes of the present applications we shall often require the f's to be *holomorphic* (with, here, G = C) and X a complex manifold--or, we may consider other related classes of functions.

So, what's a sheaf? Actually, I shan't even bother with a formal definition [which can be found in, for example, Morrow & Kodaira (1971); Gunning & Rossi (1965); Wells (1973)]. The essential point is that a sheaf is so defined that the Čech cohomology works just as well as before. In fact, a sheaf $\mathcal{S}$ defines an additive group $G_{\mathcal{U}}$ for each open set $\mathcal{U} \subset X$. For example, $G_{\mathcal{U}}$ might be the additive group of all holomorphic functions on $\mathcal{U}$ (taking X to be a complex manifold). In this case we get the sheaf, denoted $\mathcal{O}$, of *germs of holomorphic functions* on X. Slightly more generally, we might consider "twisted" holomorphic functions, i.e. functions whose values are not just ordinary complex numbers, but taken in some complex line bundle over X (think of "spin-weighted" functions, for example). An important example of such a twisted function would arise if X were taken to be *projective* twistor space PT (or a suitable portion thereof) and the functions considered were to be *homogeneous* (and holomorphic) of some fixed degree n in the twistor variable. For each open set $\mathcal{U} \subset X$ we take $G_{\mathcal{U}}$ to consist of all such twisted functions on $\mathcal{U}$, and the resulting sheaf, denoted $\mathcal{O}(n)$, is called the "sheaf of germs of holomorphic functions twisted by n" (on X). More generally we might consider functions whose values lie in some vector bundle B over X (e.g. we might consider tensor fields on X) and $G_{\mathcal{U}}$ would consist of the cross-sections of the portion of B lying above $\mathcal{U}$.

Cochains are defined as before (with $f_i \in G_{\mathcal{U}_i}$, $f_{ij} \in G_{\mathcal{U}_i \cap \mathcal{U}_j}, \ldots$) and the coboundary operator δ, just as before. Then we obtain the pth cohomology group of X, with coefficients in the sheaf $\mathcal{S}$, as

$$H^p(X,\mathcal{S}) = \binom{\text{p-cochains with}}{\text{coefficients in } \mathcal{S}} \Big/ \binom{\text{p-coboundaries with}}{\text{coefficients in } \mathcal{S}}.$$

As before, we would need to take the appropriate limit for finer and finer coverings $\{\mathcal{U}_i\}$ of X, but we can settle on one "sufficiently fine" covering if desired. Provided $\mathcal{S}$ is what is called a *coherent analytic* sheaf (and we are interested primarily in such sheaves—locally defined by n holomorphic functions factored out, if desired, by a set of s holomorphic relations), then "sufficiently fine" can be taken to mean that each of $\mathcal{U}_i$, $\mathcal{U}_i \cap \mathcal{U}_j$, $\mathcal{U}_i \cap \mathcal{U}_j \cap \mathcal{U}_k, \ldots$ is a *Stein manifold* (Gunning & Rossi 1965) (and it is sufficient just to specify that each $\mathcal{U}_i$ is Stein). In effect, a Stein manifold is a holomorphically convex open subset of C^n (or a domain of holomorphy). If X is Stein and $\mathcal{S}$ coherent, then $H^p(X,\mathcal{S}) = 0$, if $p > 0$. Note: $\mathcal{O}$ and $\mathcal{O}(n)$ *are* coherent.

Twistor functions as elements of $H^1(X,\mathcal{O}(n))$,

Let X be some suitable portion of projective twistor space PT , say some neighbourhood of a line in PT (corresponding to some neighbourhood of a point in Minkowski space), or say PT^+ , or PT^- . Suppose we can cover X with two sets $\mathcal{U}_1,\mathcal{U}_2$ (each open in X) such that every projective line L in X meets $\mathcal{U}_1 \cap \mathcal{U}_2$ in an annular region and where $\mathcal{U}_1 \cap \mathcal{U}_2$ corresponds to the domain of definition of some twistor function $f(Z^\alpha)$, homogeneous of degree n in the twistor Z^α (see Fig. C).

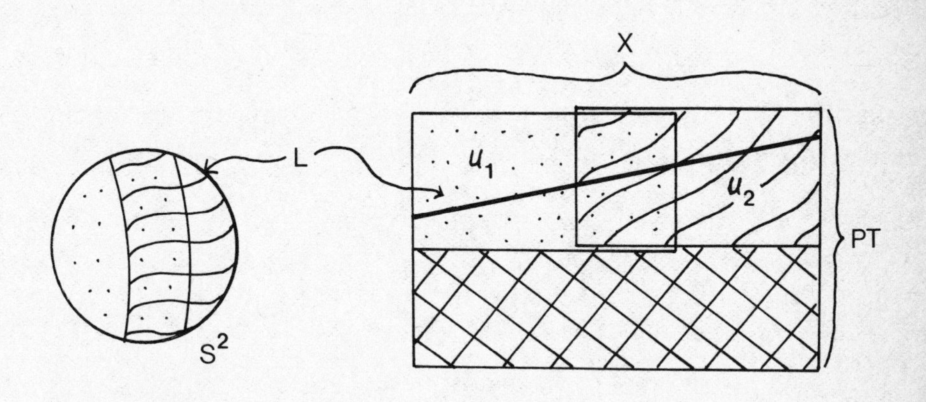

Figure C

Then $f = f_{12}$ is a twisted function on $\mathcal{U}_1 \cap \mathcal{U}_2$. There are no other $\mathcal{U}_i \cap \mathcal{U}_j$'s, so f_{12} by itself defines a 1-cochain β, with coefficients in $\mathcal{O}(n)$, for X. Clearly $\delta\beta = 0$, so β is a cocycle. The 1-coboundaries, for this covering, are functions of the form h_2-h_1, where h_2 is holomorphic on $\mathcal{U}_2$, and h_1 on $\mathcal{U}_1$. Calling $\mathcal{D} = \mathcal{U}_i \cap \mathcal{U}_2$, $\mathcal{D}^- = \mathcal{U}_1$, $D^+ = \mathcal{U}_2$, $h^- = -h_1$, and $h^+ = h_2$, we observe that the "equivalence" between twistor functions under the "gauge" freedom G that we started out with is just the normal cohomological equivalence between 1-cochains β,β' that their difference be a coboundary: $\beta' - \beta = \delta\alpha$, with $\alpha = (h_1,h_2)$. This suggests that we view the twistor function f as really defining us an element of $H^1(X,\mathcal{O}(n))$.

But this is all with respect to a particular covering of X, namely by $\{\mathcal{U}_1,\mathcal{U}_2\}$. Is this covering "fine" enough? There actually is a technical problem here. We cannot (normally) arrange that $\mathcal{U}_1$ and $\mathcal{U}_2$ are Stein manifolds (and in those exceptional cases when we can so arrange this, we would lose the invariance properties that we are striving for). It turns out, in fact, that this problem is not serious. One can show by direct construction (using the inverse twistor function--in the cases $n \leq -2$ at least, but probably in all cases) that for any *given* (analytic, positive frequency) field such a covering by two such sets $\mathcal{U}_1,\mathcal{U}_2$ is sufficient in the case $X = \overline{PT^+}$. Note that though X is invariant under SU(2,2), in this case, the *covering* is not. However, the cohomology group $H^1(X,\mathcal{O}(n))$ *is* invariant. Let us illustrate this by adding two elements of $H^1(X,\mathcal{O}(n))$ one of which is defined by a twistor function f, with respect to the covering $\{\mathcal{U}_1,\mathcal{U}_2\}$, and the other by $\hat{f}$, with respect to $\{\hat{\mathcal{U}}_1,\hat{\mathcal{U}}_2\}$, the second covering being a rotated version of the first (see Fig. D).

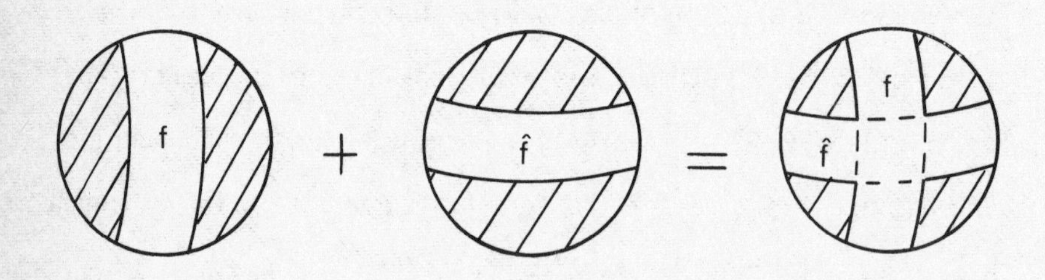

Figure D

We define a representative cochain for the sum by taking the common refinement of both coverings. (Denote $\hat{\hat{\mathcal{U}}}_1 = \mathcal{U}_1 \cap \hat{\mathcal{U}}_1, \hat{\hat{\mathcal{U}}}_2 = \mathcal{U}_1 \cap \hat{\mathcal{U}}_2, \hat{\hat{\mathcal{U}}}_3 = \mathcal{U}_2 \cap \hat{\mathcal{U}}_1$, $\hat{\hat{\mathcal{U}}}_4 = \mathcal{U}_2 \cap \hat{\mathcal{U}}_2$ to give the refined covering $\{\hat{\hat{\mathcal{U}}}_1, \hat{\hat{\mathcal{U}}}_2, \hat{\hat{\mathcal{U}}}_3, \hat{\hat{\mathcal{U}}}_4\}$.) The 1-cocycle $\beta + \hat{\beta} = \hat{\hat{\beta}}$ is

$$(\hat{\hat{f}}_{12}, \hat{\hat{f}}_{13}, \hat{\hat{f}}_{23}, \hat{\hat{f}}_{14}, \hat{\hat{f}}_{24}, \hat{\hat{f}}_{34}) = (\hat{f}, f, f-\hat{f}, f+\hat{f}, f, \hat{f}).$$

Because of the "direct construction" argument mentioned above, this 1-cocycle will be cohomologous to (i.e. differing by a coboundary from) a cocycle of the form $(0, \hat{\hat{f}}, \hat{\hat{f}}, \hat{\hat{f}}, \hat{\hat{f}}, 0)$, so we can refer it back to the original covering if desired (although with $\mathcal{U}_1$ and $\mathcal{U}_2$ perhaps reduced slightly in size).

But we need not do so if we prefer not to. We have a generalization of the concept of a twistor function, namely as a collection of functions on portions of twistor space, defining a 1-cocycle with respect to some covering. We can actually use such a cocycle *directly*, obtaining the required space-time field by means of a *branched contour integral*. I shall just illustrate this with an example. Suppose that X is covered by 3 open sets $\mathcal{U}_1, \mathcal{U}_2, \mathcal{U}_3$, where on a projective line L in X we get the picture of Fig.E.

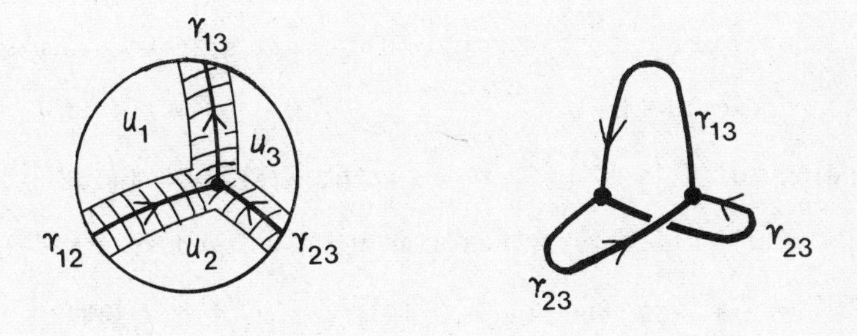

Figure E

It is a simple matter to check that the cocycle condition $f_{12} - f_{13} + f_{23} = 0$ (in $\mathcal{U}_1 \cap \mathcal{U}_2 \cap \mathcal{U}_3$) ensures that the contour's end-points can be moved without affecting the result. This easily generalizes to coverings with N open sets. The 1-cocycle β is (f_{12}, f_{13}, f_{23}). To get the field ϕ.... we perform the sum of three contour integrals with common end-points in $\mathcal{U}_1 \cap \mathcal{U}_2 \cap \mathcal{U}_3$ (see Fig.E):

$$(2\pi i)^2 \phi.... = \int_{\gamma_{12}} ....f_{12} d^2Z + \int_{\gamma_{13}} ....f_{13} d^2Z + \int_{\gamma_{23}} ....f_{23} d^2Z.$$

<u>Charge integrality in the "twisted photon"</u>

We have (Morrow & Kodaira 1971) the exact sequence

$$0 \to Z \xrightarrow{\times 2\pi i} \mathcal{O} \xrightarrow{\exp} \mathcal{O}^* \to 0,$$

where $\mathcal{O}^*$ refers to non-zero holomorphic functions taken multiplicatively, from which we derive the long exact sequence:

$$\ldots \to H^1(X,Z) \to H^1(X,\mathcal{O}) \to H^1(X,\mathcal{O}^*) \to H^2(X,Z) \to \ldots$$

$H^1(X,Z)$	$H^1(X,\mathcal{O})$	$H^1(X,\mathcal{O}^*)$	$H^2(X,Z)$
↑	↑	↑	↑
ordinary integer 1st cohomology	$\phi_{AB} = \oint \hat{\pi}_A \hat{\pi}_B f\Delta$	"twisted photon"	ordinary integer 2nd cohomology

Choose X as a region in PT corresponding to a small space-time tube surrounding a charge world-line.

Then the topology of X is $S^2 \times S^2 \times R^2$, so $H^1(X,Z) = 0$, and $H^2(X,Z) \simeq Z \oplus Z$. Thus the space $H^1(X,\mathcal{O}^*)$ effectively contains $H^1(X,\mathcal{O})$, and is strictly larger than $H^1(X,\mathcal{O})$ if the map to $Z \oplus Z$ is not simply to the zero element. The image in the first Z is always zero, but the image in the second Z is the value of the *charge*. (This much follows, for example, from examination of §3.3.) From this we see that the $H^1(X,\mathcal{O}^*)$ description only works if the charge has *integer* value (i.e. lies in Z) whereas the $H^1(X,\mathcal{O})$ description only works if the charge value is *zero*: the contour integral formula

$$\phi_{AB} = (2\pi i)^{-2} \oint \rho_X \ \hat{\pi}_A \ \hat{\pi}_B \ f(Z^\alpha) \ d^2\pi,$$

where f is homogeneous of degree zero, implies vanishing charge. In contrast to this, it may be remarked that there is no restriction on the charge value if the -4-homogeneity $f(Z^\alpha)$ description

$$\tilde{\phi}_{A'B'} = (2\pi i)^{-2} \oint \rho_X \ \pi_{A'} \ \pi_{B'} f(Z^\alpha) d^2\pi$$

is used, this corresponding to $H^1(X,\mathcal{O}(-4))$.

A finite covering for $\overline{PT^+}$

A difficulty arises, in connection with the sheaf cohomology approach if one tries to find a *fixed* covering of $\overline{PT^+}$ which is "sufficiently fine" for *all* elements of $H^1(\overline{PT^+}, \mathcal{O}(n))$ (i.e. for all analytic massless wave functions). This arises basically because PT^+ is not holomorphically (pseudo-) convex and admits no finite covering by Stein manifolds. Any covering of PT^+ by Stein manifolds must necessarily be infinite, with open sets crowding up against the boundary PN. This is reflected in the fact that while for any *fixed* analytic massless wave function, a twistor function may be found (by way of the inverse twistor function) whose domain is just $\mathcal{U}_1 \cap \mathcal{U}_2$ where $\mathcal{U}_1$ and $\mathcal{U}_2$ are open sets with $\mathcal{U}_1 \cup \mathcal{U}_2 = PT^+$, this *particular* covering can never suffice for *all* such wave functions simultaneously (which may be arbitrarily "rough" and for which the corresponding twistor function's domain may need to be arbitrarily narrow).

However, if we allow ourselves to use *closed* sets for our sheaf cohomology (and there seems no reason against this), then the covering of $\overline{PT^+}$ by the *two* closed sets, given respectively by

$$|Z^0 + Z^2| \geq |Z^1 + Z^3|, \quad Z^\alpha \bar{Z}_\alpha \geq 0$$

and by

$$|Z^0 + Z^2| \leq |Z^1 + Z^3|, \quad Z^\alpha \bar{Z}_\alpha \geq 0$$

will suffice (as the inverse twistor function shows) for *all* such wave functions.

Acknowledgment

I am grateful to George Sparling, Andrew Hodges, and Nick Woodhouse for many helpful discussions and, most particularly, to Michael Atiyah for explaining sheaf cohomology theory so clearly to us and for many insightful remarks.

REFERENCES

Gunning, R.C. & Rossi, H. 1965 *Analytic Functions of Several Complex Variables*. Prentice-Hall.

Morrow, J. & Kodaira, K. 1971 *Complex Manifolds*. Holt, Rinehart & Winston.

Penrose, R. 1975a *Twistor Theory, its Aims and Achievements*. In: *Quantum Gravity*, eds. C.J. Isham, R. Penrose & D.W. Sciama, pp. 268-407. Clarendon: Oxford.

Penrose, R. 1975b *Twistors and Particles*. In: *Quantum Theory and the Structures of Time and Space*, eds. L. Castell, M. Drieschner & C.F. von Weizsäcker, pp. 129-145.

Wells, R.O., Jr. 1973 *Differential Analysis on Complex Manifolds*. Prentice-Hall.

§2.3. TWISTOR COHOMOLOGY WITHOUT SHEAVES *by N.M.J. Woodhouse*

1. It follows from the cohomological interpretation of twistor functions that zero-rest-mass fields on Minkowski space can be represented by equivalence classes of 1-forms on twistor space. In this note, I shall show how one can go directly from these equivalence classes to the space-time fields, and back again. Going in the forward direction involves no more than a minor reinterpretation of some familiar calculations; the reverse procedure is slightly more complicated: it is based on some ideas introduced by Rawnsley (1979); Hitchin (1979); and Eastwood, Penrose & Wells (1979).

A more detailed account of the Penrose transform using forms rather than sheaf cohomology can be found in Wells (1979).

2. I shall use the following notation: M^+ is the forward tube of (complex) Minkowski space, $T^+ = \{Z^\alpha;\ Z^\alpha\bar{Z}_\alpha > 0\}$ is the corresponding (nonprojective) half twistor space, and B^+ is the bundle of primed spinors over M^+; $\sigma: B^+ \to T^+$ and $\rho: B^+ \to M^+$ are the projections of B^+ onto T^+ and M^+, given by

$$\sigma: (z^{AA'}, \pi^{B'}) \mapsto Z^\alpha = (iz^{AB'}\pi_{B'}, \pi_{A'}) \tag{1}$$

$$\rho: (z^{AA'}, \pi^{B'}) \mapsto z^{AA'} \tag{2}$$

where $z^{AA'}$, $(z^{AA'}, \pi_{B'})$, and $Z^\alpha = (\omega^A, \pi_{A'})$ are coordinates on M^+, B^+ and T^+ respectively. The (anti-)holomorphic exterior derivative is denoted ∂ ($\bar{\partial}$); thus, for example,

$$\partial(f_{\alpha\beta}\, dZ^\alpha \wedge dZ^\beta) = \partial_{[\alpha}f_{\beta\gamma]}dZ^\alpha \wedge dZ^\beta \wedge dZ^\gamma \quad \text{on } T^+ \tag{3}$$

$$\bar{\partial}(g) = \frac{\partial g}{\partial \bar{z}^{AA'}}\, d\bar{z}^{AA'} + \frac{\partial g}{\partial \bar{\pi}^{A}}\, d\bar{\pi}^{A} \quad \text{on } B^+. \tag{4}$$

The Euler vector field $Z^{\alpha} \frac{\partial}{\partial Z^{\alpha}}$ is denoted Y; θ is the 1-form $I_{\alpha\beta} Z^{\alpha} dZ^{\beta} = \pi^{A'} d\pi_{A'}$; and $\_|$ denotes the contraction between a vector field and a form. Note that $d = \partial + \bar{\partial}$, $\bar{\partial}\theta = 0$, $Y\_|\partial\theta = 2\theta$, and $Y\_|\theta = 0$.

3. Consider the set of all (0,1)-forms f on T^{+} satisfying

$$\bar{\partial} f = 0, \ \bar{Y}\_| f = 0, \text{ and } Y\_|\partial f = (-n-2)f \tag{5}$$

where n is a positive integer. That is, if $f = f^{\alpha} d\bar{Z}_{\alpha}$, then

$$\frac{\partial f^{\alpha}}{\partial \bar{Z}_{\beta}} = \frac{\partial f^{\beta}}{\partial \bar{Z}_{\alpha}}, \ \bar{Z}_{\alpha} f^{\alpha} = 0, \text{ and } Z^{\alpha} \frac{\partial f^{\beta}}{\partial Z^{\alpha}} = (-n-2) f^{\beta}. \tag{6}$$

Let H denote the set of equivalence classes of these forms under the equivalence relation $f \sim f'$ whenever $f - f' = \bar{\partial} h$ for some function $h = h(z^{\alpha}, \bar{Z}_{\beta})$ on T^{+} such that $Y\_|\partial h = (-n-2)h$. Each element of H can be identified with a cohomology class of twistor functions, homogeneous of degree $-n-2$, and therefore with a massless spin-½n field on M^{+}. Two lemmata are needed to construct this field explicitly:

Lemma 1: Let $g(Z^{\alpha})$ be a holomorphic function on T^{+}, homogeneous of degree n, let $[f] \in H$, and let $\tau = gf \wedge \theta$. Then $Y\_| d\tau = 0 = \bar{Y}\_| d\tau$.

Proof: Since $\bar{\partial}\tau = g\bar{\partial}(f \wedge \theta) = 0$,

$$\begin{aligned} Y\_| d\tau = Y\_|\partial\tau &= Y(g) f \wedge \theta + g(Y\_|\partial f) \wedge \theta + gf \wedge (Y\_|\partial\theta) \\ &= (n + (-n-2) + 2) f \wedge \theta = 0. \end{aligned} \tag{7}$$

Also, $\bar{Y}\_| d\tau = \bar{Y}\_|\partial\tau = \partial(\bar{Y}\_|\tau) = 0$. □

Lemma 2: Suppose that $k(\pi^{A'}, \bar{\pi}^{A})$ satisfies $\pi^{A'} \frac{\partial k}{\partial \pi^{A'}} = -2k$ and $\bar{\pi}^{A} \frac{\partial k}{\partial \bar{\pi}^{A}} = 0$. Then $d(k\theta) = \bar{\partial} k \wedge \theta$.

Proof: $d(k\theta) = \partial(k\theta) + \bar{\partial}(k\theta) = \partial(k\theta) + (\bar{\partial}k) \wedge \theta$ and

$$\partial(k\theta) = \frac{\partial}{\partial\pi^{A'}}[k(\pi,\bar{\pi})]d\pi^{A'} \wedge (\pi^{B'} d\pi_{B'}) + k\, d\pi^{B'} \wedge d\pi_{B'} = 0. \square \tag{8}$$

Now, given a (0,1)-form f satisfying the relations (5), and a point $z^{AA'}$ in the forward tube of Minkowski space, put

$$\psi^{A'\dots C'} = \int \pi^{A'} \dots \pi^{C'} f \wedge \theta , \tag{9}$$

the integral being taken over any real 2-surface in $L_z = \{(\omega^A, \pi_{A'}) = (iz^{AB'}\pi_{B'}, \pi_{A'})\} \subset T^+$ on which $\pi_{A'}$ varies once over almost all[1] real null directions through $z^{AA'}$. Then

(i) The integral is independent of the choice of 2-surface: by lemma 1,

$$Y \lrcorner d(\pi^{A'} \dots \pi^{C'} f \wedge \theta) = 0 = \bar{Y} \lrcorner d(\pi^{A'} \dots \pi^{C'} f \wedge \theta); \tag{10}$$

also, $Y \lrcorner (f \wedge \theta) = 0 = \bar{Y} \lrcorner (f \wedge \theta)$. Hence the integrand in (9) projects onto a well defined form on the complex projective line PL_z in the projective half twistor space PT^+; and the left hand side of (9) is equal to the integral of this projected form over PL_z (in fact, this is a slightly more satisfactory way of interpreting eqn (9)).

(ii) $\psi^{A'\dots C'}$ depends only on the equivalence class of f: for if $f = \bar{\partial}k$, then the integrand in (9) is equal to $d(\pi^{A'} \dots \pi^{B'} k\theta)$ (by lemma 2); but $\pi^{A'} \dots \pi^{B'} k\theta$ also projects onto a well defined form on PL_z, so, in this case, the integral vanishes.

(iii) $\psi^{A'\dots C'}$ satisfies the zero-rest-mass field equation: if we think of the integrand in (9) as being defined on B^+, then we have

$$\nabla_{AA'}\psi^{B'\ldots D'} = \int \frac{\partial}{\partial z^{AA'}} \lrcorner\, (\partial + \bar{\partial})(\pi^{B'} \ldots \pi^{D'} f \wedge \theta)$$

$$= \int \pi_{A'} \pi^{B'} \ldots \pi^{D'} (\frac{\partial f^{\gamma}}{\partial \omega^{A}} d\bar{Z}_{\gamma}) \wedge (\pi^{E'} d\pi_{E'}). \tag{11}$$

Contracting over A' and B' gives $\nabla_{AA'} \psi^{A'B'\ldots C'} = 0$.

(iv) $\psi^{A'\ldots C'}$ is positive frequency: a similar argument gives $\bar{\nabla}_{AA'} \psi^{B'\ldots D'} = 0$.

4. Going the other way, let $\psi^{A'\ldots C'}$ be a spin-½n solution of the zero-rest-mass field equations on M^{+}, depending holomorphically on the space-time coordinates; let $t^{AA'}$ be a constant, real time-like vector, normalized so that $t_{AA'}t^{AA'} = 2$; and let β be the (0,1)-form on B^{+} defined by

$$\beta = \frac{n+1}{2\pi i} t^{-n-2} \psi^{A'\ldots C'} t_{AA'} \bar{\pi}^{A} \ldots t_{CC'} \bar{\pi}^{C} \bar{\pi}^{E} d\bar{\pi}_{E} \tag{12}$$

where $t = t_{AA'} \bar{\pi}^{A} \pi^{A'}$. Then, by direct calculations, $\bar{\partial}\beta = 0$ and

$$\int \pi^{A'} \ldots \pi^{C'} \beta \wedge \pi^{D'} d\pi_{D'} = \psi^{A'\ldots C'} \tag{13}$$

(where the left hand side can, as before, be interpreted as an integral over the complex projective line in projective twistor space corresponding to a fixed point in M^{+}). Also, if λ^{A} is any constant spinor and X is the holomorphic vector field $X = \lambda^{A} \pi^{A'} \frac{\partial}{\partial z^{AA'}}$ on B^{+}, then

$$X \lrcorner\, \beta = 0 = \bar{X} \lrcorner\, \beta \tag{14}$$

$$\pounds_{\bar{X}} \beta = 0 \tag{15}$$

$$\pounds_{X}\beta = \bar{\partial}(X \lrcorner\, \gamma) \tag{16}$$

where γ is the (1,0)-form on B^{+}:

$$\gamma = \frac{1}{2\pi i} t^{-n-2} (\nabla_E{}^{A'} \psi^{B'\ldots D'}) t_{E'F} \bar{\pi}^F t_{AA'} \bar{\pi}^A \ldots t_{DD'} \bar{\pi}^D dz^{EE'}. \quad (17)$$

Now, as λ^A varies, the vector fields X and $\bar{X}$ span the fibres of the projection $\sigma: B^+ \to T^+$. Thus, if the right hand side of (16) vanished, then β would project onto a $\bar{\partial}$-closed (0,1)-form on T^+ satisfying (5); and, by (13), this form would lie in the equivalence class in H corresponding to $\psi^{A'\ldots C'}$. However, even though β itself does not project onto T^+, it is not hard to find a function h on B^+ such that $\beta' = \beta - \bar{\partial}h$ *is* of the form $\sigma^*(f)$ for some f satisfying (5).

Let $\nu = \gamma|_F$ be the restriction of γ to one of the fibres F of $\sigma: B^+ \to T^+$. Then

(i) ν is a holomorphic 1-form on F, since the only nonholomorphic elements in γ are the $\bar{\pi}^A$'s, and these are constant on F.

(ii) $d\nu = 0$.

To establish (ii), one introduces complex analytic coordinates ξ^A on F by writing $(z^{AA'}, \pi^{B'}) = (z_0^{AA'} + \xi^A \pi^{A'}, \pi^{B'})$ for some fixed $z_0^{AA'}$ and $\pi^{A'}$; then, in terms of these,

$$d\nu = \frac{1}{2\pi i} t^{-n-1} \pi^{F'} \nabla_{FF'} \nabla_E{}^{A'} \psi^{B'\ldots D'} t_{AA'} \bar{\pi}^A \ldots t_{DD'} \bar{\pi}^D d\xi^F \wedge d\xi^E = 0 \quad (18)$$

since $\Box\, \psi^{A'\ldots C'} = 0$.

It follows that there exists a smooth complex function h on $\bar{B}$ such that

$$X \,\lrcorner\, \gamma = X(h) \text{ and } \bar{X}(h) = 0 \quad (19)$$

for any X of the form $X = \lambda^A \pi^{A'} \frac{\partial}{\partial z^{AA'}}$. In fact, all that is necessary to construct h explicitly is to pick out a smooth *real* submanifold Σ of M^+ that intersects each α-plane in precisely one point. Then, if Λ is an α-plane and $z_0^{AA'}$ is the intersection of Λ and Σ, h can be defined at points of $\rho^{-1}(\Lambda)$ by[2]

$$h(z^{AA'}, \pi^{B'}) = \int_{z_0}^{z} \gamma \tag{20}$$

where $\pi^{A'}$ is fixed and the integral is along any path from $z_0^{AA'}$ to $z^{AA'}$ in Λ; since Λ is simply connected and the restrictions of γ to the fibres of $\sigma\colon B^+ \to T^+$ are closed, it does not matter which path is chosen (the fibres of $\sigma\colon B^+ \to T^+$ are the horizontal sections of B^+ over the α-planes in M^+).

Now consider the 1-form $\beta' = \beta - \bar{\partial}h$ on B^+. This has the following properties:

(i) $\bar{\partial}\beta' = 0$

(ii) $X \lrcorner \beta' = 0$ (since β' is of type (0,1))

(iii) $\bar{X} \lrcorner \beta' = \bar{X} \lrcorner \beta - \bar{X}(h) = \bar{X} \lrcorner \gamma = 0$

(iv)
$$\begin{aligned} \pounds_X \beta' &= \pounds_X \beta - \pounds_X \bar{\partial}h \\ &= \bar{\partial}(X(h)) - X \lrcorner \partial\bar{\partial}h \\ &= \bar{\partial}(X(h)) - \bar{\partial}(X(h)) \\ &= 0 \end{aligned}$$

(v) $\pounds_{\bar{X}} \beta' = -\bar{X} \lrcorner \partial\bar{\partial}h - \partial(\bar{X}(h)) = 0$

for any $X = \lambda^A \pi^{A'} \dfrac{\partial}{\partial z^{AA'}}$. Hence β' projects onto a $\bar{\partial}$-closed (0,1)-form f on T^+; it is clear from the homogeneities of β and γ that f satisfies (5) and, from (13), that the corresponding equivalence class in H generates $\psi^{A'\ldots C'}$.

5. An obvious possibility is to take Σ to be a negative definite real slice of M^+; for example, the set of solutions of

$$\bar{r}^{AA'} = - t^{AB'} t^{BA'} r_{BB'}. \tag{21}$$

Then $S = \rho^{-1}(\Sigma)$ is a section of the projection $\sigma: B^+ \to T^+$, and so $T^+ = S$ (as a real manifold). It follows that to construct f, we only have to find the restriction $\beta'|_S$. But knowing the values of β and γ at points of S, it is possible to write down an expression for $\beta'|_S$ *without* doing any integration.

To do this explicitly, we need to introduce some coordinates: on Σ, we can use the solutions of (21) (effectively four real coordinates); on S, we can use $(r^{AA'}, \pi^{B'}, \bar{\pi}^{B})$; and on B^+, we can use either the complex analytic coordinates $(z^{AA'}, \pi^{B'})$, or the nonholomorphic coordinates $(r^{AA'}, \xi^B, \bar{\xi}^{B'}, \pi^{C'}, \bar{\pi}^C)$, which are related to $z^{AA'}$ and $\pi^{A'}$ by

$$z^{AA'} = r^{AA'} + \xi^A \pi^{A'} \text{ and } \bar{z}^{AA'} = - t^{AB'} t^{BA'} r_{BB'} + \bar{\xi}^{A'} \bar{\pi}^A . \tag{22}$$

In these coordinates, the identification of T^+ with S (defined by restricting σ to S) is given by

$$(\omega^A, \pi_{A'}) = (i r^{AB'} \pi_{B'}, \pi_{A'}), \tag{23}$$

or, going in the other direction, by

$$r^{AA'} = -i\, t^{-1}(\omega^A t^{BA'} \bar{\pi}_B - t^{AB'} \bar{\omega}_{B'} \pi^{A'}). \tag{24}$$

Now, on S, $\xi^A = 0$ and

$$\frac{\partial h}{\partial r^{AA'}} = \frac{\partial h}{\partial \pi^{A'}} = \frac{\partial h}{\partial \bar{\pi}^A} = \frac{\partial h}{\partial \bar{\xi}^{A'}} = 0. \tag{25}$$

Hence, at points of S,

$$\bar{\partial} h = \frac{\partial h}{\partial \xi^B}\left(\frac{\partial \xi^B}{\partial \bar{z}^{AA'}} d\bar{z}^{AA'} + \frac{\partial \xi^B}{\partial \bar{\pi}^A} d\bar{\pi}^A\right). \tag{26}$$

But, from (22),

$$\frac{\partial \xi^B}{\partial \bar{z}^{AA'}} = t^{-1}\, t^B{}_{A'}\, \bar{\pi}_A \quad \text{and} \quad \frac{\partial \xi^B}{\partial \bar{\pi}^A} = 0 \tag{27}$$

on S. Also, from (19),

$$\frac{\partial h}{\partial \xi^A} = (\pi^{A'} \frac{\partial}{\partial z^{AA'}}) \_|\gamma. \tag{28}$$

Therefore

$$\begin{aligned} \bar{\partial} h|_S &= -\frac{1}{2\pi i}\, t^{-n-2} (\nabla_E{}^{A'}\, \psi^{B'\ldots C'}) t_{AA'} \bar{\pi}^A \ldots t_{CC'} \bar{\pi}^C\, \bar{\pi}^F\, t_{FE'}\, dr^{EE'} \\ &= \frac{1}{2\pi i}\, t^{-n-2} (\nabla^{EA'}\, \psi^{B'\ldots C'}) t_{AA'} \bar{\pi}^A \ldots t_{CC'} \bar{\pi}^C t_{EF'} (\bar{r}^{FF'} d\bar{\pi}_F - i d\bar{\omega}^{F'}) \end{aligned} \tag{29}$$

while $\beta|_S$ is again given by the expression on the right hand side of (12).

6. If we only know the values of $\psi^{A'\ldots B'}$ and $\nabla_{AA'}\, \psi^{B'\ldots C'}$ on the real spacelike hyperplane forming the intersection of real Minkowski space with the boundary of Σ, we can still use (12) and (29) to construct a form f on $N = \{Z^\alpha;\ Z^\alpha\, \bar{Z}_\alpha = 0\}$; and we can recover from f the values of the field $\psi^{A'\ldots C'}$ at other real points of Minkowski space using the integral formula (9). This gives yet another way of looking at the propagation of zero-rest-mass fields from their initial data.

7. Both (12) and (29) are *local* formulae: they can be applied to arbitrary open subsets of the negative definite real slice of complex Minkowski space. They thus determine not only the cohomology class on twistor space corresponding to a field in space-time, but also the precise subset of twistor space on which this class should be defined.

8. Taking a more sophisticated view, we can think of $f|_{L_Z}$ as a form on PL_Z, taking values in an Hermitian line bundle over PL_Z; and we can think

of $t_{AA'}$ as a Kähler metric on PL_z. Then, with z fixed, the right hand side of (12) can be interpreted as the unique harmonic representative of the Dolbeault cohomology class of $f|_{L_z}$.

Note, finally, that although the construction of f involves making a particular choice for Σ and $t_{AA'}$, the equivalence class of f is independent of the choices made.

I thank C.M. Patton for pointing out a mistake in an earlier version of the first half of this article.

Notes

(1) The boundary of this 2-surface, which projects onto a single point in PL_z, must be omitted; hence the qualification 'almost'.

(2) There is a slight abuse of terminology here: strictly, the integral is along a path in B^+.

References

Eastwood, M.G., Penrose, R., and Wells, R.O. 1979. *Cohomology and Massless Fields*. To be published.

Hitchin, N. 1979. *Linear Field Equations on Self-Dual Spaces*. To be published.

Rawnsley, J.H. 1979. *On the Atiyah-Singer Vanishing Theorem for Certain Cohomology Groups of Instanton Bundles*. Math. Ann. 241, 43-56.

Wells, R.O. 1979. *Complex Manifolds and Mathematical Physics*. To appear in Bulletin of the American Mathematical Society.

§2.4. MASSLESS FIELDS FROM TWISTOR FUNCTIONS *by R.S. Ward*

There is a way of deriving massless fields from twistor functions without explicitly using contour integration. It is based on a 'splitting' formula due to G.A.J. Sparling: if $g(\pi_{A'})$ is homogeneous of degree -1, with singularities in two disconnected regions S_1 and S_2, then $g = g_1 - g_2$, where

$$g_j(\pi_{A'}) = (2\pi i)^{-1} \oint_{\Gamma_j} (\xi_{B'}\, \pi^{B'})^{-1}\, g(\xi_{A'})\, \xi_{C'}\, d\xi^{C'} \tag{1}$$

for $j = 1,2$. The contours Γ_1 and Γ_2 are as indicated in Fig. A. Note

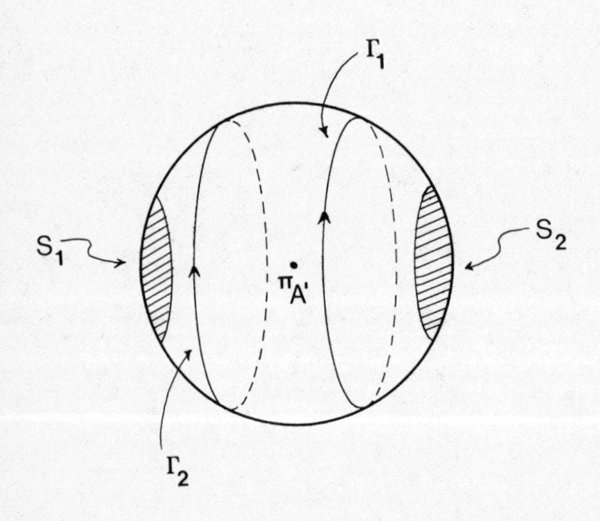

Figure A

that g_1 is holomorphic on the complement of S_2 (i.e. over the 'left hand side' of the $\pi_{A'}$-space) and g_2 over the other side.

Suppose now that $F(Z^\alpha)$ is a twistor function homogeneous of degree $n \leq -2$, and put

$$\psi_{A'\dots P'}(x,\pi_{C'}) = \pi_{A'} \dots \pi_{P'} F(i\, x^{CC'} \pi_{C'}, \pi_{C'}),$$

with (-n-1) factors of π on the right hand side.

This spinor field $\psi_{A'\dots P'}$ is homogeneous of degree -1 in $\pi_{C'}$, and so, for fixed x, it can be split:

$$\psi_{A'\dots P'}(x,\pi_{C'}) = \psi_{1A'\dots P'}(x,\pi_{C'}) - \psi_{2A'\dots P'}(x,\pi_{C'}). \tag{2}$$

Now define a spinor field $\phi_{B'\dots P'}$ by

$$\phi_{B'\dots P'} = \pi^{A'} \psi_{1A'\dots P'} .$$

I claim that $\phi_{B'\dots P'}$ is a massless free field on space-time. There are two ways of seeing this. The first is to use (1) to verify that $\phi_{B'\dots P'}$ is given by the familiar contour integral formula:

$$\phi_{B'\dots P'}(x) = \pi^{A'}[(2\pi i)^{-1} \oint (\xi_{Q'}\pi^{Q'})^{-1} \xi_{A'}\xi_{B'}\dots\xi_{P'}\rho_x F(Z^\sigma)\, \xi_{D'} d\xi^{D'}]$$

$$= (2\pi i)^{-1} \oint \xi_{B'}\dots \xi_{P'}\, \rho_x F(Z^\alpha)\, \xi_{D'}\, d\xi^{D'}.$$

The second way is to use 'globality' arguments. For example, transvecting (2) with $\pi^{A'}$ gives

$$\pi^{A'} \psi_{1A'\dots P'} = \pi^{A'} \psi_{2A'\dots P'} ; \tag{3}$$

The left hand side of (3) could only be singular in S_2, and the right hand side only in S_1, so both sides must be *globally* holomorphic. Being homogeneous of degree zero, they must therefore be independent of $\pi_{A'}$, by Liouville's theorem. Thus $\phi_{B'\dots P'}$ is a function only of x^a. Similar arguments show that $\phi_{B'\dots P'}$ is symmetric and satisfies the field equations.

The procedure can also be applied to functions $F(Z^\alpha)$ homogeneous of degree $n > -2$, as follows. Put

$$\psi_{A'A\ldots P}(x,\pi_{C'}) = \pi_{A'}\,\hat{\pi}_A \ldots \hat{\pi}_P\, F(i\,x^{CC'}\,\pi_{C'},\,\pi_{C'}),$$

split ψ into $\psi_1 - \psi_2$, and define $\phi_{A\ldots P} = \pi^{A'}\,\psi_{1A'A\ldots P}$. This field $\phi_{A\ldots P}$ is the same one as one obtains from the contour integral formula

$$\phi_{A\ldots P} = (2\pi i)^{-1} \oint \rho_x\,\hat{\pi}_A \ldots \hat{\pi}_P\, F(Z^\sigma)\,\pi_{D'}\,d\pi^{D'}.$$

All this can be restated in the language of sheaf cohomology (cf. §2.2). The basic fact is that the sheaf cohomology group $H'(CP', \mathcal{O}(-1))$ is zero. In other words, a cocycle $g_{jk}(\pi_{A'})$, homogeneous of degree -1, can be "split" into a coboundary:

$$g_{jk} = g_k - g_j.$$

When the covering of CP' consists of only two patches, so that j and k range over the values 1 and 2, g_j is given by the formula (1). For a general covering of CP', there is a generalization of (1) involving a branched contour (see §2.2). See also the second paragraph of §2.6.

§2.5 THE TWISTOR COHOMOLOGY OF LOCAL HERTZ POTENTIALS *by L.P. Hughston*

This note examines several of the well-known properties of Hertz-type potentials from a sheaf theoretic viewpoint, using twistor cohomological techniques.

I. An exact sequence for Hertz potentials. A few elementary facts and definitions will be reviewed first. If a field $\phi_A(x)$ on Minkowski space satisfies the zero rest mass [z.r.m.] equation $\nabla^{A'A}\phi_A = 0$, then a Hertz potential for ϕ_A is defined to be a solution $\psi^{A'}$ of the wave equation $\Box\psi^{A'} = 0$ such that $\phi_A = \nabla_{AA'}\psi^{A'}$. A theorem[1] can be proved to the effect that locally a $\psi^{A'}$ can always be found which satisfies the required conditions. There is some *gauge freedom* in the choice of the field $\psi^{A'}$, for if the transformation $\psi^{A'} \to \psi^{A'} + \phi^{A'}$ is made, then ϕ_A remains unchanged, providing that $\phi^{A'}$ satisfies the z.r.m. equation $\nabla_{AA'}\phi^{A'} = 0$.

This information can be synthesized neatly in a sheaf theoretic fashion. The following three sheaves are introduced: $\Phi^{A'}$ ∿ the sheaf of germs of spinor fields satisfying $\nabla_{AA'}\phi^{A'} = 0$; $\Psi^{A'}$ ∿ the sheaf of germs of spinor fields satisfying $\Box\psi^{A'} = 0$; and Φ_A ∿ the sheaf of germs of spinor fields satisfying $\nabla^{A'A}\phi_A = 0$. It's a straightforward matter to prove that the relations cited in the previous paragraph amount to the fact that the sequence

$$0 \longrightarrow \Phi^{A'} \xrightarrow{\alpha} \Psi^{A'} \xrightarrow{\beta} \Phi_A \longrightarrow 0 \qquad (1)$$

is *exact*. Here α is simply the inclusion map, whereas β is the sheaf homomorphism which induces the differential mapping $\beta^*:\psi^{A'} \to \nabla_{AA'}\psi^{A'}$ on local sections of $\Psi^{A'}$.

The exactness of the sequence is equivalent to the following set of three conditions:

(a) the map α is injective.

(b) $\mathrm{Im}(\alpha) = \mathrm{Ker}(\beta)$.

(c) the map β is surjective.

Each of these conditions is satisfied: (a), since α is simply the inclusion map; (b), since sections of $\Phi^{A'}$ satisfy the z.r.m. equation; and (c), on account of the local existence theorem for Hertz potentials, which was mentioned earlier[2].

II. <u>Sequences for higher spins</u>. A similar but somewhat more intricate result applies in the case of higher spin z.r.m. fields. For a spin-1 field ϕ_{AB} satisfying Maxwell's equations $\nabla^{A'A}\phi_{AB} = 0$, a Hertz potential is a field $\psi^{A'B'}$ which satisfies the wave equation and is such that $\phi_{AB} = \nabla_{AA'}\nabla_{BB'}\psi^{A'B'}$. In this case there also exists an "intermediate" potential $\psi^{A'}{}_{B} := \nabla_{BB'}\psi^{A'B'}$, which satisfies the equation[3] $\nabla^{BB'}\psi^{A'}{}_{B} = 0$. The gauge freedom in $\psi^{A'B'}$ is of two distinct sorts. First there is the transformation $\psi^{A'B'} \rightarrow \psi^{A'B'} + \phi^{A'B'}$, with $\nabla_{AA'}\phi^{A'B'} = 0$. Such a transformation preserves both $\psi^{A'}{}_{B}$ and ϕ_{AB}. Then there is the transformation $\psi^{A'}{}_{B} \rightarrow \psi^{A'}{}_{B} + \nabla^{A'}{}_{B}\phi$, where ϕ is a scalar subject to $\Box\phi = 0$; this is just the "usual" electromagnetic gauge transformation. Introducing appropriate sheaves for these various fields using a notation patterned in the style of the spin ½ case, the following set of exact sequences is obtained:

$$\begin{array}{ccccccccccc} 0 & \longrightarrow & \Phi^{A'B'} & \longrightarrow & \Psi^{A'B'} & \longrightarrow & \Psi^{A'}{}_{B} & \longrightarrow & 0 & & \\ & & & & & & \| & & & & \\ 0 & \longrightarrow & C & \longrightarrow & \Phi & \longrightarrow & \Psi^{A'}{}_{B} & \longrightarrow & \Phi_{AB} & \longrightarrow & 0 \end{array} \qquad (2)$$

Note that in the second of these two sequences an extra term appears, corresponding to the additional freedom $\phi \to \phi + c$, which leaves $\psi^{A'}{}_{B}$ invariant under the transformation $\psi^{A'}{}_{B} \to \psi^{A'}{}_{B} + \nabla^{A'}{}_{B}\,\phi$, where c is any complex constant[4].

III. A contour integral formula for Hertz potentials. From a twistor point of view, Hertz potentials can be obtained by means of a contour integral method. In order to produce a z.r.m. field of spin ½, a twistor function $f(Z^{\alpha})$ which is homogeneous of degree -1 ["hom(-1)"] is used in the following contour integral formula:

$$\phi_A(x) = \frac{1}{2\pi i}\oint_x \hat{\pi}_A f\, \Delta\pi ,$$

where $\hat{\pi}_A$ is the operator $-\partial/\partial\omega^A$, and $\Delta\pi = \pi_{A'}d\pi^{A'}$. A standard argument shows that ϕ_A satisfies the z.r.m. equations. To form a Hertz potential for ϕ_A, take a "spinor-valued" holomorphic function $f^{A'}(Z^{\alpha})$, which is hom(-2), and form:

$$\psi^{A'}(x) = \frac{1}{2\pi i}\oint_x f^{A'}\, \Delta\pi .$$

Differentiating, it is easy to verify that $\psi^{A'}$ is a Hertz potential for ϕ_A providing that $i\pi_{A'}f^{A'} = f$. There is some freedom available in the specification of $f^{A'}$, for the invariance of f under the transformation $f^{A'} \to f^{A'} + \pi^{A'}g$ is readily ascertainable, where $g(Z^{\alpha})$ is any holomorphic function which is hom(-3). This transformation on $f^{A'}$ induces a corresponding transformation on $\psi^{A'}$, which is given by $\psi^{A'} \to \psi^{A'} + \phi^{A'}$, where:

$$\phi^{A'}(x) = \frac{1}{2\pi i}\oint_x \pi^{A'}g\, \Delta\pi .$$

This expression for $\phi^{A'}$ is again a standard formula for a z.r.m. field, hence it follows that the freedom in $f^{A'}$ induces the proper gauge freedom available to $\psi^{A'}$.

IV. Twistor cohomology groups. In order to obtain a cohomological interpretation for the results mentioned in I, II, and III, it is necessary to introduce several sheaves on twistor space[5]:

$\mathcal{O}(n)$ ... the sheaf of germs of holomorphic functions which are hom(n).

$\mathcal{O}^{A'}(n)$ ... the sheaf of germs of primed spinor-valued holomorphic functions, hom(n).

$\mathcal{O}_A(n)$ ... the sheaf of germs of unprimed spinor-valued holomorphic functions, hom(n).

$\hat{\pi}_A\mathcal{O}(n)$ ... the sheaf of germs of unprimed spinor-valued holomorphic functions which are hom(n-1) and satisfy $\hat{\pi}^A f_A = 0$. [i.e. f_A must be of the form $\hat{\pi}_A f$, with f hom(n).] Note that $\hat{\pi}_A\ \mathcal{O}(n+1)$ is a subsheaf of $\mathcal{O}_A(n)$.

$\mathcal{O}_\pi(n)$ ... the sheaf of germs of holomorphic functions of $\pi_{A'}$ alone (i.e. functions which satisfy $\hat{\pi}_A f = 0$), which are hom(n).

Higher and mixed valence analogs of these sheaves can be formed in various ways, and will be denoted using an analogous notation.

Using these sheaves a number of exact sequences can be constructed; the most "primitive" of these sequences is

$$0 \longrightarrow \mathcal{O}(-3) \xrightarrow{\alpha} \mathcal{O}^{A'}(-2) \xrightarrow{\beta} \mathcal{O}(-1) \longrightarrow 0\ , \qquad (3)$$

where α is the injection map consisting of multiplication by $\pi^{A'}$, and β is the projection map consisting of contraction with $i\pi_{A'}$; the composition $\beta\circ\alpha = 0$ follows from the trivial spinor identity $\pi^{A'}\pi_{A'} = 0$. The idea now is to form the *exact cohomology sequence*[6] associated with the sheaf sequence above, relative to a neighbourhood L of some line in projective twistor space. Since H^n (when $n \neq 1$) vanishes, over L, for these sheaves, the only non-trivial segment of the cohomology sequence turns out to be:

$$0 \longrightarrow H^1(L,\mathcal{O}(-3)) \longrightarrow H^1(L,\mathcal{O}^{A'}(-2)) \longrightarrow H^1(L,\mathcal{O}(-1)) \longrightarrow 0 \tag{4}$$

The elements of the group $H^1(L,\mathcal{O}(-3))$ correspond to local sections of the sheaf $\Phi^{A'}$ over the region of Minkowski space corresponding to L. Similarly, the elements of $H^1(L,\mathcal{O}^{A'}(-2))$ and $H^1(L,\mathcal{O}(-1))$ correspond to local sections of the sheaves $\Psi^{A'}$ and Φ_A, respectively[7]. Letting L vary, the usual construction of a sheaf can be applied to these sections, and the original Hertz potential sequence (1) is recovered.

In the case of electromagnetism, a pair of algebraic exact sequences of sheaves of twistor functions can be constructed, following the pattern suggested by sequence (3) in the spin ½ case:

$$\begin{array}{ccccccccc} 0 \longrightarrow & \mathcal{O}(-4) & \xrightarrow{\xi} & \mathcal{O}^{A'B'}(-2) & \xrightarrow{\eta} & \mathcal{O}^{A'}(-1) & \longrightarrow 0 & & \\ & & & & & \| & & & \\ & 0 \longrightarrow & & \mathcal{O}(-2) & \xrightarrow{\alpha} & \mathcal{O}^{A'}(-1) & \xrightarrow{\beta} & \mathcal{O}(0) & \longrightarrow 0 \ . \end{array} \tag{5}$$

Here the maps α and β are defined in the same manner as before; the map ξ is multiplication by $\pi^{A'}\pi^{B'}$, and η is contraction with $i\pi_{B'}$. The exact cohomology sequences obtainable from these sheaf sequences are:

$$\begin{array}{l} 0 \longrightarrow H^1(L,\mathcal{O}(-4)) \longrightarrow H^1(L,\mathcal{O}^{A'B'}(-2)) \longrightarrow H^1(L,\mathcal{O}^{A'}(-1)) \longrightarrow 0 \\ \qquad\qquad\qquad\qquad\qquad\qquad\qquad\qquad\qquad\qquad \| \\ 0 \longrightarrow H^0(L,\mathcal{O}(0)) \longrightarrow H^1(L,\mathcal{O}(-2)) \longrightarrow H^1(L,\mathcal{O}^{A'}(-1)) \longrightarrow H^1(L,\mathcal{O}(0)) \rightarrow 0 \end{array} \tag{6}$$

Now we have a little surprise: notice the way in which the exact cohomology sequence picks up the extra term $H^0(L,\mathcal{O}(0))$, since this group does *not* vanish. In fact, $H^0(L,\mathcal{O}(0)) \simeq C$, and the extra term arising here corresponds precisely to the additional freedom which appears in diagram (2). The cohomology groups appearing in diagram (6) give rise to the spaces of local sections of the sheaves appearing in diagram (2). Essentially the same sort of analysis goes through for higher spins.

Notes

(1) See section 4 ("Zero Rest-Mass Potentials") in Penrose (1965) for a proof of the local existence theorem for Hertz-type potentials.

(2) In fact, as indicated in Penrose (1965), a sufficient condition for the existence of a Hertz potential is that the region on which the field ϕ_A is defined should have vanishing first and second homotopy groups (i.e. be simply connected and be such that any 2-sphere can be shrunk to a point.) Thus, if M is any such region then the exact sequence (1) can be modified into the following *exact sequence of space of sections* over M:

$$0 \longrightarrow \Gamma(M,\Phi^{A'}) \longrightarrow \Gamma(M,\Psi^{A'}) \longrightarrow \Gamma(M,\Phi_A) \longrightarrow 0.$$

(3) This relation implies that the potential $\psi^{A'}{}_B$ is divergence-free (a gauge condition), and is purely 'left-handed', that is $\nabla^{B(A'}\psi^{B')}{}_B = 0$.

(4) It is not difficult to see that for each spin s a diagram similar to (2) is obtained. Each such diagram is composed of a set of 2s exact sequences, and shows a "cascade" of 2s various potential fields leading down to the basic z.r.m. field. Each potential has a certain amount of gauge freedom at its disposal, and this freedom is reflected in the

structure of its accompanying exact sequence. [Cf. Penrose (1965), footnote on p.168.]

(5) Since the sheaves appearing here all involve *homogeneous* functions, they can be regarded as being defined (a) directly on twistor space, as sheaves of homogeneous functions, or (b) on projective twistor space, *as sheaves of germs of holomorphic cross-sections of certain bundles* (labeled by the integer n). For an explicit account, see Griffiths and Adams (1974), pp. 42-44. This reference also contains, incidently, on pp. 51-55 a proof of the useful result that $H^q(P^r, \mathcal{O}(n))$ vanishes, unless $q = 0$ and $n \geq 0$, or $q = r$ and $n \leq -r-1$; and that in these cases $H^0(P^r, \mathcal{O}(n))$ is isomorphic with the *space of polynomials of degree n* in the homogeneous coordinates for P^r, and $H^r(P^r, \mathcal{O}(n))$ is isomorphic with the *dual* of $H^0(P^r, \mathcal{O}(-r-n-1))$.

(6) If a sequence of sheaves $0 \longrightarrow A \xrightarrow{\alpha} B \xrightarrow{\beta} C \longrightarrow 0$ over a space X is exact, then it follows as a well-known theorem that there exists a mapping $\delta^*: H^r(X,C) \to H^{r+1}(X,A)$ called the "connecting homomorphism", and that the sequence $\cdots \xrightarrow{\delta^*} H^r(X,A) \xrightarrow{\alpha^*} H^r(X,B) \xrightarrow{\beta^*} H^r(X,C) \xrightarrow{\delta^*} H^{r+1}(X,A) \xrightarrow{\alpha^*} \cdots$ is *exact*. This is the "exact cohomology sequence". [For a proof, see Spanier (1966), chapter 4, section 5. The theorem there is given for homology, rather than cohomology, but it's not difficult to make the necessary adjustments. The proof is also given in Gunning (1966), as Theorem 1, an extremely lucid and enjoyable book, which I highly recommend.] Referring back to sequence (1) now, note that, for any region *N* of Minkowski space, the associated exact cohomology sequence contains the segment:

$$0 \longrightarrow H^0(N, \Phi^{A'}) \xrightarrow{\alpha^*} H^0(N, \Psi^{A'}) \xrightarrow{\beta^*} H^0(N, \Phi_A) \xrightarrow{\delta^*} H^1(N, \Phi^{A'}).$$

Since $H^0(N,S) \simeq \Gamma(N,S)$, it follows that $H^1(N,\Phi^{A'})$ is the "obstruction" for the existence of Hertz potentials for z.r.m. fields defined on N, i.e. any field on N which doesn't admit a Hertz potential defined globally on N will have as its *image* under δ^* some *non-vanishing element* of $H^1(N,\Phi^{A'})$.

(7) In the cases of $H^1(L,\mathcal{O}(-3))$ and $H^1(L,\mathcal{O}^{A'}(-2))$, the representative cocycles are fed directly into the basic contour integral formulas (using a 'branched contour'; see §2.2). However, in the case of $H^1(L,\mathcal{O}(-1))$ what *actually* goes into the contour formula is the image of an element of $H^1(L,\mathcal{O}(-1))$ under the operation $\hat{\pi}_A$. This gives an element of $H^1(L,\hat{\pi}_A\mathcal{O}(-1))$. In fact, $H^1(L,\mathcal{O}(-1)) \simeq H^1(L,\hat{\pi}_A\mathcal{O}(-1))$, as follows from the exact sheaf sequence $0 \to \mathcal{O}_\pi(-1) \xrightarrow{i} \mathcal{O}(-1) \xrightarrow{\hat{\pi}} \hat{\pi}_A\mathcal{O}(-1) \to 0$, using the exact cohomology sequence, and the fact that $H^r(L,\mathcal{O}_\pi(-1) \simeq 0$ for all r.

REFERENCES

Griffiths, P. & Adams, J., 1974 *Algebraic and Analytic Geometry*. Princeton University Press.

Gunning, R.C. 1966 *Lectures on Riemann Surfaces*. Princeton University Press.

Penrose, R. 1965 Proc. Roy. Soc. A284, 159-203.

Spanier, E.H. 1966 *Algebraic Topology*. McGraw-Hill.

§2.6. MASSLESS FIELDS AND SHEAF COHOMOLOGY *by R. Penrose*

In §2.2 I indicated that twistor functions are really to be regarded as providing representative cocycles for sheaf cohomology, so the twistorial representation of a massless field of "helicity $\frac{n}{2}$" is an element of $H^1(X,\mathcal{O}(-n-2))$, where X is an (open) region in PT (say PT^+) swept out by the projective lines which correspond to the points of a suitable given (open) region Y in CM (say CM^+) where the field is defined. Woodhouse (§2.3) showed how the space-time field could be obtained from the $H^1(\ldots)$ element, when the $H^1(\ldots)$ element is defined by means of Dolbeault cohomology (i.e., "anti-holomorphic" forms (Morrow & Kodaira 1971)). A method due to Ward shows how to obtain the field from a Čech cohomology cocycle f_{ij} using an extension of Sparling's "splitting" formula (cf. §2.4). Ward's method essentially replaces the "branched contour integral" described in the first section. Basically, Ward's method is as follows:

Let $f_{ij}(Z^\alpha)$ be homogeneous of degree -n-2 (with $n \geq 0$) and defined on $\mathcal{U}_i \cap \mathcal{U}_j$ (with $\{\mathcal{U}_i\}$ a covering of $X \subset PT$), $Z^\alpha = (\omega^A, \pi_{A'})$, $\omega^A = ix^{AA'}\pi_{A'}$, etc. Then $\pi_{A'} \ldots \pi_{E'} f_{ij}(Z^\alpha)$ (where π appears n+1 times) is homogeneous of degree -1 in π, so ("splitting") we get (ρ_i meaning "restriction," adopting a notational device due to Hughston, where square brackets denote skew-symmetrization):

$$\pi_{A'} \ldots \pi_{E'} f_{ij}(Z^\alpha) = \rho_{[i}\psi_{j]A'\ldots E'}(x^r,\pi_{R'}) ,$$

whence

$$\phi_{jA'\ldots D'}(x,\pi) := \pi^{E'}\psi_{jA'\ldots D'E'}$$

satisfies $\rho_{[i}\phi_{j]A'\ldots D'} = 0$. Thus, due to the sheaf property $\rho_{[i}F_{j]} = 0 \Rightarrow F_j = \rho_j G$, we have

$$\phi_{jA'\ldots D'}(x,\pi) = \rho_j \phi_{A'\ldots D'}$$

and, since $\phi_{A'\ldots D'}$ is homogeneous of degree 0 in π (and global in π), it must be a function of x only. Furthermore, $\nabla^{AA'}\phi_{A'B'\ldots D'} = 0$ readily follows. The case $n < 0$ is somewhat similar.

All this goes one way only: from twistor function to space-time field. But using the method of exact homology sequences we can readily extend this to obtain (implicitly) a version of the "inverse twistor function," which goes from space-time field to twistor function. (Compare also the Bramson-Sparling-Penrose method outlined in Penrose 1975.) The following was inspired to some extent also by Hughston (§2.5), cf. also Lerner (§2.7).

Let F be the (dual) primed spin-vector bundle (excluding the section of zero spin vectors) over CM. Thus, a point of F can be labelled $(x^a, \pi_{A'})$ (with $\pi_{A'} \neq 0$)-- except for points at infinity on CM--with x^a complex. F is also a bundle over T - {0}, the fibre over Z^α being the set of linear 2-spaces containing Z^α.

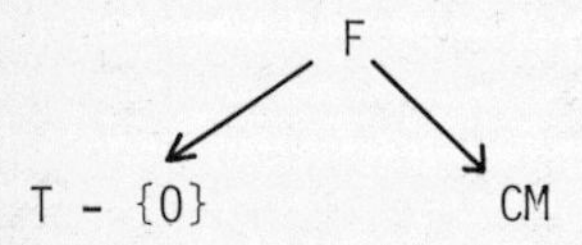

Consider the following exact sequence of sheaves over F:

$$0 \to T \xrightarrow{\times\, \pi_{A'}\cdots\pi_{E'}} Z'_{n+1} \xrightarrow{\times\, \pi^{E'}} Z'_n \to 0. \qquad (1)$$

Here Z'_n stands for the sheaf of n-index symmetric spinor fields $\phi_{A'\ldots D'}(x^r, \pi_{Q'})$ which are holomorphic in x^r and $\pi_{Q'}$ and satisfy the massless free-field equation $\nabla^{AA'}\phi_{A'B'\ldots D'} = 0$ ($\nabla_{AA'} = \frac{\partial}{\partial x^{AA'}}$, $\pi_{A'}$ being merely a passenger). The map $Z'_{n+1} \to Z'_n$ is simply

$\psi_{A'\ldots D'E'} \mapsto \phi_{A'\ldots D'} = \pi^{E'}\psi_{A'\ldots D'E'}$ and we must check that it is *onto*, i.e., that given $\phi\ldots$ we can always solve for $\psi\ldots$. One method is as follows: *First*, fix $\pi_{A'}$, and choose a primed spinor basis $o^{A'}$, $\iota^{A'}$ with $\pi^{A'} = o^{A'}$. We have $\psi_{0'\ldots 0'0'} = \phi_{0'\ldots 0'}$, $\psi_{1'0'\ldots 0'0'} = \phi_{1'0'\ldots 0'}, \ldots,$ $\psi_{1'\ldots 1'0'} = \phi_{1'\ldots 1'}$, but $\psi_{1'\ldots 1'1'}$ is fixed only by the massless field equations

$$-\frac{\partial}{\partial x_{01'}}\psi_{1'\ldots 1'1'} = \frac{\partial}{\partial x_{00'}}\psi_{1'\ldots 1'0'} \left(= \frac{\partial}{\partial x_{00'}}\phi_{1'\ldots 1'}\right)$$

and

$$-\frac{\partial}{\partial x_{11'}}\psi_{1'\ldots 1'1'} = \frac{\partial}{\partial x_{10'}}\psi_{1'\ldots 1'0'} \left(= \frac{\partial}{\partial x_{10'}}\phi_{1'\ldots 1'}\right),$$

the integrability conditions for which come simply from the massless equations on $\phi_{A'\ldots D'}$. We can ensure a unique solution locally in CM by choosing (say) $\psi_{1'\ldots 1'1'} = 0$ on an initial 2-surface $x_{11'}$ = constant, $x_{01'}$ = constant. *Second*, allow $\pi^{A'}$ to vary locally in $C^2 - \{0\}$; take $\iota^{A'}$ to depend holomorphically on $\pi^{A'}$ ($= o^{A'}$) (say $\iota^{A'} \propto$ constant). The above prescription for finding $\psi_{A'\ldots D'E'}$ ensures that it is holomorphic in $\pi_{A'}$-- though the dependence on $\pi_{A'}$ may be complicated owing to the fact that the functional dependence of $\phi\ldots$ on the components $x_{\underset{\sim}{A}0'}$ is π-dependent. (Incidentally, for simplicity, choose the *unprimed* basis *constant*.)

To see what T is we examine the kernel of the above map, i.e., find ψ's for which $\pi^{E'}\psi_{A'\ldots D'E'} = 0$ ($\pi^{E'} \neq 0$). We have (by standard lemmas) $\psi_{A'\ldots E'} = \pi_{A'}\ldots\pi_{E'}f$, and the massless equations on $\psi\ldots$ become

$\pi^{A'}\nabla_{AA'}f = 0$. Thus the dependence of f on x^a is only through the quantity $\omega^A = ix^{AA'}\pi_{A'}$ (i.e., f is constant on twistor planes in CM), so $f = f(\omega^A, \pi_{A'})$ $= f(Z^\alpha)$ is a *twistor function*. We see that T is the sheaf of twistor-type functions on F; equivalently, of holomorphic functions on T - {0}.

The short exact sequence (1) gives rise to the long exact homology sequence:

$$\ldots \to H^0(Q, Z'_{n+1}) \to H^0(Q, Z'_n) \to H^1(Q, T) \to H^1(Q, Z'_{n+1}) \to \ldots \tag{2}$$

Now, restrict Z'_n to be homogeneous of degree 0 in $\pi_{A'}$ (sheaf $Z'_n(0)$); correspondingly, Z'_{n+1} to be homogeneous of degree -1 in $\pi_{A'}$ (sheaf $Z_{n+1}(-1)$); so that T will be homogeneous of degree -n-2 in $\pi_{A'}$ (sheaf $T(-n-2)$), i.e., of degree -n-2 in Z^α. Choose $Q \subset F$ to correspond to the region of F lying above $Y \subset CM$ (so, for each $(x^a, \pi_{A'}) \in Q$ we have $(x^a, \hat{\pi}_{A'}) \in Q$ whenever $\hat{\pi}_{A'} \in C^2 - \{0\}$). Then since H^0 means "global sections" and since Z'_{n+1} is homogeneous of degree -1 in $\pi_{A'}$ we have

$$H^0(Q, Z'_{n+1}(-1)) = 0.$$

To study $H^1(Q, Z'_{n+1})$ we can use, for example, a "resolution" of Z'_{n+1} (where dependence on $\pi_{A'}$ is irrelevant)

$$0 \to Z'_{n+1} \to F_{0,n+1} \to F_{1,n} \to F_{0,n-1} \to 0, \tag{3}$$

this being exact, where $F_{p,q}$ is the sheaf of fields $\chi^{A\ldots C}_{D'\ldots F'}(x^r, \pi_{R'})$ with p symmetric unprimed indices and q symmetric primed ones (freely holomorphic), so $F_{p,q}$ is a coherent analytic sheaf, yielding $H^r(S, F_{p,q}) = 0$ whenever S is Stein and $r > 0$. The maps are:

$$\phi_{A'\ldots E'} \mapsto \chi_{A'\ldots E'} = \phi_{A'\ldots E'},$$

$$\chi_{A'B'\ldots E'} \mapsto \theta^{A}{}_{B'\ldots E'} = \nabla^{AA'}\chi_{A'B'\ldots E'},$$

$$\theta^{A}{}_{B'C'\ldots E'} \mapsto \eta_{C'\ldots E'} = \nabla^{B'}_{A}\theta^{A}{}_{B'C'\ldots E'},$$

and exactness is not hard to verify. If n=0, the sequence terminates one step sooner (and if n=-1 we also have a short exact sequence

$$0 \to Z'_0 \xrightarrow{i} F_{0,0} \xrightarrow{\Box} F_{0,0} \longrightarrow 0$$

where $\Box$ is the D'Alembertian). Note that (3) is conformally invariant, the conformal weights of $\phi\ldots$, $\chi\ldots$, $\theta\ldots$, $\eta\ldots$ being, respectively, -1, -1, -3, -4. The sequence (3) (for which $\pi_{A'}$ is irrelevant) is of interest in the study of massless fields quite independently of twistor theory. Here we just use it to derive the fact that

$$H^1(Q, Z'_{n+1}) = 0$$

whenever Y is Stein (as is the case if Y is the future- (or past-) tube in CM, or if Y is a suitable "thickening" of a portion of the real space-time M. (This follows because $H^1(Q, F_{0,n+1}(-1))$ then vanishes from the homogeneity degree -1 in $\pi_{A'}$, the fact that $H^1(CP^1, \mathcal{O}(-1)) = 0$ and suitable general theorems--for which thanks are due to R.O. Wells--and because $H^0(Q, F_{0,n+1}(-1)/Z'_n(-1))$ also vanishes--more obviously, because of -1 degree homogeneity.)

The upshot of this is that since the first and last terms in (2) both vanish (Y Stein), we have (for $n \geq 0$)

$$H^0(Q, Z'_n(0)) \cong H^1(Q, T(-n-2))$$

which, because globality in π on the left-hand side implies constancy in π, almost establishes the required isomorphism between massless fields and H^1's of twistor functions. The remaining essential problem (the non-triviality and solution of which have been pointed out to me by M. Eastwood) is that all the sheaves refer to the space F, so far, and some subtleties are involved in projecting the cohomology groups down to T. The details of this and other matters are left to a proposed later paper to be written jointly with M. Eastwood and R.O. Wells, Jr. In fact the argument can be shown to work not only for the future tube (i.e., for $X = PT^+$) but also for suitable open regions in M. (Another subtlety has been glossed over in that the (x,π) description doesn't work for points at infinity. But this is unimportant because of conformal invariance and the fact that infinity can be transformed to somewhere safely inside the singular region of the field.)

To deal with the cases $n < 0$ we need a different exact sequence:

$$0 \to \mathcal{P}_{m-2} \xrightarrow{i} \mathcal{T} \xrightarrow{\hat{\pi}_A \cdots \hat{\pi}_C} Z_{m-1} \xrightarrow{\pi^{D'} \nabla_{DD'}} Z_m \longrightarrow 0. \tag{4}$$

Here Z_m stands for the sheaf of *unprimed* massless holomorphic fields $\phi_{A\ldots D}(x^i, \pi_{R'})$ with $m = -n$ indices, and $\mathcal{P}_{m-2}$ stands for twistor functions which are polynomials in ω^A of degree at most m-2 for each fixed $\pi_{A'}$. The fact that $Z_{m-1} \to Z_m$ is onto, is basically an argument given in Penrose (1965); the kernel is this: twistor functions $\psi_{A\ldots C}(Z^\alpha)$, the massless field equations on which are $\pi^{E'} \hat{\pi}_{[D}\psi_{A]\ldots C} = 0$, yielding $\psi_{A\ldots C} = \hat{\pi}_A \cdots \hat{\pi}_C f(Z)$ as required. The short exact sequence

$$0 \to \mathcal{T}/\mathcal{P}_{m-2} \to Z_{m-1} \to Z_m \to 0$$

provides the long exact sequence

$$H^0(Q,Z_{m-1}) \to H^0(Q,Z_m) \to H^1(Q,T/P_{m-2}) \to H^1(Q,Z_{m-1}). \tag{5}$$

Choosing homogeneities -1, 0, m-2, -1, respectively, and using the same argument as above, we get (Y Stein):

$$H^0(Q,Z_m(0)) \cong H^1(Q,T(m-2)/P_{m-2}(m-2)). \tag{6}$$

To deal with the extra complication of P_{m-2} we use $0 \to P_{m-2} \to T \to T/P_{m-2} \to 0$ to derive the exact sequence

$$H^1(Q,P_k(k)) \to H^1(Q,T(k)) \to H^1(Q,T(k)/P_k(k)) \to H^2(Q,P_k(k)) \tag{7}$$

(where $k = m-2$). Now the sheaf $P_k(k)$ vanishes unless $k \geq 0$ (i.e., $m \geq 2$) and, using a suitable Kunneth formula, we derive the fact (since homogeneity in $\pi_{A'}$ for each x is non-negative) that we are concerned only with polynomial behaviour in $\pi_{A'}$ for the two end terms in (7). Consequently these terms refer, *in effect*, to the "ordinary" cohomology of Q but with coefficients which are twistor polynomials $Z^\alpha \ldots Z^\gamma P_{\alpha\ldots\gamma}$, homogeneous of degree k, i.e., equivalently, symmetric n-twistors $P_{\alpha\ldots\gamma}$. If Y describes a space-time region surrounding a "source tube," then we have $\frac{1}{6}(k+1)(k+2)(k+3) = \frac{1}{6}m(m^2-1)$ independent complex "charges," corresponding to the various independent components of $P_{\alpha\ldots\gamma}$, all of which have to *vanish* if the $H^1(Q,T)$ description is to work. This corresponds to the final map in (7) mapping to the zero element. (Note that m=2 for the anti-self-dual Maxwell case, which is consistent with a statement made in §2.2 about charge integrality). If Y is the forward tube CM^+, then the first and last terms of (7) both vanish and the required isomorphism follows (but various details remain to be worked out).

Alternative approaches to the cases $n < 0$ can be given in which a potential rather than the field is used. This avoids having to bring in P_k and $P_{\alpha\ldots\gamma}$. (Work by Ward, Hughston and Eastwood.)

Further work is in progress.

Thanks are due to M.F. Atiyah, M. Eastwood, R.O. Wells, Jr. and R.S. Ward.

References

Morrow, J. & Kodaira, K. 1971 *Complex Manifolds*. Holt, Rinehart & Winston.

Penrose, R. 1965 Proc. Roy. Soc. A284, 159-203.

Penrose, R. 1975 *Twistor Theory, its Aims and Achievements*. In: *Quantum Gravity*, eds. C.J. Isham, R. Penrose & D.W. Sciama, pp. 268-407. Clarendon: Oxford.

§2.7 THE INVERSE TWISTOR FUNCTION FOR POSITIVE FREQUENCY FIELDS *by D.E.Lerner*

A normalizable positive frequency ZRM field of helicity n/2 can be represented as

$$\phi_{A'\ldots M'}(x) = \frac{1}{2\pi i}\int_{V^+} k_{A'}\ldots k_{M'} f(k_{C'},\bar{k}_C)\exp\{-i\bar{k}_D k_{D'} x^{DD'}\} d(\Delta k \wedge \Delta\bar{k}) \qquad (1)$$

where V^+ is the forward null cone in momentum space and x ranges over the future tube M^+. The function f is in $L^2(C^2, d\mu)$ and satisfies $f(e^{i\theta}k_{C'}, e^{-i\theta}\bar{k}_C) = e^{-in\theta}f(k_{C'},\bar{k}_C)$. By integrating along the generators of V^+, we can construct a weakly $\bar{\partial}$-closed (0,1) form on the top half of twistor space; from this we can construct the element of $H^1(PT^+, \mathcal{O}(-n-2))$ which generates $\phi_{A'\ldots M'}$:

Choose an arbitrary cut of V^+ and a variable spinor $\pi_{A'}$ such that the cut is given by $\bar{\pi}_A \pi_{A'}$. A general point on V^+ can then be written as $\bar{k}_A k_{A'} = r\bar{\pi}_A \pi_{A'}$, $r > 0$; we have $d(\Delta k \wedge \Delta\bar{k}) = 2r dr \wedge \Delta\pi \wedge \Delta\bar{\pi}$, and (1) becomes

$$\phi_{A'\ldots M'}(x) = \frac{1}{2\pi i}\int_\Omega\int \pi_{A'}\ldots\pi_{M'}\int_0^\infty 2r^{(n+2)/2} f(r^{1/2}\pi_{C'}, r^{1/2}\bar{\pi}_C)$$

$$\exp\{-ir\bar{\pi}_D\pi_{D'}x^{DD'}\} dr\, \Delta\pi \wedge \Delta\bar{\pi}\,, \qquad (2)$$

or

$$\phi_{A'\ldots M'}(x) = \frac{1}{2\pi i}\int_\Omega\int \pi_{A'}\ldots\pi_{M'} F(\chi, \pi_{C'}, \bar{\pi}_C)\Delta\pi \wedge \Delta\bar{\pi}\,. \qquad (3)$$

It is easy to see that $F(\chi, \lambda\pi_{C'}, \bar{\lambda}\bar{\pi}_C) = \lambda^{-n-2}\bar{\lambda}^{-2}F(\chi, \pi_{C'}, \bar{\pi}_C)$, so the right hand side of (3) is independent of the original choice of cut. Moreover, we clearly have

$$\pi_{A'}\nabla^{AA'}F = 0; \quad \bar{\nabla}^{AA'}F = 0. \qquad (4)$$

Now define $\Psi := -F(\chi, \pi_{C'}, \bar{\pi}_C)\Delta\pi$. Because of (4), Ψ is the pull-back (to the primed spin bundle over M^+) of a (0,1) form $\Phi = -F(Z^\alpha,\bar{Z}_0,\bar{Z}_1)I^{\alpha\beta}\bar{Z}_\alpha d\bar{Z}_\beta$ defined on T^+. Since Φ is homogeneous of degree 0 in $\bar{Z}_\alpha$, it may be thought of as a distributional (0,1) form on PT^+ with values in the line bundle $L(-n-2)$.

By virtue of the specific functional form of Φ, together with the fact that $\bar{Z}_\alpha\, \partial F/\partial\bar{Z}_\alpha = -2F$ (as distributions), it follows that Φ is weakly $\bar{\partial}$-closed.

Now, by the generalized Dolbeault lemma (Serre 1955), Φ is locally $\bar{\partial}$-exact. That is, there exists an open cover $\{U_i\}$ of PT^+ and distributional cross-sections ϕ_i of $L(-n-2)|_{U_i}$ such that $\bar{\partial}\phi_i = \Phi$ in U_i.

Define $\phi_{ij} := \phi_i - \phi_j$ in each non-empty double intersection $U_i \cap U_j$. Clearly $\bar{\partial}\phi_{ij} = 0$ in $U_i \cap U_j$, so ϕ_{ij} is a weak solution to the Cauchy-Riemann equations. But any weak solution to $\bar{\partial}f = 0$ is actually holomorphic (Hörmander 1973), and thus ϕ_{ij} is a cross-section of $\mathcal{O}(-n-2)|_{U_i \cap U_j}$. Since the cocycle condition is satisfied in any non-empty triple intersection, the collection $\{\phi_{ij}\}$ determines an element of $H^1(PT^+,\mathcal{O}(-n-2))$.

It is not difficult, using Stokes' theorem, to see that

$$\phi_{A'\ldots M'}(x) = \frac{1}{2\pi i}\ \Sigma \int_{\Gamma_{ij}} \pi_{A'}\ldots\pi_{M'}\ \phi_{ij}(ix^{AA'}\pi_{A'},\pi_{A'})\Delta\pi.$$

Remarks

(1) The "general" element of $H^1(PT^+,\mathcal{O}(-n-2))$ is not obtained by the construction given above (which refers only to the Hilbert space of normalizable fields). Non-normalizable elements of $H^1(PT^+,\mathcal{O}(-n-2))$ can be constructed by applying essentially the same arguments to suitably spin-weighted distributions (and possibly even hyperfunctions) on C^2, but it is not yet clear that *all* of $H^1(PT^+,\mathcal{O}(-n-2))$ can be realized in this fashion, although it seems likely.

(2) Let $\phi_{A'...M'}$ be C^{1+} on *real* M^4. If ϕ admits a Fourier decomposition $\phi = \phi^+ + \phi^-$ (which it will if, e.g., it is continuous on I), then we can, using the above techniques, associate with ϕ an element of $H^1(PT\backslash PN, \mathcal{O}(-n-2))$.

REFERENCES

Hörmander, L. 1973 *Complex Analysis in Several Variables*, p.141. North-Holland.

Serre, J.P. 1955 Comm. Math. Helv. 29, 9-26.

§2.8 SHEAF COHOMOLOGY AND AN INVERSE TWISTOR FUNCTION *by R.S. Ward*

In §2.6, Penrose discussed massless fields and sheaf cohomology for both negative and positive homogeneity twistor functions. This note provides an alternative treatment of the positive homogeneity case.

We work in the forward tube CM^+. Let F^+ be the primed spin-bundle over CM^+, with coordinates $(x^a, \pi_{A'})$; F^+ is a bundle over both CM^+ and T^+ (see §2.6). Let $\mathcal{H}(n)$ $(n \geq -1)$ be the sheaf of holomorphic functions on F^+, homogeneous of degree n in $\pi_{A'}$. Define the differential operator $\mathcal{D}_A$ by $\mathcal{D}_A = \pi^{A'}\nabla_{AA'}$; then the subsheaf $\mathcal{O}(n) = \{f \in \mathcal{H}(n) | \mathcal{D}_A f = 0\}$ is in effect the sheaf of holomorphic functions on T^+, homogeneous of degree n. Define a third sheaf $\mathcal{Q}_A(n)$ by $\mathcal{Q}_A(n) = \{\psi_A(x^a, \pi_{A'}) | \mathcal{D}^A \psi_A = 0$, ψ_A homogeneous of degree n in $\pi_{A'}\}$. It is not difficult to check that these three sheaves fit into a short exact sequence

$$0 \to \mathcal{O}(n) \to \mathcal{H}(n) \xrightarrow{\mathcal{D}_A} \mathcal{Q}_A(n+1) \to 0.$$

This leads to a long exact sequence of cohomology groups:

$$0 \to H^0(\mathcal{O}(n)) \to H^0(\mathcal{H}(n)) \to H^0(\mathcal{Q}_A(n+1)) \to H^1(\mathcal{O}(n)) \to H^1(\mathcal{H}(n)) \to \ldots, \quad (1)$$

where the base space in each H^r is understood to be F^+. Let us investigate each of the groups in (1).

(i) $H^0(\mathcal{O}(n))$ is the space of twistor polynomials of degree n. In space-time it can be represented as

$$T_n = \{\mu_{A'\ldots C'}(x) | \nabla_{A(A'}\mu_{B'\ldots D')} = 0, \ \mu \text{ symmetric with n indices}\}.$$

(ii) $H^0(\mathcal{H}(n))$ is clearly isomorphic to

$$\Lambda_n = \{\lambda_{A'\ldots C'}(x) | \lambda \text{ holomorphic on } CM^+ \text{ and symmetric with n indices}\}.$$

(iii) An element of $H^0(\mathcal{Q}_A(n+1))$ has the form $\psi_{AA'\ldots C'}\pi^{A'}\ldots\pi^{C'}$, where $\psi_{AA'\ldots C'}(x)$ is holomorphic on CM^+, symmetric in its n+1 primed indices, and satisfies

$$\nabla_{A(A'}\psi^{A}_{B'\ldots D')} = 0. \tag{2}$$

Denote the group of such $\psi_{AA'\ldots C'}$ by Ψ_n.

(iv) $H^1(\mathcal{O}(n))$ is just $H^1(PT^+, \mathcal{O}(n))$.

(v) $H^1(\mathcal{H}(n)) = 0$.

For n = -1, take $T_n = 0$, $\Lambda_n = 0$.

Putting all this together, the sequence (1) becomes

$$0 \to T_n \xrightarrow{i} \Lambda_n \xrightarrow{\sigma_1} \Psi_n \xrightarrow{\sigma_2} H^1(PT^+, \mathcal{O}(n)) \to 0, \tag{3}$$

where i is the obvious injection, the map σ_1 is given by

$$\sigma_1 : \lambda_{A'\ldots C'} \mapsto \psi_{AA'B'\ldots D'} = \nabla_{A(A'}\lambda_{B'\ldots D')}, \tag{4}$$

and σ_2 will be described below. But first we observe that if we define Φ_n to be the group of helicity $-\tfrac{1}{2}(n+2)$ massless fields on CM^+, i.e. $\Phi_n = \{\phi_{A\ldots C}$ on CM^+, n+2 indices$|\ \nabla^{AA'}\phi_{A\ldots C} = 0\}$, then the sequence

$$0 \to T_n \xrightarrow{i} \Lambda_n \xrightarrow{\sigma_1} \Psi_n \xrightarrow{\sigma_3} \Phi_n \to 0 \tag{5}$$

is exact, the map σ_3 being defined by

$$\sigma_3 : \psi_{AA'\ldots C'} \mapsto \phi_{AB\ldots D} = \nabla^{B'}_{(B}\ldots\nabla^{D'}_{D}\ \psi_{A)B'\ldots D'}, \tag{6}$$

with n+1 derivatives on the right hand side.

[Note that the sequence (5) is conformally invariant.] Comparing (3) and (5) yields the isomorphism

$$\Phi_n \cong H^1(PT^+, \mathcal{O}(n)).$$

So we see that the map σ_2 is in effect an inverse twistor function: given a massless field $\phi_{A\ldots C}$, first find a potential $\psi_{AA'\ldots C'}$ and then apply σ_2 to give a twistor cocycle. One can describe σ_2 explicitly using a technique pioneered by G.A.J. Sparling: the construction is as follows.

Given $\psi_{AA'\ldots C'}$ satisfying (2), we want a cover $\{\mathcal{U}_j\}$ of PT^+ and twistor functions f_{jk} on $\mathcal{U}_j \cap \mathcal{U}_k$. Let $\{\mathcal{U}_j\}$ be a locally finite cover with the property that for each $\mathcal{U}_j$ there exists a point Y_j^α in PT^+ such that

$$Z^\alpha \in \mathcal{U}_j \implies \text{the line joining } Y_j^\alpha \text{ and } Z^\alpha \text{ lies entirely in } PT^+. \qquad (7)$$

Suppose now that $Z^\alpha \in \mathcal{U}_j \cap \mathcal{U}_k$. In CM^+ the setup is that Y_j^α, Y_k^α and Z^α are represented by α-planes Y_j, Y_k and Z; the points $p_j = Y_j \cap Z$ and $p_k = Y_k \cap Z$ lie in CM^+ [because of (7)]. See Fig. A. Choose some contour Γ_{jk} in Z from p_j to p_k and put

$$f_{jk}(Z) = \int_{\Gamma_{jk}} \psi_{AA'B'\ldots D'}\, \pi^{B'} \ldots \pi^{D'}\, dx^{AA'}.$$

One can now check that

(a) f_{jk} is a holomorphic 1-cocycle and hence determines an element F of $H^1(PT^+, \mathcal{O}(n))$;

(b) the choice of Γ_{jk} doesn't affect f_{jk};

(c) a gauge transformation $\psi_{AA'\ldots C'} \mapsto \psi_{AA'B'\ldots D'} + \nabla_{A(A'}\lambda_{B'\ldots D')}$ changes f_{jk} but does *not* change F.

(d) If σ_4: $H^1(PT^+, \mathcal{O}(n)) \to \Phi_n$ is the usual evaluation map, then $\sigma_4 \circ \sigma_2 = \sigma_3$.

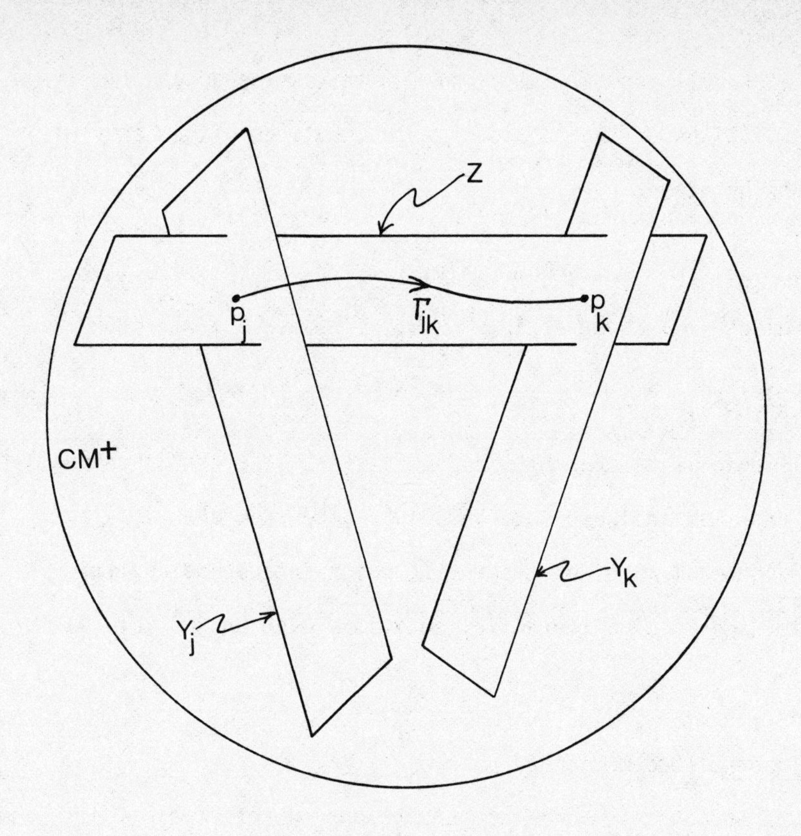

Figure A

§2.9 SOME COHOMOLOGICAL ARGUMENTS APPLICABLE TO TWISTOR THEORY

by M.G. Eastwood

(0) Introduction. Recall from §2.6 that if Z'_n denotes the space of self-dual holomorphic solutions of the massless field equations of helicity ½n on CM^+, i.e. in spinor notation

$$Z'_n = \{\phi_{A'B'\ldots D'} : CM^+ \to C \text{ holomorphic, symmetric in } A',B',\ldots,D', \text{ n indices and satisfying } \nabla^{AA'}\phi_{A'B'\ldots D'} = 0\},$$

then $Z'_n \cong H^1(PT^+, \mathcal{O}(-n-2))$.

The aim of this note is to present this result (also the corresponding isomorphism with unprimed indices [§§2.6, 2.8] and the wave equation) as special cases of a general machine (theorem 2) which interprets analytic cohomology on PT^+ in terms of differential equations over on CM^+ [cf. Wells 1979].

(1) Some Sheaf Theory [Godement 1964]

A. *Direct Image:* Suppose $\pi: E \to X$ is a continuous map of topological spaces and that S is a sheaf on E. We may construct the *direct image sheaves* $\pi^q_* S$ on X for $q \geq 0$ by means of the presheaves $V \mapsto H^q(\pi^{-1}(V), S)$; and $H^p(E,S)$ may be computed (roughly speaking) in terms of the cohomology of these direct image sheaves via the *Leray spectral sequence*

$$E_2^{p,q} = H^p(X, \pi^q_* S) \Longrightarrow H^{p+q}(E,S).$$

In (2) the conclusions which follow from this spectral sequence will be provable directly and easily without it, so we make no attempt to explain it here.

B. *Pull-back:* If $\pi: E \to X$ is a holomorphic map of complex manifolds and H is an analytic sheaf on X, then we may form the *pull-back* $\pi^* H$, an analytic sheaf on E. We shall only need the special case where H is locally free, i.e. H is isomorphic to $\mathcal{O}(V)$, the sheaf of germs of holomorphic sections of a holomorphic vector bundle V; and in this case $\pi^*\mathcal{O}(V) \cong \mathcal{O}(\pi^*V)$, where π^*V is the usual pull-back of V as a vector bundle.

C. *Resolutions:* If S is a sheaf on some topological space X, a *resolution* of S is a sequence $0 \to R^0 \to R^1 \to R^2 \to \ldots$, together with a map $\varepsilon: S \to R^0$ so that the sequence $0 \to S \to R^0 \to R^1 \to \ldots$ is exact. For shorthand we write $0 \to S \to R^{\cdot}$. If we know the cohomology of the terms of a resolution of S, then we may (roughly speaking) compute the cohomology of S via a special case of the *hypercohomology spectral sequence*, namely:

$$E_1^{p,q} = H^q(X,R^p) \Longrightarrow H^{p+q}(X,S).$$

If the resolution is acyclic, in the sense that $H^q(X,R^p) = 0$ for all $q \geq 1$, $p \geq 0$ (e.g. de Rham resolution of R, Dolbeault resolution of $\mathcal{O}$), then this gives the usual result that

$$\text{(a)} \quad H^p(X,S) \cong H^p(\Gamma(X,R^{\cdot})).$$

In (2) we will encounter a resolution with the property that $H^q(X,R^p) = 0$ for all $q \geq 2$, $p \geq 0$. In this case the spectral sequence reduces to an isomorphism $E^{00} \cong \Gamma(X,S)$ and a long exact sequence

$$0 \to E^{10} \to H^1(X,S) \to E^{01} \to E^{20} \to H^2(X,S) \to E^{11} \to E^{30} \to H^3(X,S) \to \ldots,$$

$$\text{where } E^{p,q} = H^p(H^q(X,R^{\cdot})) \equiv \frac{\ker: H^q(X,R^p) \to H^q(X,R^{p+1})}{\text{im}: H^q(X,R^{p-1}) \to H^q(X,R^p)}.$$

This can easily be proved by diagram chasing and in case $\Gamma(X,R^p) = 0$ for all $p \geq 0$ we conclude that

$$\text{(b)} \quad H^p(X,S) \cong H^{p-1}(H^1(X,R^{\cdot})).$$

Finally if $H^q(X,\mathcal{R}^p) = 0$ except for $(p,q) = (0,1)$ or $(2,0)$, then

$$\text{(c)}\quad H^1(X,S) \cong \ker: H^1(X,\mathcal{R}^0) \to \Gamma(X,\mathcal{R}^2);$$

(a), (b) & (c) are easily proved directly and are the only cases needed in (3).

(2) <u>Constructing the Machine</u>. Let CM^+, F^+ and PT^+ be as usual, with the analytic projections α and β:

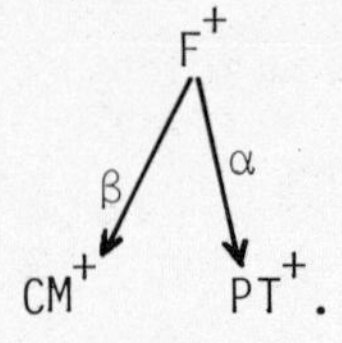

CM^+ may be regarded as a convex open subset of $C^{2\times 2}$, the space of 2×2 complex matrices $z^{AA'}$. Explicitly, under the linear isomorphism

$$(x^a + i\,y^a) = (z^a) \longmapsto \begin{bmatrix} z^0 + z^1 & z^2 + i\,z^3 \\ z^2 - i\,z^3 & z^0 - z^1 \end{bmatrix},$$

$CM^+ = \{(z^a) \text{ such that } -y^0 > ((y^1)^2 + (y^2)^2 + (y^3)^2)^{\frac{1}{2}}\} = R^4 \times \text{past cone}$. In particular CM^+ is Stein.

$F^+ \cong CM^+ \times P^1$ (holomorphically) and β is then just projection. We use the standard notation $\pi_{A'} = (\pi_{0'},\pi_{1'})$ for homogeneous coordinates on P^1 and $(\omega^A,\pi_{A'}) = (\omega^0,\omega^1,\pi_{0'},\pi_{1'})$ for homogeneous coordinates on PT and in particular PT^+. α then has the form $\alpha(z^{AA'},\pi_{A'}) = (i\,z^{AA'}\,\pi_{A'},\pi_{A'})$. To investigate α we introduce a holomorphic affine bundle $E \xrightarrow{\gamma} PT^+$ of rank 2 as follows. Let

$$\mathcal{V}^+_{0'} = \{(\omega^A,\pi_{A'}) \in PT^+ \text{ such that } \pi_{0'} \neq 0\},$$
$$\mathcal{V}^+_{1'} = \{(\omega^A,\pi_{A'}) \in PT^+ \text{ such that } \pi_{1'} \neq 0\};$$

(these cover PT^+); and define $g_{1'0'}: V^+_{1'} \cap V^+_{0'} \to$ {Invertible affine functions: $C^2 \to C^2$} by

$$g_{1'0'}(\omega^A,\pi_{A'}) \begin{bmatrix} s \\ t \end{bmatrix} = \begin{bmatrix} (\pi_{0'})^{-1}(-i\omega^0 - s\,\pi_{1'}) \\ (\pi_{0'})^{-1}(-i\omega^1 - t\,\pi_{1'}) \end{bmatrix} .$$

Use this as a transition function to construct E. Now we have

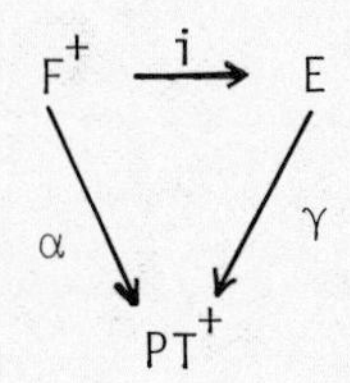

where the inclusion i is defined by

$$i\left(\begin{bmatrix} (\pi_{0'})^{-1}(-i\omega^0 - s\pi_{1'}) & s \\ (\pi_{0'})^{-1}(-i\omega^1 - t\pi_{1'}) & t \end{bmatrix}, \pi_{A'}\right) = (\omega^A,\pi_{A'}), \begin{bmatrix} s \\ t \end{bmatrix} \in V^+_{0'}\times C^2 \text{ on } F^+\big|_{V^+_{0'}},$$

$$i\left(\begin{bmatrix} u & (\pi_{1'})^{-1}(-i\omega^0 - u\pi_{0'}) \\ v & (\pi_{1'})^{-1}(-i\omega^1 - v\pi_{0'}) \end{bmatrix}, \pi_{A'}\right) = (\omega^A,\pi_{A'}), \begin{bmatrix} u \\ v \end{bmatrix} \in V^+_{1'}\times C^2 \text{ on } F^+\big|_{V^+_{1'}}.$$

The reason for the above procedure is that calculations on F^+ can be performed in coordinates on E. More precisely, we may realize $F^+\big|_{V^+_{0'}} \to V^+_{0'}$ explicitly by introducing

$$p = \omega^0/\pi_{0'},$$
$$q = \omega^1/\pi_{0'},$$
$$r = \pi_{1'}/\pi_{0'},$$

so that $V^+_{0'} \cong \{(p,q,r) \in C^3$ such that $\Phi(p,q,r) > 0\}$, where $\Phi(p,q,r) =$

$p + \bar{p} + \bar{q}r + r\bar{q}$; and then

$$\begin{array}{ccccc} F^+\big|_{V^+_{0'}} & \subset C^3 \times C^2 & \ni & \left((p,q,r), \begin{bmatrix} s \\ t \end{bmatrix}\right) \\ \downarrow & \downarrow & & \downarrow \\ V^+_{0'} & \subset C^3 & \ni & (p,q,r), \end{array}$$

where

$$\begin{bmatrix} -ip - sr & s \\ -iq - tr & t \end{bmatrix} \in CM^+ .$$

Introducing

$$G \equiv \{\begin{bmatrix} \sigma \\ \tau \end{bmatrix} = \begin{bmatrix} x_1 + i\, y_1 \\ x_2 + i\, y_2 \end{bmatrix} \in C^2 \text{ such that } -4y_2 > |\sigma|^2\},$$

we have:

<u>Lemma 1</u>:

$$\begin{array}{ccc} V^+_0 \times G & \xrightarrow[\Psi]{\simeq} & F^+\big|_{V^+_{0'}} \\ & \delta \searrow \quad \swarrow \alpha & \\ & V^+_{0'} & \end{array}$$

is commutative, where δ is projection on the first factor and

$$\Psi((p,q,r), \begin{bmatrix} \sigma \\ \tau \end{bmatrix}) = \left((p,q,r), \begin{bmatrix} \sigma - \dfrac{2\bar{r}\tau}{\Phi(p,q,r)} + i\,\bar{q} \\ \\ \dfrac{2\tau}{\Phi(p,q,r)} \end{bmatrix} \right) .$$

Proof: A messy but straightforward computation. Note that Ψ is analytic in σ and τ but only smooth in p,q and r. Thus all the fibres of α are analytically isomorphic (although $F^+ \xrightarrow{\alpha} PT^+$ is not locally trivial analytically).

Lemma 2: If $S^{open} \subset V_{0}^{+}$, is Stein then so is $\alpha^{-1}(S)$.

Proof: Using coordinates (p,q,r,s,t) as above we have:-

$$\alpha^{-1}(S) = (S \times C^2) \cap \{(p,q,r,s,t) \in C^5 \text{ s.t. } \begin{bmatrix} -ip-sr & s \\ -iq-tr & t \end{bmatrix} \in CM^+\}$$
$$= A \cap B, \text{ say.}$$

A is Stein since it's a product of such.

B is Stein since it's an inverse image of a domain of holomorphy, CM^+, under an anlytic map defined on a domain of holomorphy, C^5.

The lemma follows since an intersection of domains of holomorphy is again a domain of holomorphy. □

Corollary: If S is a coherent analytic sheaf on F^+, then $\alpha_*^q S = 0$ for all $q \geq 1$.

Proof: If $x \in V_0^+$, and S is a Stein neighbourhood of x in V_0^+, then $H^q(\alpha^{-1}(S), S) = 0$ for all $q \geq 1$ by the lemma and Cartan's theorem B. Since there are arbitrarily small Stein neighbourhoods $\alpha_*^q S_x = 0$. If $x \in V_1^+$, the argument is similar. □

Theorem 1: If S is a coherent sheaf on F^+, then

$$H^q(PT^+, \alpha_* S) \cong H^q(F^+, S) \cong \Gamma(CM^+, \beta_*^q S).$$

Proof: These isomorphisms follow by plugging information into the Leray spectral sequence of (1)A. The first uses the corollary to lemma 2 and the second uses that $H^p(CM^+, \beta_*^q S) = 0$ for all $p \geq 1$, $q \geq 0$ (which follows from theorem B plus a simple case of Grauert's direct image theorem). □

The machine is constructed by using the above theorem with specific choices of S:-

Definition: Let Ω^j_α denote the sheaf of germs of holomorphic sections of the jth exterior power of the bundle T^*_α on F^+, where T_α is the subbundle of TF^+, the holomorphic tangent bundle, consisting of these vectors tangent to the fibres of α. Thus $\mathcal{O}^0_\alpha \cong \mathcal{O}_{F^+}$ the sheaf of germs of holomorphic functions, Ω^1_α may be described by the exact sequence $0 \to \alpha^* \Omega^1_{PT^+} \to \Omega^1_{F^+} \to \Omega^1_\alpha \to 0$, and then $\Omega^2_\alpha = \Lambda^2 \Omega^1_\alpha$. Ω^j_α may be pictured more correctly using the coordinates (p,q,r,s,t) on $F^+|_{V^+_{0'}}$ as consisting of germs of holomorphic j-forms involving only the differentials ds and dt. The usual exterior derivative induces differential operators $\Omega^0_\alpha \xrightarrow{d_\alpha} \Omega^1_\alpha \xrightarrow{d_\alpha} \Omega^2_\alpha \xrightarrow{d_\alpha} 0$ ($\Omega^3_\alpha = 0$ since the fibres of α are 2-complex-dimensional). On $F^+|_{V^+_{1'}}$ we may use coordinates $(1/p, q/p, r/p, u, v)$ and on $F^+|_{V^+_{0'} \cap V^+_{1'}}$ we have

$$u = -ip - sr,$$
$$v = -iq - tr.$$

Thus

$$d_\alpha u = - r\, d_\alpha s,$$
$$d_\alpha v = - r\, d_\alpha t,$$

so that as a vector bundle Ω^1_α may be described by a transition function

$$h_{1'0'} = \begin{bmatrix} -r^{-1} & 0 \\ 0 & -r^{-1} \end{bmatrix} .$$

but $-r^{-1} = -\pi_{0'}/\pi_{1'}$, whence

$$\Omega^1_\alpha \cong \alpha^* \mathcal{O}(1)^2,$$

$$\Omega^2_\alpha \cong \alpha^* \mathcal{O}(2).$$

Notation: if K is a holomorphic vector bundle on PT^+, we will denote by K the sheaf of germs of holomorphic sections thereof. We set $S^j_K = \alpha^* K \otimes_{\mathcal{O}_{F^+}} \Omega^j_\alpha$. Thus $S^0_K \simeq \alpha^* K$ and we have the obvious inclusion $0 \to K \to \alpha_* \alpha^* K \cong \alpha_* S^0_K$.

Lemma 3: The sequence $0 \to K \to \alpha_* S^0_K \to \alpha_* S^1_K \to \alpha_* S^2_K \to 0$ is exact, i.e. $\alpha_* S^{\cdot}_K$ is a resolution of K [(1)c].

Proof: Since the statement is local we may restrict attention to a polydisc Δ in V^+_0, and suppose that K is trivial of rank 1. It will then suffice to shew that $0 \to \Gamma(\Delta, \mathcal{O}) \to \Gamma(\Delta, \alpha_* \Omega^{\cdot}_\alpha)$ is exact. We introduce the topological inverse image sheaf $\alpha^{-1}\mathcal{O}$, the sheaf of germs of holomorphic functions on F^+ constant on the fibres of α. Then our sequence becomes $0 \to \Gamma(\alpha^{-1}(\Delta), \alpha^{-1}(\mathcal{O}))$ $\to \Gamma(\alpha^{-1}(\Delta), \Omega^{\cdot}_\alpha)$. However, the sheaf sequence $0 \to \alpha^{-1}\mathcal{O} \to \Omega^{\cdot}_\alpha$ is exact. To see this, we use the usual coordinates on E and note that if we take sections over a polydisc within F^+ this is just a parametric version of the exact sequence $0 \to C \to \Gamma(P, \Omega^0) \xrightarrow{d} \Gamma(P, \Omega^1) \xrightarrow{d} \Gamma(P, \Omega^2) \to 0$ for P a polydisc in C^2 (P is a topologically trivial domain of holomorphy). Moreover by lemma 2 $\alpha^{-1}(\Delta)$ is Stein so $\Omega^{\cdot}_\alpha$ is an acyclic resolution of $\alpha^{-1}\mathcal{O}$ by theorem B. Thus $0 \to \Gamma(\alpha^{-1}(\Delta), \alpha^{-1}(\mathcal{O})) \to \Gamma(\alpha^{-1}(\Delta), \Omega^{\cdot}_\alpha)$ is exact if and only if $H^q(\alpha^{-1}(\Delta), \alpha^{-1}\mathcal{O}) = 0$ for all $q \geq 1$. This statement no longer concerns the analytic structure of F^+ and we transfer over to an analytic product $H^q(\alpha^{-1}(\Delta), \alpha^{-1}\mathcal{O}) \xleftarrow{\simeq\Psi} H^q(\delta^{-1}(\Delta), \delta^{-1}\mathcal{O})$ by lemma 1. We may compute $H^q(\delta^{-1}(\Delta), \delta^{-1}\mathcal{O})$ by using the resolution $0 \to \delta^{-1}\mathcal{O} \to \Omega^{\cdot}_\delta$ and since $\delta^{-1}(\Delta) \cong \Delta \times F$ analytically, the exactness of $D \to \Delta(\delta^{-1}(\Delta), \delta^{-1}\mathcal{O}) \to \Gamma(\delta^{-1}(\Delta), \Omega^{\cdot}_\delta)$ follows as a parametric version of $0 \to C \to \Gamma(F, \Omega^0) \xrightarrow{d} \Gamma(F, \Omega^1) \xrightarrow{d} \Gamma(F, \Omega^2) \to 0$ which is indeed exact since F is a topologically trivial domain of holomorphy. □

The machine which converts analytic cohomology on PT^+ into differential equations on CM^+ is in the form of a spectral sequence:-

Theorem 2: If K is a holomorphic vector bundle on PT^+ then there is a spectral sequence $E_1^{p,q} = T(CM^+, \beta_*^q S_K^p) \Longrightarrow H^{p+q}(PT^+, K)$ (where the differentials are induced by d_α).

Proof: Since by lemma 3 $\alpha_* S_K^{\cdot}$ is a resolution of K, we have [(1)C] a spectral sequence $E_1^{p,q} = H^q(PT^+, \alpha_* S_K^p) \Longrightarrow H^{p+q}(PT^+, K)$. But by theorem 1, $H^q(PT^+, \alpha_* S_K^p) \cong H^q(F^+, S_K^p) \cong \Gamma(CM^+, \beta_*^q S_K^p)$. □

Notes (1): $F^+ \cong M_C^+ \times P^1 = CM^+ \times [P^1\text{-north pole}] \cup CM^+ \times [P^1\text{-south pole}]$, a union of two Stein manifolds. Thus $E_1^{p,q} = 0$ for $q \geq 2$ and the spectral sequence may be interpreted as a long exact sequence as in (1)C.

(2): Regions other than CM^+, PT^+ should be amenable to these methods.

(3): One can ask - what has happened to contour integration? Actually it's lurking in β_*^1 since this involves (Serre duality) integration over P^1 [cf. §2.3].

(3) Applications. The plan is to substitute the sheaves $\mathcal{O}(n)$ into theorem 2 and see what happens. Thus we need to calculate:-

Lemma 4: Let $\odot^n C^2$ denote the nth symmetric tensor power of C^2 for $n \geq 0$ and {point} for $n \leq -1$. Then:-

$$\Gamma(CM^+,\ \beta_*^0\ S^0_{H(n)}) \cong \{\text{analytic functions} : CM^+ \to \odot^n C^2\},$$
$$\Gamma(CM^+,\ \beta_*^0\ S^1_{H(n)}) \cong \{\text{analytic functions} : CM^+ \to \odot^{n+1} C^2 \otimes_C C^2\},$$
$$\Gamma(CM^+,\ \beta_*^0\ S^2_{H(n)}) \cong \{\text{analytic functions} : CM^+ \to \odot^{n+2} C^2\},$$
$$\Gamma(CM^+,\ \beta_*^1\ S^0_{H(n)}) \cong \{\text{analytic functions} : CM^+ \to \odot^{-n-2} C^2\},$$

$$\Gamma(CM^+, \beta^1_* S^1_{H(n)}) \cong \{\text{analytic functions} : CM^+ \to \circledcirc^{-n-1}C^2 \otimes_C C^2\},$$
$$\Gamma(CM^+, \beta^1_* S^2_{H(n)}) \cong \{\text{analytic functions} : CM^+ \to \circledcirc^{-n}C^2\}.$$

Proof:

$$S^j_{H(n)} \equiv \alpha^*\mathcal{O}(n) \otimes \Omega^j_\alpha \cong \begin{cases} \alpha^*\mathcal{O}(n) \otimes \mathcal{O}_{F^+} \cong \alpha^*\mathcal{O}(n) & j = 0 \\ \alpha^*\mathcal{O}(n) \otimes \alpha^*\mathcal{O}(1)^2 \cong \alpha^*\mathcal{O}(n+1)\otimes\mathcal{O}^2, & j = 1 \\ \alpha^*\mathcal{O}(n) \otimes \alpha^*\mathcal{O}(2) \cong \alpha^*\mathcal{O}(n+2) & , j = 2. \end{cases}$$

Thus it suffices to shew:-

$$\Gamma(CM^+, \beta^0_* \alpha^*\mathcal{O}(n)) \cong \{\text{analytic functions} : CM^+ \to \circledcirc^n C^2\}$$

and $\Gamma(CM^+, \beta^1_* \alpha^*\mathcal{O}(n)) \cong \{\text{analytic functions} : CM^+ \to \circledcirc^{-n-2}C^2\}$.

Now $\alpha^*\mathcal{O}(n)$ as a bundle on F^+ is just $\mathcal{O}_{P^1}(n)$ on each P^1, when we write $F^+ \cong CM^+ \times P^1$ (and trivial in the CM^+ factor), so:-

$$\Gamma(CM^+, \beta^0_* \alpha^*\mathcal{O}(n)) \cong \{\text{analytic functions} : CM^+ \to \Gamma(P^1,\mathcal{O}(n))\}$$

and $\Gamma(P^1,\mathcal{O}(n)) \cong \{\text{homogeneous polynomials in } (\pi_{A'}) \text{ of degree } n\} \cong \circledcirc^n C^2$. Similarly $\Gamma(CM^+, \beta^1_* \alpha^*\mathcal{O}(n)) \cong \{\text{analytic functions} : CM^+ \to H^1(P^1,\mathcal{O}(n))\}$ and $H^1(P^1,\mathcal{O}(n)) \cong$ (Serre duality) $\Gamma(P^1,\mathcal{O}(-n-2)) \cong \circledcirc^{-n-2}C^2$. □

We must also compute the differentials : $E^{p,q}_1 \to E^{p+1,2}_1$ induced by d_α. This is a straightforward computation (omitted here) using the identifications made in lemma 4 and its proof. The result is a first order linear differential operator which we shall also denote by d_α. Then:-

Case 1: $n = -m-2$, $m \geq 1$. Here $E^{p,q}_1 = \Gamma(CM^+, \beta^q_* S^p_{H(n)}) = 0$ unless $q = 1$, so [(1)C(b)] $H^1(PT^+,\mathcal{O}(-m-2)) \cong \ker d_\alpha : \Gamma(CM^+, \beta^1_* S^0_{H(n)}) \to \Gamma(CM^+, \beta^1_* S^1_{H(n)}) \cong$ $\ker d_\alpha$: {analytic functions: $CM^+ \to \circledcirc^m C^2\} \to$ {analytic functions $CM^+ \to \circledcirc^{m-1}C^2 \otimes C^2\}$.
d_α in this case turns out to be $\nabla^{AA'}$, so $H^1(PT^+,\mathcal{O}(-m-2)) \cong Z'_m$. □

Case 2: $n = -2$. Here only $E_1^{0,1}$ and $E_1^{2,0}$ are non-zero, so [(1)C(c)] $H^1(PT^+, \mathcal{O}(-2)) \cong \ker: \Gamma(CM^+, \beta_*^1 S^0_{H(n)}) \to \Gamma(CM^+, \beta_*^0 S^2_{H(n)})$. This is a map on the second level of the spectral sequence and so will be a second order differential operator (just on analytic functions). In this way we obtain the wave equation. □

Case 3: $n \geq -1$. Here $E_1^{p,q} = 0$ unless $q = 0$, so using lemma 4 [and (1)C(a)] $H^1(PT^+, \mathcal{O}(n)) \cong H^1(E_1^{\cdot 0})$

$$\cong \frac{\ker d_\alpha : \{\text{analytic functions: } CM^+ \to \mathcal{O}^{n+1} C^2 \otimes C^2\} \to \{\text{analytic functions:} CM^+ \to \mathcal{O}^{n+2} C^2\}}{\text{im } d_\alpha : \{\text{analytic functions: } CM^+ \to \mathcal{O}^n C^2\} \to \{\text{analytic functions:} CM^+ \to \mathcal{O}^{n+1} C^2 \otimes C^2\}}.$$

To interpret this as massless fields (anti-self-dual helicity $\frac{1}{2}n$), we use diagram chasing with the exact sequence

$$\Gamma(CM^+, \Omega^0) \xrightarrow{d} \Gamma(CM^+, \Omega^1) \xrightarrow{d} \Gamma(CM^+, \Omega^2) \xrightarrow{d} \Gamma(CM^+, \Omega^3) \xrightarrow{d} \Gamma(CM^+, \Omega^4).$$

REFERENCES

Godement, R. 1964 *Topologie Algébrique et Théorie des Faisceaux;* Chapter II §4. Paris: Hermann.

Wells, R.O. 1979 *Complex Manifolds and Mathematical Physics;* §9. To appear in Bull. Amer. Math. Soc.

§2.10 MASSLESS FIELDS AND TOPOLOGY *by M.G. Eastwood*

Representing solutions of the zero-rest-mass field equations on CM^+ in terms of $H^1(PT^+ \mathcal{O}(n))$ or $H^1(PT^{*-}, \mathcal{O}(n))$ has been described in various ways [§§2.2, 2.6, 2.8 etc.]. For regions other than CM^+ the matter is less well understood. The following is a little speculative and *assumes* that the representation as described in §2.9 works for any Stein region X in CM. We describe here the case of the anti-self-dual Maxwell equations.

Let PX denote the region in PT corresponding to X in CM (identifying CM with $Gr(2,C^4)$, PX is just those lines within the planes in X) and similarily let PX* denote the corresponding region in PT*.

Let $\Omega^p(X)$ denote the space of holomorphic p-forms on X and recall that the 2-forms split: $\Omega^2(X) = \Omega^2_+(X) \oplus \Omega^2_-(X)$ where $\Omega^2_+(X)$ is the space of self-dual 2-forms and $\Omega^2_-(X)$ the space of anti-self-dual 2-forms. Then trivially we have the following commutative diagram:

$$\begin{array}{ccccccccccccc}
 & & 0 & & 0 & & 0 & & 0 & & 0 & & \\
 & & \uparrow & & \uparrow & & \uparrow & & \uparrow & & \uparrow & & \\
0 & \to & \Omega^0(X) & \xrightarrow{d} & \Omega^1(X) & \xrightarrow{d_+} & \Omega^2_+(X) & \to & 0 & \to & 0 & \to & 0 \\
 & & \uparrow & & \uparrow & & \uparrow{\scriptstyle \text{projection}} & & \uparrow & & \uparrow & & \\
0 & \to & \Omega^0(X) & \xrightarrow{d} & \Omega^1(X) & \xrightarrow{d} & \Omega^2(X) & \xrightarrow{d} & \Omega^3(X) & \xrightarrow{d} & \Omega^4(X) & \to & 0 \\
 & & \uparrow & & \uparrow & & \uparrow{\scriptstyle \text{inclusion}} & & \uparrow & & \uparrow & & \\
0 & \to & 0 & \to & 0 & \to & \Omega^2_-(X) & \xrightarrow{d} & \Omega^3(X) & \xrightarrow{d} & \Omega^4(X) & \to & 0 \\
 & & \uparrow & & \uparrow & & \uparrow & & \uparrow & & \uparrow & & \\
 & & 0 & & 0 & & 0 & & 0 & & 0 & &
\end{array}$$

Then it turns out that the first row of this diagram has $H^*(PX,\mathcal{O})$ as cohomology ($H^0(PX,\mathcal{O}) \cong \ker d: \Omega^0(X) \to \Omega^1(X)$ etc.) and that $H^*(PX^*,\mathcal{O}(-4))$ is the cohomology of the last row ($H^1(PX^*, \mathcal{O}(-4)) \cong \ker d: \Omega^2_-(X) \to \Omega^3(X)$ etc.). Since X is Stein, the middle row has $H^*(X,C)$ as cohomology. All the columns of the diagram are exact, so by standard homological reasoning we obtain $H^0(X,C) \cong H^0(PX,\mathcal{O})$ and an exact sequence

$$0 \to H^1(X,C) \to H^1(PX,\mathcal{O}) \to H^1(PX^*,\mathcal{O}(-4)) \to H^2(X,C) \to H^2(PX,\mathcal{O}) \to \dots$$

In case $X = CM^+$ the isomorphism just confirms that there are no non-constant holomorphic functions on PT^+ and since X is topologically trivial we obtain the usual twistor transform isomorphism

$$H^1(PT^+,\mathcal{O}) \xrightarrow{\cong} H^1(PT^{*-}, \mathcal{O}(-4)).$$

$H^2(PT^+,\mathcal{O}) = 0$ by convexity arguments or from arguing as in §2.6. (Thanks here to Lane Hughston.)

For other X it would appear that $H^1(PX^*, \mathcal{O}(-4))$ represents zero-rest-mass fields directly whereas $H^1(PX,\mathcal{O})$ gives a "Hertz potential" representation of those fields which can be so represented. In other words in the sequence $H^1(PX,\mathcal{O}) \to H^1(PX^*,\mathcal{O}(-4)) \to H^2(X,C)$ the last map should be called "charge" (Thanks here to L.P. Hughston, R. Penrose, & R.S. Ward).

For the neutrino equations the topology of X does not enter at all and for higher helicity the above method will give rise to a spectral sequence (which will degenerate to the twistor transform for $X = CM^+$).

§2.11 FURTHER REMARKS ON MASSLESS FIELDS AND COHOMOLOGY *by L.P. Hughston*

The purpose of this note is to describe a short exact sequence from which it is possible to derive, in a straightforward way, the connection between massless fields and twistor cohomology for both positive *and* negative helicities. The sequence to be described is, in fact, a special case of the sequence (1) in §2.6. It is suggested there, however, that to deal with the *negative* helicity cases we need a *different* exact sequence, viz. sequence (4) in §2.6. We shall demonstrate here that, remarkably, sequence (1) suffices for *all* helicities.

Consider the following short exact sequence of sheaves:

$$0 \to \mathcal{O}(-2S-2) \xrightarrow{\pi^{A'}} \Phi^{A'}(-2S-1) \xrightarrow{\pi_{A'}} \Phi(-2S) \to 0. \tag{1}$$

Here $\Phi^{A'}(-2S-1)$ is the sheaf of functions $\phi^{A'}(x,\pi)$ satisfying $\nabla_{AA'}\phi^{A'}(x,\pi) = 0$, homogeneous of degree $-2S-1$ in π. Similarly, $\Phi(-2S)$ is the sheaf of functions $\phi(x,\pi)$ satisfying $\Box\phi(x,\pi) = 0$, homogeneous of degree $-2S$ in π. By $\mathcal{O}(-2S-2)$ we denote the sheaf of holomorphic twistor functions, homogeneous of degree $-2S-2$. The helicity S can assume positive or negative half-integral values. We restrict x to lie in a (holomorphically and geometrically) convex and trivial region Q of complex Minkowski space (letting Q denote also the corresponding region in twistor space).

<u>Theorem</u>. *Letting S vary we obtain from the cohomology of sequence (1) above the isomorphism* $H^1(Q,\mathcal{O}(-2S-2)) \simeq H^0(Q,Z_S)$ *where* Z_S *is the sheaf of germs of solutions to the zero rest mass equations for helicity S.*

We shall simply outline the proof here by way of illustration with several examples. We require the long exact cohomology sequence obtained from sequence (1):

$$0 \to H^0 \mathcal{O}(-2S-2) \to H^0\ \Phi^{A'}(-2S-1) \to H^0\ \Phi(-2S) \to H^1 \mathcal{O}(-2S-2)$$

$$\to\ H^1\ \Phi^{A'}(-2S-1) \to H^1 \Phi(-2S) \to H^2 \mathcal{O}(-2S-2) \to H^2\ \Phi^{A'}(-2S-1) \to \ldots \quad (2)$$

For the evaluation of some of these groups we need the following useful resolutions (cf. §2.6):

$$0 \to \Phi^{A'}(-2S-1) \to F^{A'}(-2S-1) \xrightarrow{\nabla_{A'A}} F_A(-2S-1) \to 0, \quad (3)$$

$$0 \to \Phi(-2S) \longrightarrow F(-2S) \xrightarrow{\Box} F(-2S) \longrightarrow 0\ , \quad (4)$$

where $F^{A'}$, F_A, and F are (coherent) sheaves of (unrestricted) holomorphic functions of x and π, homogeneous in π of the designated degrees. (In what follows, Q is suppressed.)

S = 0 Case. From sequence (2) we obtain the segment:

$$H^0\ \Phi^{A'}(-1) \to H^0\ \Phi(0) \to H^1 \mathcal{O}(-2) \to H^1\ \Phi^{A'}(-1). \quad (5)$$

Using the cohomology of resolution (3), together with the fact that Q is Stein and the F's are coherent, we obtain $H^0\ \Phi^{A'}(-1) = 0$ and $H^1\ \Phi^{A'}(-1) = 0$, and the desired isomorphism between $H^1 \mathcal{O}(-2)$ and $H^0 Z_0$ follows immediately. Note that we must use resolution (4) in order to deduce that $H^0 \Phi(0) \simeq H^0 Z_0$.

S = 1/2 Case. From sequence (2) we obtain the segment:

$$H^0\ \Phi(-1) \to H^1 \mathcal{O}(-3) \to H^1\ \Phi^{A'}(-2) \to H^1\ \Phi(-1). \quad (6)$$

Using resolution (4) we obtain that $H^0\ \Phi(-1)$ and $H^1\ \Phi(-1)$ both vanish. Using resolution (3) we obtain that $H^1\ \Phi^{A'}(-2)$ is the space of spinor fields $\phi^{A'}(x)$ satisfying $\nabla_{AA'}\ \phi^{A'} = 0$, and we have the desired isomorphism.

S = -1/2 Case. From sequence (2) we obtain the segment:

$$H^0\mathcal{O}(-1) \to H^0\ \Phi^{A'}(0) \to H^0\ \Phi(1) \to H^1\mathcal{O}(-1) \to H^1\ \Phi^{A'}(0). \tag{7}$$

Using resolution (3) we deduce $H^1\ \Phi^{A'}(0) = 0$. We know that $H^0\mathcal{O}(-1)$ must vanish, since it vanishes when restricted down to any line. The group $H^0\ \Phi^{A'}(0)$ consists, by (3), of fields $\phi^{A'}$ satisfying $\nabla_{AA'}\ \phi^{A'} = 0$. The group $H^0\ \Phi(1)$ is, using resolution (4), the space of polynomials of degree one in π satisfying the wave equation, i.e. functions of the form $\psi^{A'}\ \pi_{A'}$ with $\Box\psi^{A'} = 0$. Accordingly, we can identify $H^0\ \Phi(1)$ with the space of *spinor* solutions of the wave equation. This shows us that $H^1\mathcal{O}(-1)$ is given by spinor solutions of the wave equation, modulo solutions of the $s = \frac{1}{2}$ massless equations. Thus, $H^1\mathcal{O}(-1)$ is (as desired) the space of solutions of the $s = \frac{1}{2}$ massless equations (see, e.g., equation (1), and the attendant remarks, in §2.5.)

Higher spins, for both positive and negative helicity fields, can be treated similarly.

§2.12 LOCAL H^1's AND PROPAGATION *by R. Penrose*

We have now become accustomed to the fact that, at least for a region $R \subset \mathbb{C}M$ that is appropriately convex (e.g. $R = \mathbb{C}M^+$, the forward tube), *massless free fields* of helicity s in R are represented, in twistor space, by the elements of the *sheaf cohomology group* $H^1(\mathcal{Q},\mathcal{O}(-2s-2))$, where $\mathcal{Q} \subset \mathbb{P}T$ is the region swept out by all the lines in $\mathbb{P}T$ that correspond to the points of R (e.g. $\mathcal{Q} = \mathbb{P}T^+$ when $R = \mathbb{C}M^+$, giving the positive frequency fields). There is an asymmetry involved in this correspondence in that $\mathcal{Q}$ is an extended region swept out by lines, whereas R can be a local region of $\mathbb{C}M$, say a small neighbourhood of a point R of $\mathbb{C}M$. However (as was pointed out to me by C.D. Hill), non-vanishing H^1's can arise also for *local* regions $\mathcal{Q}$ of $\mathbb{P}T$. What is the space-time interpretation of these local H^1's? I shall attempt to give here a partial answer to this question.

Consider, first, a somewhat more general situation, where an n-complex-dimensional Stein manifold S is divided into two open regions S^+ and S^- by a smooth (2n-1)-real-dimensional hypersurface N whose Levi form has constant signature (p,q), with $p+q = n-1$. Thus, S^+ has "q degrees of convexity" at N and S^- has "p degrees of convexity" at N (so that S^+ would be Stein if $p = 0$, and S^- Stein if $p = n-1$). Assume $pq > 0$. Then (Andreotti & Hill 1972; Hill 1971) there is a tangential Dolbeault complex for N, whose cohomology, which I here write $H^r(N,\mathcal{D})$ (= "$H^{0,r}$"), with $r = 0,1,2,\ldots$, is given by $H^0(N,\mathcal{D}) \cong H^0(S,\mathcal{O})$, $H^p(N,\mathcal{D}) \cong H^p(S^+,\mathcal{O})$, $H^q(N,\mathcal{D}) \cong H^q(S^-,\mathcal{O})$, provided that $p \neq q$; and all other $H^r(N,\mathcal{D})$'s vanish. This is assuming that we are concerned with hyperfunction* cohomology on N. Had we used, say, just C^∞ (or even distribution)

* A hyperfunction (Sato 1959, 1960) is a generalization of a distribution which is, in an appropriate sense the "dual" concept to that of a real analytic function. (Recall that a distribution is the dual concept to that of a C^∞ function.) The boundary values of holomorphic functions are always hyperfunctions. More generally, hyperfunctions represent *jumps* between such boundary values.

cohomology on N then the above isomorphisms would have to be replaced by inclusions. In the case p = q the only modification is that now we have $H^p(N,\mathcal{D}) \cong H^p(S^-,\mathcal{O}) \oplus H^p(S^+,\mathcal{O})$, so that, in effect, $H^p(N,\mathcal{D})$ naturally decomposes into a negative and positive "frequency" part. In each non-zero case, the space $H^r(N,\mathcal{D})$ (and $H^r(S^{\pm},\mathcal{O})$) is *infinite-dimensional*. The Stein manifold S can be chosen to be an arbitrarily small (pseudo-convex) neighbourhood of a given point P of a *given* analytic real hypersurface N in a complex manifold. In this sense, these ∞-dimensional H^r's may be regarded as defined *locally* on N.

We are concerned with the case N = PN, so p = q = 1, n = 3. Thus, we have local H^1's living on PN which have, in some sense, canonically defined positive and negative "frequency" parts. What do they mean in space-time terms? A point to bear in mind is that since non-analytic cohomology (in fact hyperfunction cohomology) on N is involved, the space-time interpretation must refer to *non-analytic* behaviour in the space-time fields. My contention is that these local H^1's refer to *propagation* of massless fields. Until now, a weakness of the twistor description of fields has been that such questions (e.g. domains of dependence etc.) have been almost totally ignored. We now have the potentiality to remedy this situation. But I can, as yet, give only two partial attempts at interpretation.

The problem is, of course, that $S \subset$ PT, being Stein, can contain no projective line in PT (since a line is compact and of positive dimension). Thus the "normal" interpretation of $H^1(S^{\pm},\mathcal{O}(n))$ in terms of massless fields in CM must break down. One way to get some feeling for $H^1(S^{\pm},\mathcal{O}(n))$ is to "approximate" S by taking, instead, a local neighbourhood L, of a point P $\in$ PN, where $L = A \cap B$, the open sets $A,B \subset$ PT being of the "normal" type, i.e. swept out by lines in PT corresponding to the points of two disjoint sets $\hat{A}$ and $\hat{B}$ (hats denote corresponding regions in CM). See Fig. A. Since $P \in A \cap B$, we have $\hat{P}$ passing through the two regions $\hat{A}$ and $\hat{B}$. Here $\hat{P}$ is an α-plane in CM, but

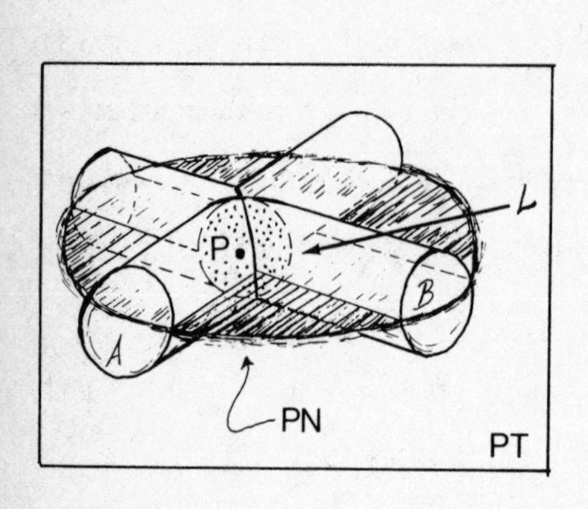

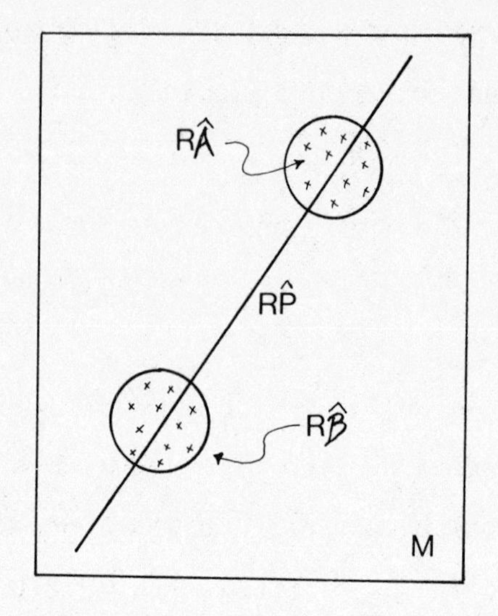

Figure A

we can equally well think in terms of R$\hat{P}$, the corresponding *null line* in M, where $\hat{A}$ and $\hat{B}$ determine subsets of M through which the null line R$\hat{P}$ passes. These two real subsets are denoted R$\hat{A}$ ($= \hat{A} \cap M$) and R$\hat{B}$ ($= \hat{B} \cap M$) (cf. Fig.A). The subset L of PN cannot be Stein since its boundary is not pseudo-convex (because the boundaries of A and B are not). But L can be *contained* in a Stein manifold S (at least if A and B are "narrow" enough, i.e. if $\hat{A}$ and $\hat{B}$ are small enough). Thus, if we can interpret $H^1(L^{\pm}, \mathcal{O})$, we shall have some sort of *approximation* to $H^1(S^{\pm}, \mathcal{O})$, where $L^{\pm} = L \cap PT^{+}$, provided that the contributions to $H^1(S^{\pm}, \mathcal{O}(n))$ from the parts of the boundary not on PN can be ignored. By application of Mayer-Vietoris sequences etc. to this situation we find that an essential part of $H^1(L \cap PN, \mathcal{D}(-2s-2))$ describes something like: massless free fields that exist in the union $\hat{RA} \cup \hat{RB}$ *modulo* those that can be extended to solutions of the field equations in the whole space M (the fields being hyperfunctional and of helicity s). This is an expression of the *influence* that fields in $\hat{RA}$ have on fields in $\hat{RB}$ (and vice-versa). We know, after all, that if the field in $\hat{RA}$ has a discontinuity or non-analyticity along a null geodesic (say $\hat{RP}$) that intersects $\hat{RB}$ as well as $\hat{RA}$, then this non-analyticity must show up also in $\hat{RB}$.

This seems to give an approximate interpretation of the local H^1's on PN in the neighbourhood of a point P of PN. As our sets A, B get narrower and narrower about P, so also do $\hat{RA}$ and $\hat{RB}$ get smaller and smaller. We may consider that in the limiting situation we are concerned simply with *propagation* along the null line $\hat{RP}$. Thus, $H^1(S^+) \oplus H^1(S^-)$, where S is small and Stein surrounding P, describes, in some sense, the *influence* that propagates along $\hat{RP}$.

There is another, more clear cut, way of investigating these local H^1's, as suggested to me by H. Grauert. Let S be a Stein neighbourhood of a point $P \in PN$, with $S^{\pm} = S \cap PT^{\pm}$, $N = S \cap PN$. To construct a non-zero element of $H^1(S^+)$ we find some analytic surface Σ in S whose intersection Σ^+ with PT^+ is Stein. (Pseudo-convexity at P is sufficient.) We can choose a simple situation, in fact, in which Σ is just the intersection of S with a plane, and if this is given by $A_\alpha Z^\alpha = 0$, where $A_\alpha \bar{A}^\alpha \geq 0$, then Σ^+ will indeed be Stein. Now choose a holomorphic function f (or twisted function f, a section of $\mathcal{O}(-2s-1)$) on Σ^+ which does not extend meromorphically beyond P into Σ^- (possible since Σ^+ is Stein). Take a Stein covering $\{U_i\}$ of S^+ (necessarily infinite since $\cup_i U_i = S^+$, which is not pseudo-convex) and let f_i be holomorphic in U_i with $f_i = f$ on Σ^+. Define $F_{ij} = (f_i - f_j)(A_\alpha Z^\alpha)^{-1}$. Then F_{ij} is holomorphic (section of $\mathcal{O}(-2s-2)$), since the f_i's agree on Σ^+, and the cocycle condition is satisfied. Now $\{F_{ij}\}$ cannot be a coboundary $F_{ij} = G_i - G_j$, because $G_i - f_i(A_\alpha Z^\alpha)^{-1}$ would then be global and meromorphic on S^+, but not extendible meromorphically beyond P - and this is impossible, since S^+ is Levi (+,-) at P.

The *twistor function* F_{ij}, having a simple pole near P, defines a *null massless free field* (Penrose 1969), taking $s \geq 1$, defined in the neighbourhood of $\hat{RP}$ in M, and which is non-analytic (possibly hyperfunctional) along $\hat{RP}$. Thus, again, we see that $H^1(S^{\pm})$ has to do with the propagation of non-analytic behaviour along $\hat{RP}$.

REFERENCES

Andreotti, A. & Hill, C.D. 1972 Ann. Scuola Norm. Sup. Pisa C1 Sci. 26, 325 & 747.

Hill, C.D. 1971 In: *Partial Differential Equations*. Proc. Sympos. Pure Math., Vol. 23, p.135. Amer. Math. Soc: Providence, R.I.

Penrose, R. 1969 J. Math. Phys. 10, 38-39.

Sato, M. 1959 J. Fac. Sci. Univ. Tokyo, Sect. I 8, 139-193.

Sato, M. 1960 J. Fac. Sci. Univ. Tokyo, Sect. I 8, 387-437.

§2.13 TWISTORS AS HELICITY RAISING OPERATORS *by R. Penrose*

Twistors have a bewildering number of different-looking interpretations, e.g. Robinson congruences, spinning massless particles (kinematics), α-planes in complexified space-time,... . In most of these there is no evident connection between the complex linearity of twistor space and quantum complex linear superposition (and, indeed, in most, the twistor phase is not even represented), whereas it has been one of the primary stated motivations behind twistor theory that twistors should provide an essential link between the complex linearity of quantum theory and the geometry of space-time. There is, however, one interpretation of a twistor, namely as a *helicity raising operator*, where the relation between twistor linearity and quantum linear superposition becomes quite apparent. The same remark applies to the interpretation of a dual twistor as a *helicity lowering operator*.

Let $f_{ij} = f_{ij}(Z^{\alpha}) = -f_{ji}$, $\rho_{[i}f_{jk]} = 0$ be a Čech cocycle for some covering $\{U_i\}$ of T^+, so that if f_{ij} has homogeneity $-n-2$ it is a twistor wave function (not necessarily normalizable) for a free massless particle of helicity $\frac{n}{2}$. Now consider a twistor P^{α} and a dual twistor Q_{α}. These provide new twistor wave functions $(P^{\alpha}\,\partial/\partial Z^{\alpha})\,f_{ij}$ and $(Q_{\alpha}Z^{\alpha})f_{ij}$ for particles of helicity $\frac{n}{2}+\frac{1}{2}$ and $\frac{n}{2}-\frac{1}{2}$, respectively. (The operations obviously apply to the cohomology classes and not just to the representatives.) Quantum linear superposition refers to the linear structure of the f's (i.e. states f_{ij} and g_{ij} superpose linearly to give $\lambda f_{ij} + \mu g_{ij}$; $\lambda,\mu \in C$), and this clearly relates to the linearity in P^{α} and Q_{α}. If the (complex) space-time field description of f_{ij} is $\psi_{\underbrace{A'B'\ldots L'}_{n}}$ $(n \geq 0)$ or $\phi_{\underbrace{AB\ldots L}_{-n}}$ $(n \leq 0)$, then that of $(P^{\alpha}\partial/\partial Z^{\alpha})f_{ij}$ is

$$-P^M\nabla_{MM'}\psi_{A'B'\ldots L'} - (n+1)P_{(M'}\psi_{A'B'\ldots L')} (n\geq 0) \text{ or } -P^L\phi_{AB\ldots KL} (n\leq -1),$$

respectively, and that of $(Q_\alpha Z^\alpha)f_{ij}$ is

$$Q^{L'}\psi_{A'B'\ldots K'L'} (n\geq 1) \text{ or } -iQ^{M'}\nabla_{MM'} - (n-1)Q_{(M}\psi_{AB\ldots L)} (n\leq 0),$$

respectively, where the pair of spinor fields $(P^A, P_{A'})$ represents P^α and the pair of spinor fields $(Q_A, Q^{A'})$ represents Q_α, with $\nabla_{AA'}P^B = -i\varepsilon_A{}^B P_{A'}$, $\nabla_{AA'}Q^{B'} = i\varepsilon_{A'}{}^{B'}Q_A$ (see Penrose 1975 and §2.14).

We investigate the kernel and quotient of these maps

$$\Phi_n \xrightarrow{P^\alpha} \Phi_{n+1}, \quad \Phi_n \xrightarrow{Q_\alpha} \Phi_{n-1},$$

where Φ_n stands for the space of helicity $\frac{n}{2}$ massless wave functions (not necessarily normalizable), i.e. equivalently, $\Phi_n = H^1(PT^+, \mathcal{O}(-n-2))$. We use the short exact sequences of sheaves over PT:

$$0 \to \mathcal{O}^P(-n-2) \to \mathcal{O}(-n-2) \xrightarrow{P^\alpha\partial/\partial Z^\alpha} \mathcal{O}(-n-3) \to 0,$$

$$0 \to \mathcal{O}(-n-2) \xrightarrow{Q_\alpha Z^\alpha} \mathcal{O}(-n-1) \to \mathcal{O}_Q(-n-1) \to 0,$$

where $\mathcal{O}^P(-n-2)$ refers to homogeneous degree $-n-2$ functions which are constant along the linear 2-spaces in T through P^α, and where $\mathcal{O}_Q(-n-1)$ refers to homogeneous degree $-n-1$ functions on the 3-space $Q_\alpha Z^\alpha = 0$, so in each case these describe twisted holomorphic functions on a CP^2. The corresponding long exact sequences are (where in each case the space under consideration is PT^+)

$$\ldots \to H^1(\mathcal{O}^P(-n-2)) \to \Phi_n \xrightarrow{P^\alpha} \Phi_{n+1} \to H^2(\mathcal{O}^P(-n-2)) \to \ldots$$

and

$$0 \to H^0(\mathcal{O}(-n-2)) \to H\ (\mathcal{O}(-n-1)) \to H^0(\mathcal{O}_Q(-n-1)) \to \Phi_n \xrightarrow{Q_\alpha} \Phi_{n-1} \to$$

$$\to \quad H^1(\mathcal{O}_Q(-n-1)) \to H^2(\mathcal{O}(-n-2)) \to \ldots .$$

The second sequence is easier to handle owing to the fact that $\mathcal{O}_Q$ refers to a subspace, whereas $\mathcal{O}^P$ refers to a factor space. There are three cases to consider, $Q_\alpha \bar{Q}^\alpha \gtreqless 0$. We have $H^p(\mathcal{O}_Q(m)) = H^p(Q^+,\mathcal{O}(m))$ where $Q^+ = Q \cap PT^+$, with $Q \subset PT$ given by $Q_\alpha Z^\alpha = 0$. When $Q_\alpha \bar{Q}^\alpha \geq 0$, Q^+ is Stein, so $H^1(\mathcal{O}_Q(-n-1))$ $= 0$ while $H^0(\mathcal{O}_Q(-n-1))$ is an infinite-dimensional space — given by the functional freedom of one holomorphic function of two variables, whence $\Phi_n \xrightarrow{Q_\alpha} \Phi_{n-1}$ is surjective:

$$Q_\alpha \bar{Q}^\alpha \geq 0\text{: } 0 \to C^{\infty^2} \to \Phi_n \xrightarrow{Q_\alpha} \Phi_{n-1} \to 0 \qquad \text{(exact)},$$

where C^{∞^2} symbolically stands for the freedom of one holomorphic function of two variables. When $Q_\alpha \bar{Q}^\alpha < 0$, then $Q^- = Q \cap PT^-$ is Stein whereas Q^+ has pseudoconcave boundary. Holomorphic functions on Q^+ extend to the whole of Q, so $H^0(Q^+,\mathcal{O}(m)) \doteq H^0(Q,\mathcal{O}(m))$. This is non-vanishing only if $m \geq 0$ (polynomials of degree m) and then we have

$$\begin{array}{l} 0 \to H^0(PT,\ \mathcal{O}(m-1)) \to H^0(PT,\ \mathcal{O}(m)) \to H^0(Q,\ \mathcal{O}(m)) \to 0 \\ \qquad\qquad\qquad\qquad\qquad\qquad\qquad\qquad\qquad\qquad \text{(exact)} \\ 0 \to H^0(PT^+,\mathcal{O}(m-1)) \to H^0(PT^+,\mathcal{O}(m)) \to H^0(Q^+,\mathcal{O}(m)) \underbrace{\to 0}_{\text{implied}} \end{array}$$

whence (since $H^2(\mathcal{O}(-n-2)) = 0$ also), $\Phi_n \xrightarrow{Q_\alpha} \Phi_{n-1}$ is injective, with

$$Q_\alpha \bar{Q}^\alpha < 0\text{: } 0 \to \Phi_n \xrightarrow{Q_\alpha} \Phi_{n-1} \to C^{\infty^2} \to 0 \qquad \text{(exact)},$$

the functional freedom in $H^1(\mathcal{O}_Q(-n-1))$ being the same as in $H^0(\mathcal{O}_Q(-n-1))$ because of an argument involving Serre duality, and the relation between the cohomology of Q^+ and $\bar{Q}^-$.

The results for $\Phi_n \xrightarrow{P^\alpha} \Phi_{n-1}$ are similar and follow from the above by complex conjugation.

$$P^\alpha \bar{P}_\alpha > 0: \quad 0 \to \Phi_n \xrightarrow{P^\alpha} \Phi_{n+1} \to C^{\infty^2} \to 0 \qquad \text{(exact)},$$

$$P^\alpha \bar{P}_\alpha \leq 0: \quad 0 \to C^{\infty^2} \to \Phi_n \xrightarrow{P^\alpha} \Phi_{n+1} \to 0 \qquad \text{(exact)}.$$

Question: How does one characterize these maps *intrinsically*, so as to provide a "definition" of a twistor?

Acknowledgement: Thanks are due to Mike Eastwood.

REFERENCE

Penrose, R. 1975 *Twistor Theory, its Aims and Achievements.* In: *Quantum Gravity*, eds. C.J. Isham, R. Penrose & D.W. Sciama; pp. 267-407. Clarendon: Oxford.

§2.14 ON RAISING AND LOWERING HELICITY *by M.G. Eastwood*

Suppose Y is an open subset of CM and let X denote the corresponding region in PT ($X = \bigcup_{y \in Y} L_y$). If Y satisfies mild topological restrictions [(a) $Y \cap \hat{Z}$ connected and $H^1(Y \cap \hat{Z}, C) = 0$ for all α-planes $\hat{Z}$ will do if $n \geq -1$, and (a) together with (b) $H^1(Y,C) = H^2(Y,C) = 0$ will do if $n \leq -2$], then $H^1(X, \mathcal{O}(-n-2))$ is isomorphic to the holomorphic, helicity n/2, massless fields on Y. We suppose from now on that Y is such a region.

A dual twistor $Q_\alpha = (\eta_A, \xi^{A'}) \in T^*$ may be regarded (essentially by definition) as a section $Q \equiv Q_\alpha Z^\alpha \in \Gamma(PT, \mathcal{O}(1))$. Multiplication by Q induces a sheaf homomorphism $Q: \mathcal{O}(k) \to \mathcal{O}(k+1)$ and hence a map on cohomology $Q: H^1(X, \mathcal{O}(k)) \to H^1(X, \mathcal{O}(k+1))$. This must correspond to some operation on massless fields on Y which we now identify:

Let $S = Y \times P^1$ (a subset of the primed spin bundle). As usual we have

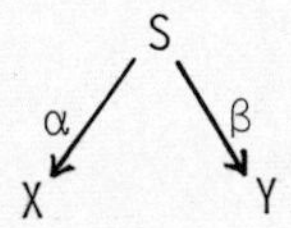

and $H^1(X, \mathcal{O}(k))$ may be identified by using the exact sequence

$$0 \to \alpha^{-1}\mathcal{O}(k) \to \mathcal{O}(k) \xrightarrow{\pi_{A'}\nabla_A^{A'}} \mathcal{O}_A(k+1) \xrightarrow{\pi_{A'}\nabla^{AA'}} \mathcal{O}(k+2) \to 0. \qquad (1)$$

Case 1: $k = -n-2$, $n \geq 1$: In this case (1) gives

$$H^1(S, \alpha^{-1}\mathcal{O}(-n-2)) \cong \ker \nabla^{AA'} : H^1(S, \mathcal{O}(-n-2)) \to H^1(S, \mathcal{O}(-n-1))$$

$$\underbrace{\phi_{A'B'\ldots D'}}_{n \text{ indices}} \longmapsto \nabla^{AA'}\phi_{A'B'\ldots D'}$$

Thus to see what effect Q has on fields we just compute the effect of α^*Q on $H^1(S, \mathcal{O}(-n-2))$. In the usual coordinates $(x^{AA'}, \pi_{A'})$ on S, $\alpha^*Q(x,\pi) = \eta_A(ix^{AA'}\pi_{A'}) + \xi^{A'}\pi_{A'} = \zeta^{A'}\pi_{A'}$ where $\zeta^{A'} = \xi^{A'} + ix^{AA'}\eta_A$. Thus

$$H^1(S,\mathcal{O}(-n-2)) \xrightarrow{\alpha^*Q} H^1(S,\mathcal{O}(-n-1))$$

$$\phi_{A'B'\ldots D'} \mapsto \zeta^{A'}\phi_{A'B'\ldots D'}$$

and so this is what happens on fields. This is called "lowering helicity" [see §2.13]. It is clear from the dual twistor equation $\nabla^{B(B'}\zeta^{A')} = 0$ that $(Q\phi)_{B'\ldots D'} = \zeta^{A'}\phi_{A'B'\ldots D'}$ satisfies the field equations $\nabla^{BB'}(Q\phi)_{B'\ldots D'} = 0$.

Case 2: k = -2: In this case a similar argument shews

$$Q: H^1(X,\mathcal{O}(-2)) \to H^1(X,\mathcal{O}(-1))$$
$$\phi \mapsto \zeta^{A'}\nabla_{AA'}\phi + \frac{1}{2}\phi\nabla_{AA'}\zeta^{A'} \quad .$$

Case 3: k = n -2, n ≥ 1:

$$Q: H^1(X,\mathcal{O}(n-2)) \to H^1(X,\mathcal{O}(n-1))$$
$$\phi_A^{B'\ldots D'} \mapsto \zeta^{(E'}\phi_A^{B'\ldots D')} \text{ on potentials}$$

or equivalently $\psi_{AB\ldots D} \mapsto \zeta^{E'}\nabla_{E'(E}\psi_{A\ldots D)} + \frac{1}{2}(n+1)\,\psi_{(A\ldots D}\nabla_{E)E'}\zeta^{E'}$ on fields.

Note that this last formula agrees with case 2 when n = 0.

The dual operation of "raising helicity" also arises naturally in twistor theory. A twistor $P^\alpha = (\sigma^A, \tau_{A'}) \in \mathbb{T}$ may be regarded as a differential operator $P \equiv P^\alpha \frac{\partial}{\partial Z^\alpha} : \mathcal{O}(k) \to \mathcal{O}(k-1)$ and hence gives rise to a map on cohomology which may be interpreted as a map on fields. Calculating the effect on fields may be achieved as follows:

On S define a differential operator $\partial_{E'}: \mathcal{O}(k) \to \mathcal{O}_{E'}(k)$ by $\partial_{E'} = \pi_{E'}\tau_{F'}\frac{\partial}{\partial\pi_{F'}} - i\theta^E\nabla_{EE'}$ where $\theta^E = \sigma^E - ix^{EE'}\tau_{E'}$. It is straightforward to check that

$$\begin{array}{ccc} 0 \to \alpha^{-1}\mathcal{O}(k) & \to & \mathcal{O}(k) \\ \alpha^{-1}\pi_{E'}T\downarrow & & \downarrow \partial_{E'} \\ 0 \to \alpha^{-1}\mathcal{O}_{E'}(k) & \to & \mathcal{O}_{E'}(k) \end{array} \qquad \text{commutes.}$$

Thus for $k = -n-2$, $n \geq 0$, we may compute (by 1) the effect of P on fields by applying $\partial_{E'}$ to $H^1(S,\mathcal{O}(-n-2))$ (and using case 1 of lowering spin with $\zeta^{A'} = \delta^{A'}_{E'}$ to remove the effect of $\pi_{E'}$). The answer is

$$iP: H^1(X,\mathcal{O}(-n-2)) \to H^1(X,\mathcal{O}(-n-3))$$

$$\psi_{A'B'\ldots D'} \mapsto \frac{(n+1)}{2}\tau_{(E'}\psi_{A'B'\ldots D')} + \theta^E \nabla_{E(E'}\psi_{A'\ldots D')}$$

$$= \theta^E \nabla_{E(E'}\psi_{A'\ldots D')} + \frac{(n+1)}{2} \psi_{(A'\ldots D'} \nabla_{E')E}\theta^E$$

which is just the dual version of case 3 of lowering spin. For $k = n-2$, $n \geq 1$, we obtain the dual version of case 1, namely

$$iP: H^1(X,\mathcal{O}(n-2)) \to H^1(X,\mathcal{O}(n-3))$$

$$\phi_{AB\ldots D} \mapsto \theta^A\phi_{AB\ldots D}.$$

These operations are called "raising helicity". Field equations follow from $\nabla^{A'(A}\theta^{E)} = 0$.

Penrose in §2.13 shews how to calculate the kernel and cokernel of a helicity raising or lowering operation. We indicate here how to investigate a similar question when lowering helicity (raising is similar) with respect to a line L in PT rather than a plane Q. A line may be defined by a pair of planes $L_{A\alpha}Z^\alpha = 0$ and lowering helicity with both of them gives a pair of fields. Up to taking linear combinations this operation depends only on the line L and not on the particular choice of defining functions.

Similarily we can take a pair of fields and use L^A to lower helicity and contract to form a single field. This is a very natural operation from the twistor theory point of view since L must correspond to a point in CM. Indeed, apart from certain constants, it is easy to check that the value of the field at this point is unchanged by raising and lowering helicity based on the corresponding line. This enables us to regard the integral formulae for producing a field from a twistor function as first adjusting the helicity to zero and then producing the field. To investigate further we use the exact sequence

$$0 \to \mathcal{O}(m) \xrightarrow{L_A} \mathcal{O}_A(m+1) \xrightarrow{L^A} \mathcal{O}(m+2) \to \mathcal{O}_L(m+2) \to 0 \tag{2}$$

(which should be compared with a similar sequence in §2.15). There are a number of different cases depending on the value of m and whether (1) $L \cap X = \phi$, (2) $L \cap X \neq \phi$ by $L \not\subset X$, (3) $L \subset X$. (1), (2) and (3) have obvious interpretations in CM. In case (2) L is Stein and so in all cases $H^p(L,\mathcal{O}_L(m+2))$ may be computed. $H^0(X,\mathcal{O}(m))$ may be interpreted as solutions of a multi-twistor equation.

The conclusions which follow from (2) are easy to derive. For example, if $m = -3$, $H^2(X,\mathcal{O}(k)) = 0$ (e.g. $X = PT^+$), and either (1) $L \cap X = \phi$ or (3) $L \subset X$ then we obtain the rather interesting exact sequence:

$$0 \to H^1(X,\mathcal{O}(-3)) \to H^1(X,\mathcal{O}_A(-2)) \to H^1(X,\mathcal{O}(-1)) \to 0$$

R.H. Neutrino ↦ Pair spin-0 field ↦ L.H. Neutrino.

§2.15 MASSLESS FIELDS BASED ON A LINE *by M.G. Eastwood and L.P. Hughston*

Preliminary Remarks. Twistor methods can be used very effectively to gain information concerning certain aspects of the structure of singularities arising in solutions of the zero rest-mass equations in Minkowski space. In this note we shall be investigating massless fields that have the property of being based on a single line in P^3: that is to say, fields which are well-defined throughout compactified complex Minkowski space (CM) except on a single complex null-cone. The vertex of this null-cone corresponds to the line L in P^3 that we have singled out, and the remaining points on the cone correspond to lines in P^3 that meet L.

The simplest class of massless fields based on L consists of the so-called 'elementary states'. These can be described as follows: Let P_α and Q_α be a pair of distinct planes containing L, and let $p_{A'}(x)$ and $q_{A'}(x)$ be the solutions of the dual twistor equation that are naturally associated with P_α and Q_α (i.e. $p_{A'} = 0$ is the β-plane in CM corresponding, via the Klein representation, to $P_\alpha Z^\alpha = 0$ in P^3). Then the field $\phi(x) = 1/(p.q)$ satisfies the wave equation, and is the elementary state (of spin zero) associated with L. One can check that $\phi(x)$ is independent of the choice of P_α and Q_α in that the only freedom available to $\phi(x)$ is scaling with an arbitrary complex number. Moreover, one can verify that $p.q = 0$ defines the complex null-cone whose vertex is the line L.

A more complicated solution of the wave equation, singular only at $p.q = 0$, is given by the formula

$$\phi(x) = \frac{p.r + q.s}{(p.q)^2} \quad , \tag{1}$$

where $r_{A'}(x)$ and $s_{A'}(x)$ are arbitrary solutions of the dual twistor equation. What is the dimensionality of this family of solutions? The functions $r_{A'}$

and $s_{A'}$ each contain four complex parameters. However, $\phi(x)$ is invariant under the following transformations:

$$\left.\begin{aligned} r &\longrightarrow r + \alpha p + \gamma q \\ s &\longrightarrow s + \beta q + \gamma p \end{aligned}\right\},$$

where α, β and γ are complex parameters. Thus the dimensionality of the family (1) above is *five*. How do we know that (1) represents the entire family of solutions of the wave equation with a 'second-order' singularity on the null-cone associated with L? In what follows we shall indicate how such questions can be handled systematically. Two rather distinct approaches to the problem will be presented. First we shall boldly blow the line in P^3 corresponding to the vertex of the null-cone on which we would like the field to be singular up.

Blowing Up the Line. A very neat way of classifying massless fields based on a line (motivated by a suggestion due to Penrose) is to regard these fields as elements of the group $H^1(V,\mathcal{O}_V(a,b))$, where V is P^3 with the line L blown up, and $\mathcal{O}_V(a,b)$ is a sheaf of germs of twisted holomorphic functions on V. The twist parameters a and b, which will be described below, carry the information of the helicity and the singularity type of the associated fields.

The manifold V can be characterized as follows. The line L is given by the equation $L_{A\alpha}Z^\alpha = 0$, where $L_{1\alpha} = P_\alpha$ and $L_{2\alpha} = Q_\alpha$. The spinor index on $L_{A\alpha}$ represents the P^1 of planes containing L. Now we let λ^A be homogeneous coordinates on P^1 and consider the submanifold $V \subset P^3 \times P^1$ defined by $\lambda^A L_{A\alpha}Z^\alpha = 0$. Note that when $L_{A\alpha}Z^\alpha \neq 0$, λ_A is fixed on V by $\lambda_A \sim L_{A\alpha}Z^\alpha$. On the other hand, when $L_{A\alpha}Z^\alpha = 0$, the spinor λ_A can take on any value; thus the line L is blown up in V to a $P^1 \times P^1$, with a separate copy of L for each distinct plane through L in the original space.

Associated with the blown-up space V we have the following exact sequence:

$$0 \longrightarrow \mathcal{O}_{P^3\times P^1}(a-1,b-1) \xrightarrow{\lambda^A L_{A\alpha} Z^\alpha} \mathcal{O}_{P^3\times P^1}(a,b) \xrightarrow{\rho_V} \mathcal{O}_V(a,b) \longrightarrow 0.$$

This sequence can be regarded as a definition for the sheaf $\mathcal{O}_V(a,b)$, whence we see the origin of the twist parameters a and b. In order for $H^1(V,\mathcal{O}_V(a,b))$ to be non-vanishing we must have at least $a \geq 0$ and $b \leq -2$. With this assumption a simple calculation shews that the sequence

$$0 \to H^0(V,\mathcal{O}_V(a,b)) \to H^1(P^3\times P^1,\mathcal{O}(a-1,b-1)) \to H^1(P^3\times P^1,\mathcal{O}(a,b)) \to H^1(V,\mathcal{O}_V(a,b)) \to 0 \tag{2}$$

is exact. Using this sequence we can deduce various properties of the group $H^1(V,\mathcal{O}_V(a,b))$.

Let us consider, for example, the case $a = 0$, $b = -2$. The cohomology sequence (2) above tells us that $H^1(V,\mathcal{O}_V(0,-2)) \simeq C$, as is appropriate for elementary states of spin 0. In this case we have an isomorphism $H^1(P^3\times P^1, \mathcal{O}(0,-2)) \xrightarrow{\simeq} H^1(V,\mathcal{O}_V(0,-2))$. Now, a representative in the first of these groups must have the form

$$\frac{1}{(P_A\lambda^A)(Q_B\lambda^B)} , \tag{3}$$

where P_A and Q_B are constant spinors. Away from $L_{A\alpha}Z^\alpha = 0$ we must have $\lambda_A \sim L_{A\alpha}Z^\alpha$ on V; thus, away from $L_{A\alpha}Z^\alpha = 0$, the representative (3) above restricts under the action of ρ_V to become

$$\frac{1}{(P_\alpha Z^\alpha)(Q_\beta Z^\beta)} , \tag{4}$$

where $P_\alpha = P_A L^A_\alpha$ and $Q_\alpha = Q_A L^A_\alpha$. This is the canonical form for a spin-zero elementary state based on the line L, so we see that there is indeed a natural correspondence between elements of $H^1(V, \mathcal{O}_V(0,-2))$ and spin-zero elementary states based on L. Similar reasoning holds for other values of a and b. We find that the helicity of the field associated with an element of $H^1(V, \mathcal{O}_V(a,b))$ is $-(a+b+2)/2$. The number $\ell := -b-1$ is a convenient measure for the order of the singularity.

As for dimensionalities, with the assumptions $a \geq 0$ and $b \leq -2$, we have:

$$\begin{cases} \rho := \dim_C H^1(P^3 \times P^1, \mathcal{O}(a,b)) = -(b+1)(a+1)(a+2)(a+3)/6 \\ \sigma := \dim_C H^1(P^3 \times P^1, \mathcal{O}(a-1,b-1)) = -ba(a+1)(a+2)/6 \\ \tau := \dim_C H^0(V, \mathcal{O}_V(a,b)) = \begin{cases} 0 & \text{if } a+b < 0 \\ (a+b+1)(a+b+2)(a+b+3)/6 & \text{if } a+b \geq -3, \end{cases} \end{cases}$$

whence, by the rule of alternating dimensionalities summing to zero, we obtain

$$\dim_C H^1(V, \mathcal{O}_V(a,b)) = \rho - \sigma + \tau . \tag{5}$$

In particular, a short calculation shews that $\dim_C H^1(V, \mathcal{O}_V(1,-3)) = 5$, corresponding to the set of massless spin-zero fields with a second order singularity on the chosen cone. This figure agrees with the informal calculation we made earlier, in connection with formula (1). In order to derive (1) directly, we note that a general representative in the group $H^1(P^3 \times P^1, \mathcal{O}(1,-3))$ takes the canonical form

$$\frac{\lambda^A T_{A\alpha} Z^\alpha}{(P_A \lambda^A)^2 (Q_B \lambda^B)^2} ,$$

where $T_{A\alpha}$ is arbitrary. Writing $P^A T_{A\alpha} = R_\alpha$ and $Q^A T_{A\alpha} = S_\alpha$, this expression reduces to

$$\frac{R_\alpha Z^\alpha}{(P_\beta Z^\beta)(Q_\beta Z^\beta)^2} + \frac{S_\alpha Z^\alpha}{(P_\beta Z^\beta)^2(Q_\beta Z^\beta)} \tag{6}$$

when restricted down to V, away from the blown up version of L. This is the standard 'twistor function' for the field in formula (1). For other values of a and b one proceeds analogously.

Annihilating Cohomology Classes: The production of a massless field from an element of $H^1(V,\mathcal{O}_V(a,b))$ may be described more abstractly as follows. The blow-up V comes equipped with a projection map $\pi\colon V \to P^3$ such that $V - \pi^{-1}(L) \to P^3 - L$ is an isomorphism. Under this isomorphism it is easy to check that $\mathcal{O}_V(a,b)|_{V-\pi^{-1}(L)} \simeq \mathcal{O}(a+b)|_{P^3-L}$ and therefore we have a homomorphism

$$H^1(V,\mathcal{O}_V(a,b)) \to H^1(V-\pi^{-1}(L)),\ \mathcal{O}_V(a,b)) \xrightarrow{\simeq} H^1(P^3-L,\mathcal{O}(a+b)).$$

An element of $H^1(P^3-L,\mathcal{O}(a+b))$ then gives rise to a massless field (of helicity $-(a+b+2)/2$) by one of the standard methods. Therefore we ask: What is the image of $H^1(V,\mathcal{O}_V(a,b))$ in $H^1(P^3-L,\mathcal{O}(a+b))$? In this section we will characterize this image in a rather natural way.

We first observe that we can actually calculate $H^1(P^3-L,\mathcal{O}(a+b))$ quite explicitly. To do this we choose homogeneous coordinates (x_0,x_1,x_2,x_3) on P^3 such that $L = \{x \text{ s.t. } x_2 = x_3 = 0\}$. Then $\mathcal{V}_2 := \{x_2 \neq 0\}$ and $\mathcal{V}_3 := \{x_3 \neq 0\}$ form a cover of P^3-L by Stein sets. From the Mayer-Vietoris sequence we conclude that

$$\Gamma(\mathcal{V}_2,\mathcal{O}(m)) \oplus \Gamma(\mathcal{V}_3,\mathcal{O}(m)) \to \Gamma(\mathcal{V}_2 \cap \mathcal{V}_3,\mathcal{O}(m)) \to H^1(P^3-L,\mathcal{O}(m)) \to 0$$

is exact and this enables us to calculate $H^1(P^3-L,\mathcal{O}(m))$ since we may identify $\Gamma(V_2,\mathcal{O}(m))$ etc. by means of Laurent series. For example:

$$\Gamma(V_2,\mathcal{O}(m)) \simeq \{ \sum_{0\le r,s,t} c_{rst} \frac{x_0^r\, x_1^s\, x_3^t}{x_2^{r+s+t-m}} \} \text{ (suitably convergent).}$$

The calculation of $H^1(P^3-L,\mathcal{O}(m))$ is simply a matter of comparing coefficients and the answer is:

$$H^1(P^3-L,\mathcal{O}(m)) \simeq \{ \sum_{0<j,k} \frac{A_{jk}(x_0,x_1)}{x_2^{\,j}\, x_3^{\,k}} \} ,$$

where $A_{jk}(x_0,x_1)$ is a homogeneous polynomial of degree $j+k+m$. If, instead of choosing coordinates on P^3, we choose planes P_α, Q_α with intersection L as before then this may be rewritten

$$H^1(P^3-L,\mathcal{O}(m)) \simeq \{ \sum_{0<j,k} \frac{A_{jk}}{(P.Z)^j(Q.Z)^k} \} , \tag{7}$$

where $A_{jk} \in \Gamma(L,\mathcal{O}(j+k+m))$ and P.Z means $P_\alpha Z^\alpha$, etc..

It is clear from this representation that there is a filtration

$$H^1_1(P^3-L,\mathcal{O}(m)) \subseteq H^1_2(P^3-L,\mathcal{O}(m)) \subseteq \ldots \subseteq H^1(P^3-L,\mathcal{O}(m)), \text{ where}$$

$$H^1_\ell(P^3-L,\mathcal{O}(m)) \simeq \{ \sum_{\substack{0<j,k \\ j+k\le\ell+1}} \frac{A_{jk}}{(P.Z)^j(Q.Z)^k} \} .$$

Moreover, this fitration agrees with that described in the previous section. If m = -2 then A_{11} is a constant so $H^1_1(P^3-L,\mathcal{O}(-2)) \simeq \{A_{11}/(P.Z)(Q.Z)\}$ i.e. the elementary states as indentified by (4). Similarly the twistor function (6) yields the cohomology classes in $H^1_2(P^3-L,\mathcal{O}(m))$.

We can easily compute $\dim_C H^1_\ell(P^3-L,\mathcal{O}(m))$. To do this we observe that the dimension of the space of A_{jk} $(= \dim_C\Gamma(L,\mathcal{O}(j+k+m)))$ is $\underline{j+k+m+1}$ (where

$\underline{r} := \max(r,0)$) and that there are i possibilities for (j,k) if $j+k = i+1$ and $0 < j,k$. Thus:

$$\dim_C H^1_\ell(P^3-L,\mathcal{O}(m)) = \sum_{i=1}^{\ell} \underline{(i+m+2)}i. \tag{8}$$

For example $\dim_C H^1_\ell(P^3-L,\mathcal{O}(m)) = \sum_{i=1}^{\ell} i^2 = \ell(\ell+1)(2\ell+1)/6$ so that

$\dim_C H^1_1(P^3-L,\mathcal{O}(-2)) = 1$ and $\dim H^1_2(P^3-L,\mathcal{O}(-2)) = 5$. This agrees with the figures obtained by blowing up.

The above method of describing $H^1_\ell(P^3-L,\mathcal{O}(m))$ depends on choosing coordinates whereas blowing up L is an invariant notion. Thus there should be a co-ordinate-free way of characterizing $H^1_\ell(P^3-L,\mathcal{O}(m))$ more directly. $H^1(P^3-L,\mathcal{O}(\cdot)) := \bigoplus_m H^1(P^3-L,\mathcal{O}(m))$ may be considered as a graded module over the graded ring $\bigoplus_m \Gamma(P^3,\mathcal{O}(m))$ ($\simeq C[Z^\alpha]$). For example if $f \in \Gamma(P^3,\mathcal{O}(1))$ then multiplication by f induces a map of sheaves $\mathcal{O}(m) \xrightarrow{\times f} \mathcal{O}(m+1)$ and hence a map on cohomology $H^1(P^3-L,\mathcal{O}(m)) \xrightarrow{\times f} H^1(P^3-L,\mathcal{O}(m+1))$. It is easy to identify this module structure in terms of the isomorphism (6). In particular,

$$(P.Z)^r(Q.Z)^s \sum_{0<j,k} A_{jk}/(P.Z)^j(Q.Z)^k = \sum_{0<j,k} A_{j+r,k+s}/(P.Z)^j(Q.Z)^k$$

and hence $H^1_\ell(P^3-L,\mathcal{O}(m))$ is exactly the submodule of $H^1(P^3-L,\mathcal{O}(m))$ consisting of those elements annihilated by $(P.Z)^r(Q.Z)^s$ for $r+s = \ell$. But now we observe that P.Z, Q.Z generate the homogeneous ideal I(L) consisting of those polynomials which vanish on L and hence we may characterize $H^1_\ell(P^3-L,\mathcal{O}(m))$ in a manifestly invariant manner as

$$H^1_\ell(P^3-L,\mathcal{O}(m)) = \{\text{elements of } H^1(P^3-L,\mathcal{O}(m)) \text{ annihilated by } I(L)^\ell\}.$$

We note that this definition makes sense for any subvariety (instead of L) of any open subset of P^3 and can therefore be used to investigate massless fields based thereon. This method and the corresponding 'blowing-up' procedure will be taken up elsewhere [see §2.16 for the *twisted cubic*].

We may calculate $H^1_\ell(P^3-L,\mathcal{O}(m))$ without using (7). To do this set $L_A = L_{A\alpha}Z^\alpha$ and consider the sequence of sheaves

$$0 \to \mathcal{O}(m) \xrightarrow{L_A\ldots L_D} \mathcal{O}_{\underbrace{A\ldots D}_{\ell}}(m+\ell) \xrightarrow{L_A} \mathcal{O}_{B\ldots D}(m+\ell+1) \to 0$$

where $\mathcal{O}_{A\ldots D}(k)$ is symmetric in the spinor indices. It is clear that this is exact on P^3-L. To interpret the corresponding cohomology sequence we observe that $\Gamma(P^3-L,\mathcal{O}(k)) \simeq \Gamma(P^3,\mathcal{O}(k))$ by the Riemann removable singularities theorem and that $L_A\ldots L_D$ generate $I(L)^\ell$. Thus we obtain the exact sequence

$$0 \to \Gamma(P^3,\mathcal{O}(m)) \to \Gamma(P^3,\mathcal{O}_{A\ldots D}(m+\ell)) \to \Gamma(P^3,\mathcal{O}_{B\ldots D}(m+\ell+1)) \to H^1_\ell(P^3-L,\mathcal{O}(m)) \to 0$$

which enables us to calculate $H^1_\ell(P^3-L,\mathcal{O}(m))$ as before. In particular, $\dim_C H^1_\ell(P^3-L,\mathcal{O}(m))$ is easily determined since $\dim_C\Gamma(P^3,\mathcal{O}(k)) = (k+1)(k+2)(k+3)/6$. A straightforward calculation gives

$$\dim_C H^1_\ell(P^3-L,\mathcal{O}(m)) = \begin{cases} 0 & \text{if } m \leq -\ell-2 \\ [\ell(\ell+1)(2\ell+3m+7)-(m+1)(m+2)(m+3)]/6 & \text{if } -\ell-1\leq m\leq -1 \\ \ell(\ell+1)(2\ell+3m+7)/6 & \text{if } m \geq -3 \end{cases} \tag{9}$$

The two parameters m,ℓ are related to the parameters a,b involved in blowing up by $m = a+b$ and $\ell = -b-1$. Thus we have the natural identification

$$H^1_\ell(P^3-L,\mathcal{O}(m)) \simeq H^1(V,\mathcal{O}_V(m+\ell+1,\ -\ell-1)).$$

An elementary calculation establishes the equivalence of the dimensionality formulae (5), (8), and (9). For example (9) gives

$$\dim_C H^1_\ell(P^3-L,\mathcal{O}(-2)) = \ell(\ell+1)(2\ell+1)/6, \text{ as before.}$$

Finally we remark that since the effect on $H^1(P^3-L,\mathcal{O}(m))$ of multiplication by L_A may be interpreted on fields as 'lowering helicity' [see §§2.13 & 2.14], it is possible, therefore, to give a direct space-time description of $H^1_\ell(P^3-L,\mathcal{O}(m))$ in terms of fields being annihilated by various helicity lowering operators. The details will appear elsewhere.

Gratitude is expressed to S.A. Huggett and R. Penrose for useful discussion.

§2.16 MASSLESS FIELDS BASED ON A TWISTED CUBIC *by M.G. Eastwood, L.P. Hughston, and T.R. Hurd*

The twisted cubic is the simplest non-planar curve one comes across in projective geometry. The aim here is to study zero rest-mass fields in Minkowski space that are based on a twisted cubic curve in projective twistor space[1]. To begin, we shall review some of the well-known elementary properties of the twisted cubic[2]. Using (W,X,Y,Z) as homogeneous coordinates on P^3, the twisted cubic curve Q is given by the locus (t^3, t^2u, tu^2, u^3) with $(t,u) \in P^1$. It is the intersection of the three quadric surfaces Q_i defined by:

$$Q_1 := WY-X^2 = 0, \quad Q_2 := WZ-XY = 0, \quad Q_3 := XZ-Y^2 = 0 . \tag{1}$$

In fact, any quadric containing the curve Q can be expressed as a linear combination of Q_1, Q_2, and Q_3; thus projectively there is a two-parameter family of such quadrics. Any two quadrics in this net intersect on a residual line as well as on Q. The residual line is necessarily a chord (possibly a tangent) of the cubic. Through any point in P^3-Q there passes a unique chord of the curve. An interesting and important feature of the three quadrics Q_i is the identity

$$\frac{aW+bX}{Q_1Q_2} - \frac{aX+bY}{Q_1Q_3} + \frac{aY+bZ}{Q_2Q_3} = 0 \;, \tag{2}$$

which holds for all values of a and b.

To proceed spinorially, let $\xi_{ABC} = \xi_{(ABC)}$ be homogeneous coordinates on P^3. Then[3]

$$Q_{CD} := \xi_{AB(C}\,\xi^{AB}_{D)} = 0 \tag{3}$$

defines the twisted cubic. The general solution to equation (3) is given by $\xi_{ABC} = \psi_A \psi_B \psi_C$ with ψ_A arbitrary, and we see therefore that the twisted cubic can be regarded as the image of the Veronese mapping from P^1 to P^3.

Let us introduce a spinor basis (o_A, ι_A) with $o_A \iota^A = 1$ and put $W = \xi_{000}$, $X = \xi_{001}$, $Y = \xi_{011}$, and $Z = \xi_{111}$. We recover the parametrization mentioned above if we put $(\psi_0, \psi_1) = (t,u)$. A short calculation establishes that the three quadrics Q_i are given by the formula $Q_i = \frac{1}{2} Q_{AB}\alpha_i^{AB}$, where α_i^{AB} is defined by

$$\alpha_1^{AB} = o^A o^B, \quad \alpha_2^{AB} = 2o^{(A}\iota^{B)}, \quad \alpha_3^{AB} = \iota^A \iota^B .$$

The cubic identity (2) above can be expressed more concisely as the following basic result:

PROPOSITION 1. For any symmetric spinor ξ^{ABC}, one has the identity

$$Q_{AB}\, \xi^{ABC} = 0 , \tag{4}$$

with Q_{AB} defined as in equation (3).

Proof. Transvect the expression $\xi^{ABC}\, \xi^{DE}_{A}\, \xi^{F}_{EB}$ with the elementary spinor identity $\varepsilon_{[CD}\varepsilon_{F]G} = 0$. □

One obtains (2) directly by hitting (4) with $\theta_C = (a,b)$. An important fact to note is:

PROPOSITION 2. For any symmetric spinors ξ_{ABC} and η_{AB},

$$Q_{AB}\, \eta^{AB} = 0 \Longleftrightarrow \begin{cases} Q_{AB} = 0 , & \textit{or} \\ \eta^{AB} = \xi^{ABC} \lambda_C & \text{for some } \lambda_C. \end{cases}$$

Proof. First we shall establish the fact mentioned earlier that through any point ξ^{ABC} in P^3-Q there exists a unique chord of Q: If $\xi^{ABC} \in P^3$-Q then $Q_{AB} \neq 0$ and $Q_{AB} = o_{(A}\iota_{B)}$ for some o^A, ι^A. When $o_A \iota^A \neq 0$ observe that *Proposition 1* implies

$$\xi^{ABC} = a\, o^A o^B o^C + b\, \iota^A \iota^B \iota^C \tag{5}$$

for some $(a,b) \in P^1$. Thus ξ^{ABC} lies on the join of the two points $o^A o^B o^C$ and $\iota^A \iota^B \iota^C$ on Q. On the other hand, if $Q_{AB} = o_A o_B$ then *Proposition 1* implies $\xi^{ABC} \; o^{(A} o^B \beta^{C)}$ for some β^C, and one sees that ξ^{ABC} lies on the line tangent to Q at the point $o^A o^B o^C$. See Figure A.

If $Q_{AB} = o_{(A}\iota_{B)}$ with $o_A \iota^A \neq 0$, then $Q_{AB}\, \eta^{AB} = 0$ implies that $\eta^{AB} = c o^A o^B + d \iota^A \iota^B$ for some (c,d). From (5) it follows that $\eta^{AB} = \xi^{ABC} \lambda_C$ where $\lambda_A = ad\, o_A - bc\, \iota_A$.

If $Q_{AB} = o_A o_B$ then $\eta^{AB} = o^{(A}\gamma^{B)}$ for some γ^B. A short calculation verifies that $\eta^{AB} = \xi^{ABC} \lambda_C$ with $\lambda_C = 2(\beta_B \gamma^B) o_C + o_B \gamma^B \beta_C$, where β^A is given, as above, by $\xi^{ABC} = o^{(A} o^B \beta^{C)}$. □

With this information at hand, we proceed to "blow up" the twisted cubic. There is a two-dimensional net of quadrics containing Q, and we propose to blow up the curve in such a way as to obtain a separate copy of the curve for each of these quadrics. Our blown-up space V will be a smooth variety in $P^3 \times P^2$. Using (ξ^{ABC}, η^{AB}) as coordinates on $P^3 \times P^2$, we define V by the equation $Q_{AB}\, \eta^{AB} = 0$. *Proposition 2* tells us that away from the locus $\tilde{Q}$ defined by $Q_{AB} = 0$, η^{AB} must have the form $\xi^{ABC} \lambda_C$. This means $V - \tilde{Q} \simeq (P^3 - Q) \times P^1$. When $Q_{AB} = 0$, η^{AB} is arbitrary and therefore $\tilde{Q} \simeq Q \times P^2$.

If $\pi: V \to P^3$ is the natural projection, the preimage $\pi^{-1}(\xi)$ of any point $\xi \in P^3$ can be identified with the system of all quadrics containing both ξ and Q. Alternatively, $\pi^{-1}(\xi)$ can be identified naturally with the set of all the lines in P^3 intersecting both ξ and Q.

The key short exact sequence for the blow-up is:

$$0 \to \mathcal{O}_{P^3\times P^2}(a-2,b-1) \xrightarrow{Q_{AB}\eta^{AB}} \mathcal{O}_{P^3\times P^2}(a,b) \xrightarrow{\rho_V} \mathcal{O}_V(a,b) \to 0, \qquad (6)$$

where the map ρ_V is the restriction to v. Away from $\tilde{Q}$ this map is given by replacing η^{AB} with $\xi^{ABC}\lambda_C$.

We want to obtain elements of $H^1(P^3-Q,\mathcal{O}(m))$. This group corresponds to the set of all zero rest-mass fields based on the twisted cubic. Note that we have the homomorphism

$$H^i(V,\mathcal{O}_V(a,b)) \longrightarrow H^i((P^3-Q)\times P^1,\ \mathcal{O}(a+b,b)) \qquad (7)$$

obtained by restricting to $V-\tilde{Q} \simeq (P^3-Q)\times P^1$. Inspection of (6) and (7) indicates that nothing of special interest arises, for our purposes, unless $a \geq 0$, $b \leq -3$, and $i = 2$. With these restrictions, the cohomology sequence associated with (6) gives:

$$0 \to H^1(V,\mathcal{O}_V(a,b)) \to H^2(P^3\times P^2,\mathcal{O}(a-2,b-1)) \to H^2(P^3\times P^2,\mathcal{O}(a,b)) \to H^2(V,\mathcal{O}_V(a,b)) \to 0 \qquad (8)$$

In general all four main terms of the sequence are non-vanishing. When $a = 0$ or $a = 1$, however, things simplify and we get the isomorphism $H^2(P^3\times P^2,\mathcal{O}(a,b)) \xrightarrow{\simeq} H^2(V,\mathcal{O}_V(a,b))$.

We propose to examine the case $a = 0$, $b = -3$ in some detail. This is the simplest case, and corresponds directly to the case originally analyzed by Penrose. Associated with each element of $H^2(V,\mathcal{O}_V(0,-3)) \simeq C$ there is some element of $H^2((P^3-Q)\times P^1,\ \mathcal{O}(-3,-3)) \simeq H^1(P^3-Q,\mathcal{O}^A(-3))$. Thus the group $H^2(V,\mathcal{O}_V(0,-3))$ consists of a one-parameter family of pairs of positive helicity solutions of the neutrino equation. In order to study $H^2(V, \mathcal{O}_V(0,-3))$ more closely it is convenient to define $U_i := \{\eta^{AB} : \eta^{AB}\alpha_{iAB} \neq 0,\ i = 1,2,3\}$. This is an open Stein cover of P^2. Representative cocycles of the group $H^2(P^3 \times P^2, \mathcal{O}(0,-3))$, when restricted to $V-\tilde{Q}$, look like:

$$f = \frac{k}{(\xi^{ABC}\lambda_C \underset{1}{\alpha}_{AB})(\xi^{ABC}\lambda_C \underset{2}{\alpha}_{AB})(\xi^{ABC}\lambda_C \underset{3}{\alpha}_{AB})},\ k \in C. \tag{9}$$

We shall now proceed to evaluate the pair of zero rest mass fields associated with the cocycle (9). This pair is completely determined up to the choice of the constant k.

It is convenient at this point to transform to standard twistor coordinates Z^α via the relation $\xi^{PQR} = \Xi_\alpha^{PQR} Z^\alpha$. The precise choice of Ξ_α^{PQR} fixes the location of the curve in twistor space(4). [Throughout the remainder of this calculation we shall use letters from the middle of the alphabet (P,Q,R,...) for the twisted cubic indices, and letters from the beginning of the alphabet (A,B,C,...) for ordinary Minkowski space spinor indices. It should be noted that these two systems of indices are quite distinct.] As usual, for Z^α restricted to the space-time point $x^{AA'}$ we shall write: $\rho_x Z^\alpha = (ix^{AA'}\pi_{A'}, \pi_{A'})$. Define the spinor fields $L_i^{PA'}(x)$ and $Q_i^{A'B'}(x)$ by the formulae:

$$\rho_x\ \xi^{PQR}\alpha_{iQR} = L_i^{PA'}(x)\pi_{A'},\ \rho_x Q_i = Q_i^{A'B'}(x)\pi_{A'}\pi_{B'}\ . \tag{10}$$

These fields satisfy dual twistor equations:

$$\nabla^{B(B'}L_i^{A')P} = 0,\quad \nabla^{C(C'}Q_i^{A'B')} = 0\ .$$

Using various algebraic properties of α_{iPQ} we deduce:

$$Q_1^{A'B'} = 2\underset{2}{L}^{P(A'}\underset{1}{L}^{B')}{}_P,\ Q_2^{A'B'} = \underset{3}{L}^{P(A'}\underset{1}{L}^{B')}{}_P,\ Q_3^{A'B'} = 2\underset{3}{L}^{P(A'}\underset{2}{L}^{B')}{}_P\ , \tag{11}$$

while the cubic identity (2) becomes:

$$\underset{1}{Q}^{(A'B'}\underset{3}{L}^{C')P} - 2\underset{2}{Q}^{(A'B'}\underset{2}{L}^{C')P} + \underset{3}{Q}^{(A'B'}\underset{1}{L}^{C')P} = 0 .$$

Restricting the cocycle f to the space-time point $x^{AA'}$ using (10) we get:

$$\rho_x f = \frac{k}{(\underset{1}{L}^{PA'}\lambda_P\pi_{A'})(\underset{2}{L}^{PA'}\lambda_P\pi_{A'})(\underset{3}{L}^{PA'}\lambda_P\pi_{A'})} ,$$

which is a representative cocycle on $P^1 \times P^1$. The associated pair of spacetime fields is:

$$\phi_{PA'}(x) = \oint \lambda_P\pi_{A'}\rho_x f \; \Delta\pi ,$$

and this integral can be evaluated to give:

$$\phi_{PA'}(x) = \frac{k\ell_{PA'}(x)}{\ell_{PA'}(x)\ell^{PA'}(x)} , \qquad (12)$$

where:

$$\ell_{PA'}(x) := \varepsilon_{PA'\,QB'\,RC'\,SD'}\,\underset{1}{L}^{QB'}\,\underset{2}{L}^{RC'}\,\underset{3}{L}^{SD'} .$$

Another expression for the denominator in (12) can be obtained by using formulae (11):

$$\ell_{PA'}\,\ell^{PA'} = \underset{1}{Q}_{A'B'}\,\underset{2}{Q}^{A'}_{C'}\,\underset{3}{Q}^{B'C'} .$$

It is straightforward to see that if the space-time point $x^{AA'}$ corresponds in twistor space to any line which meets the twisted cubic, then the denominator in (12) vanishes.

As with the case of a line in P^3 [see §2.15], the cohomology $H^1(P^3\text{-}Q,\mathcal{O}(m))$

is amenable to direct study. Although it is possible to cover P^3-Q by two Stein sets (P^3-H_2 and P^3-H_3 for suitable quadric and cubic hypersurfaces H_2, H_3), it is perhaps more natural to use the cover $U_i = P^3 - \{Q_i = 0\}$, as did Penrose in his original considerations of the twisted cubic. The calculation would require the identification of $\Gamma(U_i, \mathcal{O}(m))$ etc. and this could be achieved by using the Veronese $P^3 \to P^9$ to straighten the hypersurface $Q_i = 0$ and then using Laurent series as in the preceeding article. It is, however, possible to generalize the more indirect approach of §2.15 to calculate $H^1_\ell(P^3\text{-}Q, \mathcal{O}(m))$, where:

$$H^1_\ell(P^3\text{-}Q,\mathcal{O}(m)) := \{\text{Those elements of } H^1(P^3\text{-}Q,\mathcal{O}(m)) \text{ annihilated by } I(Q)^\ell\}.$$

Since I(Q) is generated by Q_{AB},

$$H^1_1(P^3\text{-}Q,\mathcal{O}(m)) = \ker Q_{AB}: H^1(P^3\text{-}Q,\mathcal{O}(m)) \to H^1(P^3\text{-}Q,\mathcal{O}_{AB}(m+2)).$$

The multiplication map $Q_{AB}:\mathcal{O}(m) \to \mathcal{O}_{AB}(m+2)$ is part of the Koszul complex

$$0 \to \mathcal{O}(m) \xrightarrow{Q_{AB}} \mathcal{O}_{AB}(m+2) \xrightarrow{Q^A_{(C}} \mathcal{O}_{BC}(m+4) \xrightarrow{Q^{BC}} \mathcal{O}(m+6) \to 0\ , \tag{13}$$

which is easily seen to be exact[5] on P^3-Q. Now given any exact sequence of sheaves $0 \to K^0 \to K^1 \to K^2 \to K^3 \to \ \dots$ it is easy to show that

$$\ker : H^1(K^0) \to H^1(K^1) \simeq \frac{\ker\colon \Gamma(K^2) \to \Gamma(K^3)}{\text{im}\colon \ \Gamma(K^1) \to \Gamma(K^2)}$$

and this, together with the Riemann removable singularities theorem, implies:

$$H^1_1(P^3\text{-}Q,\mathcal{O}(m)) \simeq \frac{\ker Q^{BC}\colon \Gamma(P^3,\mathcal{O}_{BC}(m+4)) \to \Gamma(P^3,\mathcal{O}(m+6))}{\text{im } Q^A_{(C}\ \colon \Gamma(P^3,\mathcal{O}_{AB}(m+2)) \to \Gamma(P^3,\mathcal{O}_{BC}(m+4))} \tag{14}$$

The right-hand side of this isomorphism is easily identified as it just involves linear algebra. Moreover, a straightforward untangling of isomorphisms shows that an element f_{BC} of $\ker Q^{BC}: \Gamma(P^3,\mathcal{O}_{BC}(m+4)) \to \Gamma(P^3,\mathcal{O}(m+6))$ is explicitly represented in $H^1(P^3-Q,\mathcal{O}(m))$ by the cocycle

$$f_{jk} = \alpha_j^{BD}\, \alpha_{kD}^{C}\, f_{BC} \,/\, Q_j Q_k \in \Gamma(U_{jk},\mathcal{O}(m)) \;.$$

In the case m = -3 we conclude $H_1^1(P^3-Q,\mathcal{O}(-3)) \simeq \ker Q^{BC}:\Gamma(P^3,\mathcal{O}_{BC}(1)) \to \Gamma(P^3,\mathcal{O}(3))$. By *Proposition 2* $\ker Q^{BC} = \{\lambda^P\, \xi_{BCP} : \lambda^P$ a constant spinor$\}$. Thus we conclude that $\dim H_1^1(P^3-Q,\mathcal{O}(-3)) = 2$. These two parameters correspond to taking arbitrary linear combinations of the two fields appearing in (12). The corresponding representative cocycle in $H^1(P^3-Q,\mathcal{O}(-3))$ is

$$f_{jk} = \alpha_j^{BD}\, \alpha_{kD}^{C}\, \lambda^P\, \xi_{BCP} \,/\, Q_j Q_k \;,$$

which can be recognized immediately as the cocycle originally obtained by Penrose.

Finally, we indicate a method of calculating $H_\ell^1(P^3-Q,\mathcal{O}(m))$, and in particular its dimension, which avoids having to explicitly study the linear maps Q^{BC} and $Q^A_{(C}$ in (14). The key observation is that if L is a line (or indeed any 1-dimensional subvariety) in P^3 then the exact sequence (13) on $P^3-(Q \cup L)$ gives rise (cf. 14) to an isomorphism

$$\ker Q_{AB} : H^1(P^3-(Q\cup L),\mathcal{O}(m)) \to H^1(P^3-(Q\cup L),\, \mathcal{O}_{AB}(m+2))$$

$$\simeq \frac{\ker Q^{BC}:\, \Gamma(P^3,\mathcal{O}_{BC}(m+4)) \to \Gamma(P^3,\mathcal{O}(m+6))}{\operatorname{im} Q^A_{(C} :\, \Gamma(P^3,\mathcal{O}_{AB}(m+2)) \to \Gamma(P^3,\mathcal{O}_{BC}(m+4))} \;,$$

which is the right hand side of (14). In particular, we may take L to be the residual intersection of two quadrics containing Q, say Q_{A0}. Then $Q \cup L$

is defined by the two equations $Q_{AO} = 0$. Indeed, by the Nullstellensatz, these polynomials generate $I(Q \cup L)$ and the exact sequence

$$0 \to \mathcal{O}(m) \xrightarrow{Q_{AO}} \mathcal{O}_A(m+2) \xrightarrow{Q_0^A} \mathcal{O}(m+4) \to 0$$

on $Q \cup L$ gives

$$H_1^1(P^3-(Q \cup L), \mathcal{O}(m)) \simeq \text{coker}: \Gamma(P^3, \mathcal{O}_A(m+2)) \xrightarrow{Q_0^A} \Gamma(P^3, \mathcal{O}(m+4)). \qquad (15)$$

By definition, every element of $H_1^1(P^3-(Q \cup L), \mathcal{O}(m))$ is annihilated by every element of $I(Q \cup L)$, but conversely it is not hard to show that the only homogeneous polynomials which annihilate any elements of the group $H_1^1(P^3-Q \cup L), \mathcal{O}(m))$ are those which vanish on $Q \cup L$. Now $H_1^1(P^3-Q, \mathcal{O}(m))$ consists of those elements of $H_1^1(P^3-(Q \cup L), \mathcal{O}(m))$ annihilated by Q_{11}. However, Q_{11} vanishes on Q but not on L so $f \in \Gamma(P^3, \mathcal{O}(m+4))$ represents via (15) an element of $H_1^1(P^3-Q, \mathcal{O}(m))$ if and only if f vanishes on L. Thus, from (15) we have

$$H_1^1(P^3-Q, \mathcal{O}(m)) \simeq \frac{\text{ker}: \Gamma(P^3, \mathcal{O}(m+4)) \to \Gamma(L, \mathcal{O}_L(m+4))}{\text{im}: \Gamma(P^3, \mathcal{O}_A(m+2)) \to \Gamma(P^3, \mathcal{O}(m+4))}. \qquad (16)$$

Hence we can read off:

$$\dim H_1^1(P^3-Q, \mathcal{O}(m)) = \dim \Gamma(P^3, \mathcal{O}(m+4)) - \dim \Gamma(P^3, \mathcal{O}_A(m+2)) + \dim \Gamma(P^3, \mathcal{O}(m)) - \dim \Gamma(L, \mathcal{O}_L(m+4)).$$

In particular, $\dim H_1^1(P^3-Q, \mathcal{O}(m)) = 0$ for $m \leq -4$. We find, as before, that $\dim H_1^1(P^3-Q, \mathcal{O}(-3)) = 4-2 = 2$. For $m = -2$ we obtain $\dim H_1^1(P^3-Q, \mathcal{O}(-2)) = 10-2-3 = 5$; the field in this case, as derived using sequence (8), is

$$\phi(x) = r^{A'} L_{A'P} \lambda^{P} / L_{A'P} L^{A'P} , \tag{17}$$

where $r^{A'}(x)$ is an arbitrary solution of the dual twistor equation, λ^P is an arbitrary constant spinor, and $L_{A'P}$ is given as before. Since $r^{A'}$ carries four parameters and λ^P two, it is evident that $\phi(x)$ does indeed depend on precisely five parameters. Other values of m can be treated similarly.

We have only discussed the case $\ell = 1$ above, but (13) may be modified to cope with the case $\ell > 1$ as in §2.15. $H^1_\ell(P^3 - Q, \mathcal{O}(m))$ may be interpreted in space-time by using $Q_i^{A'B'}$ to lower helicity. All of this theory can be modified so as to apply to arbitrary curves in P^3, as will be described elsewhere.

Notes

(1) The problem of constructing solutions of the zero rest-mass equations based on an algebraic curve in P^3 arose originally out of considerations involving the explicit construction of instantons (see, e.g., Atiyah & Ward 1977; Hartshorne 1978). The problem is, of course, of considerable interest in its own right. The twisted cubic curve was first examined in this light by Penrose (see §5.9).

(2) For a standard and very readable exposition of the geometry of the twisted cubic curve, see Semple & Kneebone 1952, chapter XII.

(3) Cf. Hughston 1979, pp. 123-124.

(4) Projectively Ξ_α^{PQR} has 15 components. However, it is easy to see that Ξ_α^{PQR} and $\ell_S^P \ell_T^Q \ell_U^R \Xi_\alpha^{STU}$ with $\ell_B^A \in SL(2,C)$ define the same curve. Thus we conclude that P^3 contains a 12-dimensional family of twisted cubics, any two of which are equivalent under automorphisms of P^3.

(5) Here we use the fact that $\eta_{BC}\, Q^{BC} = 0$ and $Q^{BC} \neq 0$ implies, for any Q^{BC}, that $\eta_{BC} = Q^{A}_{(C}\, \lambda_{B)A}$ for some symmetric λ_{AB}.

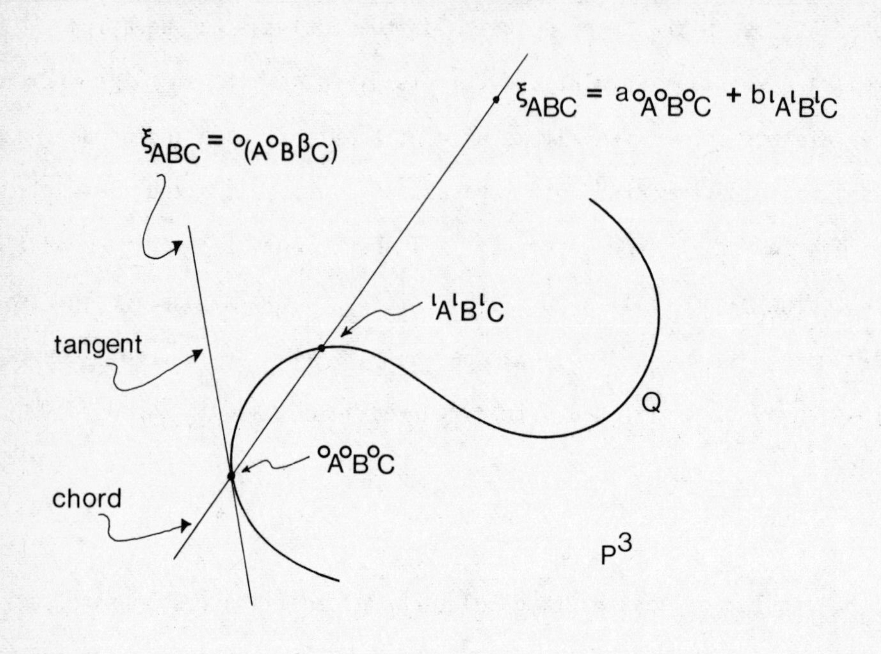

Figure A *A twisted cubic curve*

REFERENCES

Atiyah, M.F. & Ward, R.S. 1977 Commun. Math. Phys. 55, 117-124.

Hartshorne, R. 1978 Commun. Math. Phys. 59, 1-15.

Hughston, L.P. 1979 *Twistors and Particles*. Springer Lecture Notes in Physics, Volume 97.

Semple, J.G. & Kneebone, G.T. 1952 *Algebraic Projective Geometry*. Clarendon Press: Oxford.

3 Curved twistor spaces

§3.1. INTRODUCTION *by R.S. Ward*

The theory of twistors described in chapter 1 is relevant only to flat space-time; can it be generalized to a curved space-time background? The theory of curved twistor spaces arises out of this problem. Despite considerable progress, no completely satisfactory generalization exists; in this section we shall look at some of the different approaches to the problem. For a more detailed review, the reader is referred to Penrose & Ward (1979). We shall also review a topic which, although it does not involve curved space-time, is closely related: the use of curved twistor space techniques in self-dual gauge theories.

(a) Twistors in curved space-time

Let M be a curved space-time; we would like to define twistors in M. One possibility is to have a separate twistor space T_x for each point x of M; in effect, such *local twistors* are constructed in the tangent space at x, which of course is a copy of Minkowski space-time. If we move along a curve in the space-time, there is a natural way of propagating a local twistor along the curve: this is known as *local twistor transport*. In general, this transport is not integrable (i.e. if we propagate a local twistor around a closed loop, then it may not return to its original value). See Penrose & MacCallum (1972); Penrose (1975); Dighton (1974) for more details about this theory. The main disadvantage of local twistors is that they are local; it appears that for this reason the theory does not inherit many of the attractive features of flat-space twistor theory, such as the

Kerr theorem and the contour integral formula for massless fields.

A more global approach is based on the correspondence between null twistors and null geodesics. Let PN' be the space of null geodesics in M. In general PN' might fail to have a Hausdorff manifold structure, because of the occurrence of pairs of conjugate points on the null geodesics in M. Let us for simplicity avoid this difficulty by taking M to be not the whole space-time, but a suitable subregion of the space-time. The following question now arises: Is PN' naturally embedded in a three-complex-dimensional projective twistor space PT'? To throw some light on this question, recall the Kerr theorem: shear-free congruences of null geodesics in Minkowski space-time correspond to the intersection with PN of a holomorphic surface in PT (cf. chapter 1). However, if a shear-free congruence enters a region of space-time where there is conformal curvature, then it picks up shear (Penrose 1968a). In a general curved space-time, therefore, there are *no* shear-free congruences (although there are congruences which are shear-free 'for an instant', e.g. where they intersect some spacelike hypersurface). Thus the Kerr theorem suggests that the complex structure of twistor space is destroyed by conformal curvature. An investigation of what happens when null geodesics pass through an impulsive plane-fronted gravitational wave reveals this phenomenon explicitly (Penrose 1968b; Penrose & MacCallum 1972).

Thus it appears that the above question has to be answered in the negative. However, there are ways of avoiding the difficulties. One of these is to fix a point on each null geodesic, so that the shearing effect of gravitation no longer matters; this leads to the theory of *hypersurface twistors*. Another way is to consider complex space-times in which the right-handed (or left-handed) half of the conformal curvature vanishes, so that there exist "half"-shear-free null congruences. This leads to the *nonlinear graviton* construction. These two theories will now be described.

(b) Hypersurface Twistors

Details of the theories of hypersurface twistors and asymptotic twistors may be found in Penrose & MacCallum (1972); Penrose (1975); Ko, Newman & Penrose (1977); Penrose & Ward (1979); Ward (1977b). The essential idea is that one picks a point on each null geodesic by taking its intersection with a fixed hypersurface S. The theory is simplest if S is taken to be spacelike and to intersect each null geodesic exactly once. But one may also consider the case when S is null, even though one then has to think of how to deal with the null geodesics lying entirely in S. In particular, in an asymptotically flat space-time (Hawking & Ellis 1973) one can take $S = \mathscr{I}^+$ or $S = \mathscr{I}^-$; the hypersurface twistors one then obtains are called *asymptotic twistors*.

The precise definition of hypersurface twistors will not be given here. Briefly, a hypersurface twistor is a certain type of complex curve in the complexification of the hypersurface S. The space PT(S) of hypersurface twistors with respect to S is a three-dimensional complex manifold containing PN' as a real hypersurface. The complex structure of PT(S) depends on S and contains some (but not all) of the information about the space-time metric on S. It is an unsolved problem to determine exactly how much information about the metric *is* contained in the structure of PT(S).

Remarks. (1) The Kerr theorem is replaced by a hypersurface Kerr theorem: an analytic congruence of null geodesics which is shear-free *at S* corresponds in the twistor picture to the intersection with PN' of a holomorphic 2-surface in PT(S).

(2) Since hypersurface twistors are localized to the hypersurface S, one cannot expect there to be a contour integral formula which solves the massless field equations on the curved space-time background. In any event, the existence in curved space-time of ordinary massless fields of helicity greater than one is severely constrained by the Buchdahl conditions (Penrose 1968a).

(3) If S is null, then there is a naturally defined scalar product $\Pi(Z,\bar{Z})$ on PT(S); it is the generalization of the flat-space scalar product $Z^\alpha \bar{Z}_\alpha$. The definition for the case $S = \mathscr{I}$ is given in Penrose & MacCallum (1972) and the definition for the case of a general null hypersurface is essentially the same; see Penrose & Ward (1979). A twistor Z in PT(S) is null (i.e. lies in PN' and so corresponds to a null geodesic) if and only if $\Pi(Z,\bar{Z}) = 0$. The scalar product in turn leads to a Kähler structure on PT(S): the Kähler form is

$$\frac{\partial^2 \Pi}{\partial Z^\alpha \, \partial \bar{Z}_\beta} \, dZ^\alpha \wedge d\bar{Z}_\beta \quad . \tag{1}$$

The Kähler curvature is related to the space-time curvature; see Ko, Newman & Penrose (1977).

(4) Hypersurface twistor spaces have been constructed in some specific space-times, for example in plane-fronted waves $[ds^2 = 2\,du\,dv - dx^2 - dy^2 + 2\,h(v,x,y)\,dv^2]$ and in Schwarzschild space-time $[ds^2 = (1 - 2mr^{-1})\,dv^2 + 2\,dv\,dr - r^2(d\theta^2 + \sin^2\theta\,d\phi^2)]$, taking S in both cases to be the null hypersurface v = constant (Ward 1977b). In both cases, PT(S) turns out to be flat, i.e. to be an open submanifold of CP^3. If one takes two hypersurfaces S_1 (given by $v = v_1$) and S_2 (given by $v = v_2$), one can relate the twistors in $PT(S_1)$ to those in $PT(S_2)$ as follows. There is a natural map from the

null twistors in $PT(S_1)$ to those in $PT(S_2)$, since null twistors in either $PT(S_1)$ or $PT(S_2)$ correspond to null geodesics in the space-time. Now extend this map to the whole space $PT(S_1)$; one then obtains a canonical transformation, preserving the 2-form (1). The map is non-holomorphic and illustrates the way in which the gravitational curvature between the surfaces S_1 and S_2 "shifts" the complex structure of twistor space (cf. Penrose 1968b, p.99).

(c) The Nonlinear Graviton

The curvature tensor of a space-time is made up of two parts: the Weyl conformal curvature tensor C_{abcd} and the Ricci tensor R_{ab}. The Weyl tensor can be split up into its self-dual and anti-self-dual parts:

$$C_{abcd} = C^{+}_{abcd} + C^{-}_{abcd} ,$$
$$C^{\pm}_{abcd} = \tfrac{1}{2}(C_{abcd} \mp \tfrac{1}{2} i \varepsilon_{ab}{}^{ef} C_{efcd}).$$

C^{+}_{abcd} and C^{-}_{abcd} are complex conjugates of each other. However, if we complexify the space-time by allowing the coordinates to become complex and analytically extending the space-time metric (assuming it to be real analytic to begin with), then R_{ab} and C_{abcd} become complex tensors; C^{+}_{abcd} and C^{-}_{abcd} are then no longer complex conjugates. More generally, we can consider complex-Riemannian 4-spaces which are not necessarily the complexifications of real space-times. Such a complex space is said to be *right-flat* (or anti-self-dual) if

$$C^{+}_{abcd} = 0, \quad R_{ab} = 0.$$

It is clear from the above discussion that a right-flat space cannot be the complexification of a real space-time, unless it is flat. However, it is possible for a complex space, or a real space with signature ++++ or ++--, to be right-flat without being flat. Such spaces have arisen in two different areas of general relativity:

(1) The theory of *H*-space (Newman 1976; Hansen, Newman, Penrose & Tod 1978; Ko, Ludvigsen, Newman & Tod 1978). *H*-space is a complex right-flat space which is naturally associated with the $\mathscr{I}^+$ or $\mathscr{I}^-$ of an asymptotically flat space-time. The theory of *H*-space is closely related to asymptotic twistor theory.

(2) Gravitational instantons and the approach to quantum gravity which involves functional integration in positive definite 4-space (Hawking 1977; Hawking 1978).

H-space and asymptotic twistor theory is closely associated with Penrose's (1976) *nonlinear graviton* construction. Penrose proved essentially the following result.

Theorem. There is a one-to-one correspondence between

(i) right-flat spaces M; and

(ii) curved twistor spaces *PT*, together with a certain pair of differential forms on *PT*.

For proofs and further details, see Penrose (1976); Curtis, Lerner & Miller (1978b); Atiyah, Hitchin & Singer (1978).

The construction tells one how to go back and forth between (i) and (ii). It is a generalization of the flat-space correspondence (cf. chapter 1)

α-plane in space-time $\leftrightarrow$ point in PT,

point in space-time $\leftrightarrow$ line in PT.

The concept of an α-plane is generalized to that of an α-*surface* in the curved space-time M. The condition for a three-parameter family of α-surfaces to exist is precisely $C^+_{abcd} = 0$. The lines in PT generalize to compact holomorphic curves (satisfying certain topological conditions) in *PT*.

Thus given a right-flat space M, we can take the corresponding curved twistor space PT to be the space of α-surfaces in M. Conversely, given PT, we take M to be the space of compact holomorphic curves in PT. This correspondence as it stands is between

(i) complex 4-manifolds M with conformal metric satisfying $C^{+}_{abcd} = 0$; and

(ii) curved twistor spaces PT.

The extra structure on M represented by having a right-flat metric (as opposed to just a conformal metric) corresponds in PT to having the two differential forms mentioned in Penrose's theorem above.

In principle the construction can be used to find all right-flat metrics. In practice, only certain special cases and classes have so far proved tractable. One of the first examples was the right-flat analogue of the Schwarzschild solution (see §3.4); this space is also known as the anti-self-dual Taub-NUT space (Hawking 1977). For more recent progress on this problem, see Curtis, Lerner & Miller (1978a); Ward (1978); Hitchin (1979); Tod & Ward (1979).

(d) Outlook

Our original problem was that of defining twistors in curved space-time. Clearly neither hypersurface twistors nor the nonlinear graviton can be regarded as a completely satisfactory solution to this problem. The former theory suffers as a result of being localized to a particular hypersurface; the latter applies only to right-flat spaces. What one really needs is something which is analogous to the nonlinear graviton, but which applies to a general space-time. Some preliminary ideas aimed at this have been put forward (see §3.9), but the problem is as yet unsolved.

(e) Self-Dual Gauge Theories

The gravitational field is usually thought of in terms of geometry, and this geometric interpretation is a crucial feature of the nonlinear graviton construction described above. There is a large class of field theories which have a neat geometric interpretation; they are called *gauge theories*. From the mathematical point of view, gauge theories are described in terms of connections on principal bundles or vector bundles; see Wu & Yang (1975) and Cho (1975).

Let G be a Lie group (the *gauge group*) and let B be a principal G-bundle (or its associated vector bundle) over space-time M. Let A be a connection on B; A is called the *gauge potential*. The corresponding *gauge field* is the curvature F of the connection. If we choose some section of B, to act as a basis, then A and F may be represented (respectively) as a 1-form and a 2-form on M, taking values in the Lie algebra of G. The form F is then given by

$$F = dA + [A,A]. \qquad (2)$$

The Bianchi identities $dF + [A,F] = 0$ follow from (2). The *Yang-Mills equations* are

$$d{*F} + [A,{*F}] = 0,$$

where ${*F}_{ab} = \frac{1}{2}\varepsilon_{abcd}\, F^{cd}$ is the dual of F. A Yang-Mills field is a curvature form F which satisfies the Yang-Mills equations. If the gauge group G is U(1), then the Yang-Mills theory is the same as Maxwell theory. The case G = SU(2) gives the theory originally introduced by Yang and Mills.

We are interested here in special types of Yang-Mills field, namely those which are self-dual or anti-self-dual (i.e. ${*F} = i\,F$ or ${*F} = -i\,F$). Clearly it follows from the Bianchi identities that any self-dual or anti-self-dual curvature automatically satisfies the Yang-Mills equations. There is a correspondence between anti-self-dual ("left-handed") Yang-Mills fields and certain holomorphic vector bundles over projective twistor space: this is described in §3.5. This is preceded by a discussion of the Maxwell case in §3.2. In §3.3 a study is made of a particular Maxwell field, namely the anti-self-dual part of the Coulomb field. The Maxwell construction can also be looked at from the point of view of asymptotic twistor theory and H-space theory: see §3.6.

More details about the twistor technique for solving the anti-self-dual Yang-Mills equations may be found in Atiyah, Hitchin, Drinfeld & Manin (1978); Atiyah, Hitchin & Singer (1978); Atiyah & Ward (1977); Christ, Weinberg & Stanton (1978); Corrigan, Fairlie, Templeton & Goddard (1978); Corrigan, Fairlie, Yates & Goddard (1978); Drinfeld & Manin (1978); Hartshorne (1978); Ward (1977a).

The problem of dealing with Yang-Mills fields which are not self-dual or anti-self-dual, using twistor methods, is still unsolved. There have been two approaches to the problem. One of these may be found in Witten (1978) and Isenberg, Yasskin & Green (1978); the other (which thus far applies only to the Maxwell case) in §3.7 and §3.8.

REFERENCES

Atiyah, M.F., Hitchin, N.J., Drinfeld, V.G. & Manin, Yu. I. 1978 Phys. Lett 65A, 185-187.

Atiyah, M.F., Hitchin, N.J. & Singer, I.M. 1978 Proc. Roy. Soc. A362, 425-461.

Atiyah, M.F. & Ward, R.S. 1977 Comm. Math. Phys. 55, 117-124.

Cho, Y.M. 1975 J. Math. Phys. 16, 2029-2035.

Christ, N.H., Weinberg, E.J. & Stanton, N.K. 1978 Phys. Rev. D18, 2013-2025.

Corrigan, E.F., Fairlie, D.B., Templeton, S. & Goddard, P. 1978 Nucl. Phys. B140, 31-44.

Corrigan, E.F., Fairlie, D.B., Yates, R.G. & Goddard, P. 1978 Comm. Math. Phys. 58, 223-240.

Curtis, W.D., Lerner, D.E. & Miller, F.R. 1978a J. Math. Phys. 19, 2024-2027.

Curtis, W.D., Lerner, D.E. & Miller, F.R. 1978b *Some remarks on the non-linear graviton*. Gen. Rel. Grav., to appear.

Dighton, K. 1974 Int. Jour. Theor. Phys. 11, 31-43.

Drinfeld, V.G. & Manin, Yu.I. 1978 Comm. Math. Phys. 63, 177-192.

Hansen, R.O., Newman, E.T., Penrose, R. & Tod, K.P. 1978 Proc. Roy. Soc. A363, 445-468.

Hartshorne, R. 1978 Comm. Math. Phys. 59, 1-15.

Hawking, S.W. 1977 Phys. Lett. 60A, 81-83.

Hawking, S.W. 1978 Phys. Rev. D18, 1747-1753.

Hawking, S.W. & Ellis, G.F.R. 1973 *The Large Scale Structure of Space-Time*. Cambridge: University Press.

Hitchin, N.J. 1979 Proc. Camb. Phil. Soc. 85, 465-476.

Isenberg, J., Yasskin, P.B. & Green, P.S. 1978 *Non-self-dual Gauge Fields*. Phys. Lett. 78B, 462-464.

Ko, M., Ludvigsen, M., Newman, E.T. & Tod, K.P. 1978 *The Theory of H-Space*. Phys. Reports, to appear.

Ko, M., Newman, E.T. & Penrose, R. 1977 J. Math. Phys. 18, 58-64.

Newman, E.T. 1976 Gen. Rel. Grav. 7, 107-111.

Penrose, R. 1968a *The Structure of Space-Time*. In: *Battelle Rencontres*, eds. C.M. DeWitt & J.A. Wheeler, pp. 121-235. Benjamin: New York.

Penrose, R. 1968b Int. Jour. Theor. Phys. 1, 61-99.

Penrose, R. 1975 *Twistor Theory, its Aims and Achievements.* In: *Quantum Gravity,* eds. C.J. Isham, R. Penrose & D.W. Sciama, pp. 268-407. Oxford: University Press.

Penrose, R. 1976 Gen. Rel. Grav. 7, 31-52.

Penrose, R. & MacCallum, M.A.H. 1972 Phys. Reports 6, 241-315.

Penrose, R. & Ward, R.S. 1979 *Twistors for Flat and Curved Space-Time.*[*]

Tod, K.P. & Ward, R.S. 1979 *Self-dual Metrics with Self-dual Killing Vectors.* To appear in Proc. Roy. Soc.

Ward, R.S. 1977a Phys. Lett. 61A, 81-82.

Ward, R.S. 1977b *Curved Twistor Spaces.* D.Phil. Thesis: Oxford.

Ward, R.S. 1978 Proc. Roy. Soc. A363, 289-295.

Witten, E. 1978 Phys. Lett. 77B, 394-395.

Wu, T.T. & Yang, C.N. 1975 Phys. Rev. D12, 3845-3857.

---

* To appear in an Einstein centennial volume, eds. P.G. Bergmann, J.N. Goldberg & A.P. Held.

§3.2. THE TWISTED PHOTON: MASSLESS FIELDS AS BUNDLES *by R.S. Ward*

The complex structure of projective twistor space PT determines, and is determined by, the conformal structure of Minkowski space-time. By deforming the *non*-projective space T while preserving PT, one can code information about massless free fields on space-time into the global complex structure of non-projective twistor space.

Let B denote the primed spin-bundle over space-time and let $(x^a, \pi_{A'})$ be coordinates on B. The x^a are coordinates on space-time and the $\pi_{A'}$ are coordinates on primed spin-space. Let Υ denote the Euler vector field $\pi_{A'} \frac{\partial}{\partial \pi_{A'}}$ on B. If a function f on B satisfies $\Upsilon f = 0$, then f is homogeneous of degree zero in $\pi_{A'}$ and hence is defined on the projective spin-bundle. Consider now the two-dimensional distribution on B which is spanned by the two vector fields $\{\pi^{A'} \nabla_{AA'}, A = 0,1\}$. The space of integral surfaces of this distribution is precisely the non-projective twistor space T; each integral surface in fact projects down to an α-plane in space-time. In order to deform T without changing PT, we replace the distribution $\{\pi^{A'} \nabla_{AA'}\}$ by $\{\pi^{A'} \nabla_{AA} - \psi_A \Upsilon\}$, where ψ_o and ψ_1 are two functions on B. Frobenius' theorem tells us that this new distribution is integrable if and only if

$$\pi^{A'} \nabla_{AA'} \psi^A - \psi_A \quad \Upsilon \psi^A = 0 \tag{1}$$

for all values of $\pi^{A'}$. So if (1) is satisfied, then there exists a four-dimensional space T' of integral surfaces, and T' is a holomorphic bundle over the (undeformed) projective twistor space PT. T' is a deformation of the "flat" twistor space T.

Let us take ψ_A to have the form

$$\psi_A(x,\pi_{A'}) = i\, \Phi_A^{A'\ldots C'}(x)\, \pi_{A'} \ldots \pi_{C'}.$$

Then equation (1) becomes

$$\nabla^{A(A'}\, \Phi_A^{B'\ldots D')} = 0. \tag{2}$$

The field $\Phi_A^{A'\ldots C'}$ is a potential for a massless free field

$$\phi_{AB\ldots D} = \nabla^{B'}_{(B} \cdots \nabla^{D'}_{D}\, \Phi_{A)B'\ldots D'}\,.$$

The field equations $\nabla^{AA'}\, \phi_{AB\ldots D} = 0$ follow automatically from (2). The information of the field $\phi_{AB\ldots D}$ is contained in the deformed twistor space T'. The details of the construction are not yet well understood, except in the case of spin one (i.e. Maxwell theory). We shall now examine this case in more detail.

Suppose, therefore, that $\psi_A = i\, \Phi_A^{A'}(x)\pi_{A'}$, with

$$\nabla^{A(A'}\, \Phi_A^{B')} = 0. \tag{3}$$

In this case, T' is a principal fibre bundle over PT, with group C* (the multiplicative group of non-zero complex numbers). In the usual way, one can also think of it as a line bundle. (In the case of spin not equal to one, T' is not a line bundle: it is some other kind of bundle.) It should be pointed out that we are using PT to denote not the whole of complex projective 3-space, but that portion of it which corresponds to the region of space-time under consideration. One should think of PT as the neighbourhood of a line in CP^3.

There is a close connection between the sheaf cohomological treatment of Maxwell fields described in §§2.2 and 2.8 and the standard sheaf cohomological treatment of line bundles (for an exposition of the latter topic, see e.g.

R.C. Gunning, "Lectures on Riemann Surfaces", Princeton 1966). Briefly, the correspondence is as follows. We have the long exact sequence

$$\ldots \to H^1(PT,Z) \to H^1(PT, \mathcal{O}) \to H^1(PT, \mathcal{O}^*) \to H^2(PT,Z) \to \ldots \tag{4}$$

Assuming PT to have topology $R^4 \times S^2$, the group $H^1(PT,Z)$ vanishes. $H^1(PT, \mathcal{O})$ is isomorphic to the space of left-handed Maxwell fields $\{\phi_{AB} | \nabla^{AA'} \phi_{AB} = 0\}$. $H^1(PT, \mathcal{O}^*)$ is the space of line bundles over PT. $H^2(PT,Z) \cong Z$ is the space of possible Chern classes of such bundles. It is not difficult to convince oneself that the bundles T' constructed above all have Chern class zero. Putting all this together, we conclude from the sequence (4) that the space of left-handed Maxwell fields is isomorphic to the space of deformed line bundles T'.

One can realize this correspondence as follows. Cover T with the two coordinate patches U_1 (with coordinates Z_1^α) and U_2 (with coordinates Z_2^α). The flat twistor space T is obtained by patching U_1 and U_2 together according to $Z_2^\alpha = Z_1^\alpha$. The deformed space T' is obtained by using a patching relation

$$Z_2^\alpha = \exp\{i\ f(Z^\beta)\}\ Z_1^\alpha ,$$

where $f(Z^\beta)$ is a holomorphic function on $U_1 \cap U_2$, homogeneous of degree zero in Z^α. We know from the above discussion that T' corresponds to a Maxwell field ϕ_{AB}. In fact, it is not difficult to show that ϕ_{AB} is given by

$$\phi_{AB} = (2\pi i)^{-1} \oint \rho_x\ \hat{\pi}_A\ \hat{\pi}_B\ f(Z)\ \pi_{C'}\ d\pi^{C'}$$

(cf. §2.1).

There is yet another way of looking at the construction. A twistor determines an α-plane, together with a primed spinor $\pi_{A'}$ propagated over the α-plane. The usual propagation law is parallel propagation, i.e.

$\pi^{A'} \nabla_{AA'} \pi_{B'} = 0$. But in the deformed case we take the propagation equation to be

$$\pi^{A'} (\nabla_{AA'} + i\, \Phi_{AA'})\, \pi_{B'} = 0.$$

The integrability condition for this equation is precisely (3).

Acknowledgement

I should like to thank G.A.J. Sparling and R. Penrose for useful discussions and suggestions.

§3.3. THE TWISTOR QUADRILLE: A LINE BUNDLE BASED ON THE COULOMB FIELD

by R. Penrose and G.A.J. Sparling

The aim of this article is to show explicitly how to construct a twistor space which encapsulates the global structure of a left-handed (i.e. anti-self-dual) Coulomb field. The construction dovetails in neatly with an analogous construction for the Schwarzschild-like graviton (see §3.4). We carry out the construction of §3.2 for a line bundle over part of projective twistor space incorporating the Coulomb field in any suitable sufficiently small four-dimensional neighbourhood (with topology C^4), of complexified Minkowski space, on which the field is non-singular. We then try to extend this line bundle to handle the Coulomb field in a complex four-dimensional neighbourhood of a two-dimensional real sphere, surrounding the source singularity. We find that our construction breaks down unless a certain integrality condition holds; the construction forces electric charge to come in fixed multiples of one basic unit. This is closely related to the famous Dirac argument (which uses magnetic monopoles) but is not quite the same.

The construction goes as follows. We first recall the building of the line bundle over that portion of projective twistor space which contains all the projective twistors which pass through points belonging to a neighbourhood with topology C^4 in complex Minkowski space. This portion may be regarded as a certain neighbourhood U, say, of a line in projective twistor space. As usual we split the neighbourhood into two overlapping regions: U_1, with projective coordinates $\{Z_1^\alpha\}$, and U_2, with projective coordinates $\{Z_2^\alpha\}$, such that the overlap $U_1 \cap U_2$ contains an annular region of each compact projective line in U. We also use coordinates Z_i^α for that part of $C^4-\{0\}$ which projects down to U_i under the canonical map $C^4-\{0\} \to CP^3$, i=1,2.

We may assume that the projective coordinates $\{Z_1^\alpha\}$ for a point on $U_1 \cap U_2$ are the same as $\{Z_2^\alpha\}$. Then if we identified Z_1^α with Z_2^α on the overlap region we would obtain just the usual non-projective twistors which project to U, as a principal bundle of a line bundle over U. However here we identify the point with coordinates Z^α in the U_2 chart with the point with coordinates $\exp\{f(Z^\beta)\}\ Z^\alpha$ in the U_1 chart (i.e. $Z_1^\alpha = Z^\alpha$ corresponds to $Z_2^\alpha = \exp\{f(Z^\beta)\}Z^\alpha$), where we take

$$f(Z^\alpha) = \log\left(\frac{Q}{Z^2\ Z^3}\right) \quad ,$$

$$Q = Z^1\ Z^2 - Z^0\ Z^3.$$

So our overlap region has to be such that it avoids the singularities $Q = 0$, $Z^2 = 0$ or $Z^3 = 0$, appropriately. This function f produces a Coulomb field when it is substituted into the familiar contour integral formula.

To study the overlap region more carefully, we picture the singularities of the twistor function in Minkowski space. We may restrict our attention to a real two-sphere surrounding the charge, as the picture will be essentially unchanged in a sufficiently small complex neighbourhood of that sphere. We may assume that the sphere lies in a t=constant plane with centre on the charge wordline in a frame in which the charge is at rest. At each point $x^{AA'}$ of the two-sphere there are (generically) four spinors $\pi_{A'}$, defined up to proportionality, such that if $Z^\alpha = (i\ x^{AA'}\ \pi_{A'}, \pi_{A'})$, then $f(Z^\alpha)$ is singular. We project the null vectors $\bar{\pi}^A\ \pi^{A'}$ into the surface t=constant and obtain the picture in Fig. A, the flag-ended lines representing the singular directions at each point. The flags A and B at each point correspond to the singularities $Z^2 = 0$ and $Z^3 = 0$ respectively. The flags O and I correspond to the singularities $Q = 0$, O pointing outward from and I pointing inward to the

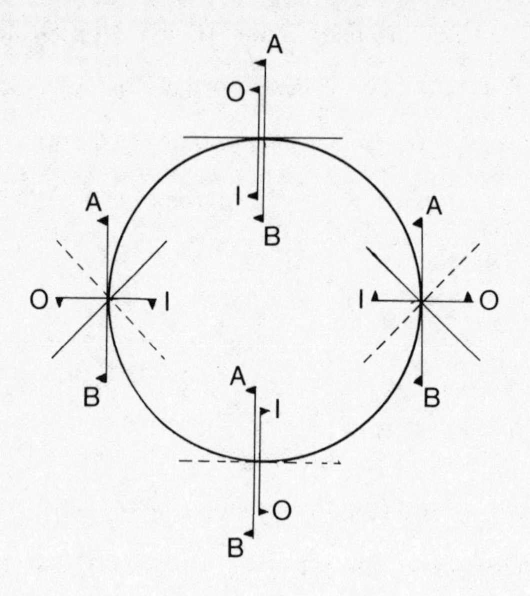

Figure A

centre. For simplicity and with no real loss of generality we may assume that A coincides with O at the north pole, and B with I. Then the whole picture is rotationally symmetric about the north-south axis. We portray the overlap region at various points of the sphere by drawing a solid line to represent the orientation, in the sphere of directions at each point, of the central great circle of the overlap region, for the sphere's worth of twistors that represent the given point in twistor space. (In this construction the overlap region may always be taken to consist of an annular neighbourhood of a great circle of the sphere of directions, for each point.) At the north pole the annular region may be taken to be equatorial as indicated in Fig. A. As we move away from the north pole, O moves away from

A and the annular region gradually has to tip up to allow for this. Eventually when we reach the south pole, the region becomes pinched. We may define the region U_1 by saying that for each point on the partial sphere in which a neighbourhood of the south pole is deleted, we have a partial sphere of directions for the twistors through those points, this partial sphere being tipped up at the point x on the equator (given the tipping up at x, we may work it out everywhere else by continuity). This is illustrated in Fig. B, where the sets U_i and their intersections are represented in the

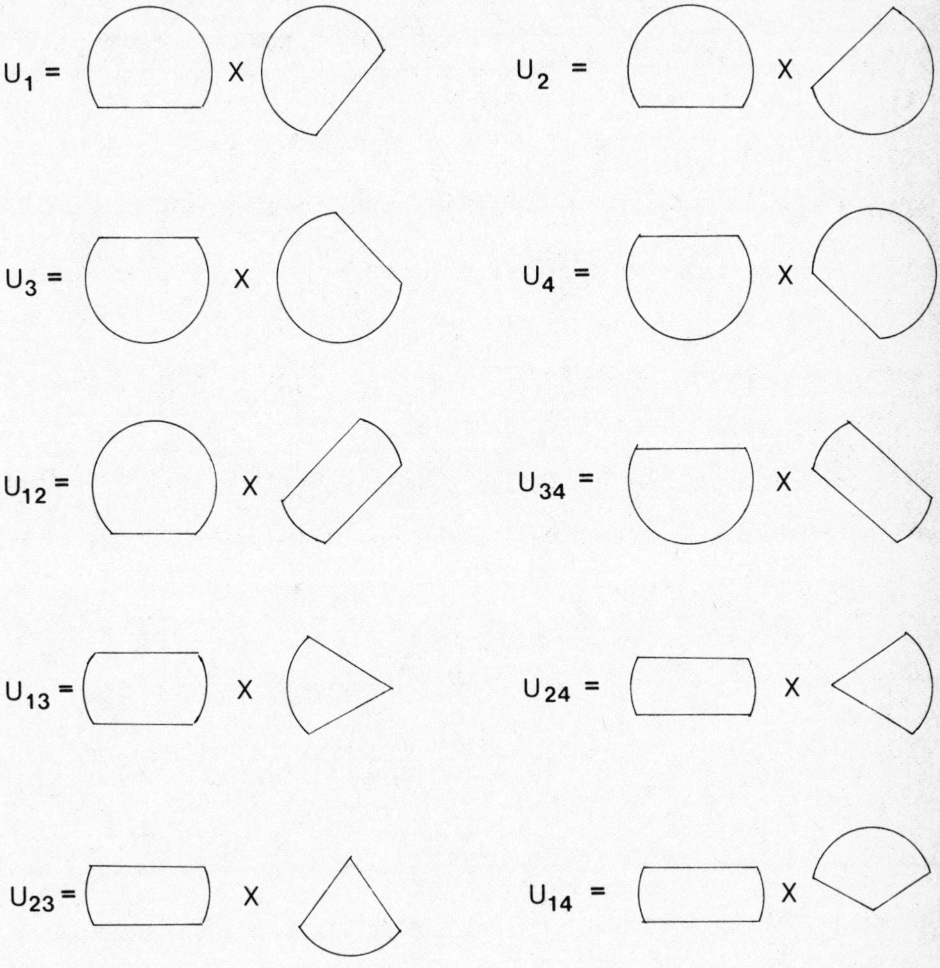

U_1 = X U_2 = X

U_3 = X U_4 = X

U_{12} = X U_{34} = X

U_{13} = X U_{24} = X

U_{23} = X U_{14} = X

Figure B

form U_i = (partial sphere of points) × (allowed directions at a point x on the equator). The transition function to be used on the overlap region is $c_{12}(Z^\alpha) = \exp\{f(Z^\alpha)\} = Q\,(Z^2\,Z^3)^{-1}$.

This finishes the construction as far as the northern hemisphere is concerned and produces a twistor space carrying the structure of the Coulomb field in a small neighbourhood of that hemisphere. However we could equally well start from the south pole, using regions U_3 and U_4 as indicated with a dotted line in Fig. A and illustrated in Fig. B. The coordinates Z_3^α in U_3 and Z_4^α in U_4 are patched together according to $Z_3^\alpha = c_{34}(Z^\beta)\,Z_4^\alpha$ on the overlap region $U_{34} = U_3 \cap U_4$, where $c_{34}(Z^\alpha) = Q\,(Z^2\,Z^3)^{-1}$. Now can we patch all the patches together to give a line bundle corresponding to a neighbourhood of the whole two-sphere? We *can* do so if we can find transition functions $c_{ij}(Z^\alpha)$ defined on $U_{ij} = U_i \cap U_j$ such that $c_{ij}\,c_{jk} = c_{ik}$ on U_{ijk} = $U_i \cap U_j \cap U_k$, for all i,j,k (with $c_{ii} = 1$, $c_{ij} = c_{ji}{}^{-1}$), identifying $Z_j^\alpha = Z^\alpha$ in U_j with $Z_i^\alpha = c_{ij}(Z^\beta)\,Z^\alpha$ in U_i.

Now consider, for instance, the relation $c_{12} = c_{14}\,c_{42}$ on U_{124}, so $Q(Z^2\,Z^3)^{-1} = c_{14}\,c_{42}$. Now U_{14} contains the direction $Z^2 = 0$, so c_{14} cannot be singular there. However U_{14} does not contain directions where $Q = 0$ or $Z^3 = 0$. Also U_{24} contains a direction where $Q = 0$ but not where $Z^2 = 0$ or $Z^3 = 0$. So we may solve for

$$c_{14}(Z) = Q\,(Z^3)^{-2} \quad \text{and} \quad c_{42} = Z^3\,(Z^2)^{-1}.$$

Next consider $c_{12} = c_{13}\,c_{32}$ on U_{123}. A similar analysis yields

$$c_{32} = Q\,(Z^2)^{-2} \quad \text{and} \quad c_{13} = Z^2\,(Z^3)^{-1}.$$

Then we have $c_{32}\,c_{34} = c_{34}$ on U_{234} and $c_{31}\,c_{14} = c_{34}$ on U_{134}, so we have a complete solution. The four singularities are just right that they can dance out of trouble. The solution is unique up to multiplication by an overall constant. To see the integrality of charge showing up, we look at the effect of replacing $f(Z)$ by $\lambda f(Z)$ for some $\lambda \neq 0$, $\lambda \in C$. Then $c_{ij}(Z)$ becomes $(c_{ij}(Z))^{\lambda}$ and again this is the only possible solution up to an overall constant. But then $(c_{13})^{\lambda}$ is not single-valued on U_{13} nor is $(c_{24})^{\lambda}$ on U_{24}, unless λ is integral. So the construction fails to work globally around the two-sphere unless λ is integral. Although we have not proved here that no other construction will work, this follows from a sheaf-cohomological discussion (see §2.2).

§3.4. THE SCHWARZSCHILD GRAVITON *by G.A.J. Sparling*

We use Penrose's "Nonlinear Graviton" construction to construct the right-flat analogue of the Schwarzschild (or Taub-NUT) space-time. We begin with two coordinate patches U_1 (with coordinates Z_1^α) and U_2 (with coordinates Z_2^α). The twistor function

$$f(Z^\alpha) = Q \log\{Q(Z^2 Z^3)^{-1}\},$$

where $\quad Q = Z^1 Z^2 - Z^0 Z^3,$

produces the linearized Schwarzschild solution when substituted into the standard contour integral formula. This suggests that we should patch U_1 and U_2 as follows: $Z_1^\alpha = Z^\alpha = (\omega^A, \pi_{A'})$ is identified with $Z_2^\alpha = \tilde{Z}^\alpha = (\tilde{\omega}^A, \tilde{\pi}_{A'})$, where

$$\tilde{\omega}^A = \omega^A + \lambda\varepsilon^{AB} \frac{\partial f}{\partial \omega^B},$$

$$\tilde{\pi}_{A'} = \pi_{A'},$$

or equivalently

$$\tilde{Z}^\alpha = Z^\alpha + \lambda P^\alpha \log\{Q (Z^2 Z^3)^{-1}\},$$

$$P^\alpha = (Z^2, Z^3, 0, 0).$$

Under $Z^\alpha \mapsto Z^\alpha + \mu P^\alpha$, we have $\tilde{Z}^\alpha \mapsto \tilde{Z}^\alpha + \mu P^\alpha$, so we may identify Z_1^α with $Z_1^\alpha + 2\pi i \lambda P^\alpha$ and Z_2^α with $Z_2^\alpha + 2\pi i \lambda P^\alpha$ and ignore henceforth the fact that the logarithm may not be single-valued.

Our patches U_1, U_2, U_3, U_4 will be almost the same as we used for the construction of the Coulomb twistor space in §3.3. We again use a neighbourhood of the product of a real two-sphere and a partial sphere of directions at each point; this time, however, we use a complex three-dimensional neighbourhood of the real two-sphere, lying in the plane t=constant, and then spread the neighbourhood by time translation, $t \mapsto t + 2\pi\kappa\lambda$, together with parallel translation of the partial sphere of directions, where κ occupies a closed strip in its Argand plane bounded by the straight lines $\mathrm{Re}\,\kappa = 0$ and $\mathrm{Re}\,\kappa = 1$. Note that Q, Z^2 and Z^3 are invariant under time translation. Each partial sphere of directions is made as large as possible, so omits just two directions (generically) and each partial sphere of points is made as large as possible, so omits just one point, the north or south pole, the south for U_1 and U_2, the north for U_3 and U_4.

Then the patching goes as follows (compare with the patching in §3.3).

$Z_1^\alpha = Z^\alpha$ is identified with $Z_2^\alpha = Z^\alpha + \lambda P^\alpha \log\{Z^2 Z^3 Q^{-1}\}$ on U_{12},
with $Z_3^\alpha = Z^\alpha + \lambda P^\alpha \log\{Z^3 (Z^2)^{-1}\}$ on U_{13},
with $Z_4^\alpha = Z^\alpha + \lambda P^\alpha \log\{(Z^3)^2 Q^{-1}\}$ on U_{14}.

$Z_2^\alpha = Z^\alpha$ is identified with $Z_1^\alpha = Z^\alpha + \lambda P^\alpha \log\{Q(Z^2 Z^3)^{-1}\}$ on U_{12},
with $Z_3^\alpha = Z^\alpha + \lambda P^\alpha \log\{Q(Z^2)^{-2}\}$ on U_{23},
with $Z_4^\alpha = Z^\alpha + \lambda P^\alpha \log\{Z^3(Z^2)^{-1}\}$ on U_{24}.

$Z_3^\alpha = Z^\alpha$ is identified with $Z_1^\alpha = Z^\alpha + \lambda P^\alpha \log\{Z^2(Z^3)^{-1}\}$ on U_{13},
with $Z_2^\alpha = Z^\alpha + \lambda P^\alpha \log\{(Z^2)^2 Q^{-1}\}$ on U_{23},
with $Z_4^\alpha = Z^\alpha + \lambda P^\alpha \log\{Z^2 Z^3 Q^{-1}\}$ on U_{34}.

$Z_4^\alpha = Z^\alpha$ is identified with $Z_1^\alpha = Z^\alpha + \lambda P^\alpha \log\{Q (Z^3)^{-2}\}$ on U_{14},
with $Z_2^\alpha = Z^\alpha + \lambda P^\alpha \log\{Z^2 (Z^3)^{-1}\}$ on U_{24},
with $Z_3^\alpha = Z^\alpha + \lambda P^\alpha \log\{Q (Z^2 Z^3)^{-1}\}$ on U_{34}.

It is easy to verify that these relations are consistent on U_{123}, U_{124}, U_{134} and U_{234}, using the fact that P^α, Q, Z^2 and Z^3 are all preserved under the identifications. This completes the global definition of the twistor space.

There is a four-parameter group of symmetries, given by:

$$Z_1^\alpha \mapsto Z_1^\beta T_\beta^\alpha + \lambda P^\beta T_\beta^\alpha \log \{Z^3 (Z^\gamma T_\gamma^3)^{-1}\}, \tag{1}$$
$$Z_2^\alpha \mapsto Z_2^\beta T_\beta^\alpha + \lambda P^\beta T_\beta^\alpha \log \{Z^\gamma T_\gamma^2 (Z^2)^{-1}\}, \tag{2}$$
$$Z_3^\alpha \mapsto Z_3^\beta T_\beta^\alpha + \lambda P^\beta T_\beta^\alpha \log \{Z^2 (Z^\gamma T_\gamma^2)^{-1}\}, \tag{3}$$
$$Z_4^\alpha \mapsto Z_4^\beta T_\beta^\alpha + \lambda P^\beta T_\beta^\alpha \log \{Z^\gamma T_\gamma^3 (Z^3)^{-1}\}, \tag{4}$$

where T_β^α satisfies

$$Q(Z^\beta T_\beta^\alpha) = Q(Z^\alpha), \text{ for all } Z^\alpha,$$
$$T_\gamma^\alpha T_\delta^\beta I_{\alpha\beta} = I_{\gamma\delta},$$
$$T_\gamma^\alpha T_\delta^\beta I^{\gamma\delta} = I^{\alpha\beta},$$

$I_{\alpha\beta}$ and $I^{\alpha\beta}$ being the "infinity twistors" (see Chapter 1). Note that it is possible that $\log (Z^\alpha T_\alpha^3)$ is singular in the U_1 region for suitable Z^α and T_α^β, so (1) breaks down. This is just because the point Z^α moves outside the U_1 region, so cannot be described by coordinate transformations in the U_1 coordinates only. However the transformation can still be described in terms of suitable coordinates.

We note that the twistor space possesses a four-parameter set of globally defined functions homogeneous of degree two, namely the set $\{\alpha\, Q(Z) + S^{A'B'} \pi_{A'} \pi_{B'}\}$, where α and $S^{A'B'}$ are arbitrary constants. Every holomorphic curve representing a point of the space-time then lies entirely on a "quadric" of equation $\alpha Q + S^{A'B'} \pi_{A'} \pi_{B'} = 0$ for some α, $S^{A'B'}$. Each symmetry of the space leaves each quadric invariant.

We next describe the holomorphic curves lying in $U_1 \cup U_2$ representing the points of the space-time. Instead of the coordinates ω^A, introduce the new coordinates $\omega_{A'}$ defined by $\omega^0 = \omega_{0'}$, $\omega^1 = \omega_{1'}$. The patching identifies $Z_2^\alpha = (\omega_{A'}, \pi_{B'})$ with $Z_1^\alpha = (\tilde{\omega}_{A'}, \pi_{B'})$, where

$$\tilde{\omega}_{A'} = \omega_{A'} + \lambda \pi_{A'} \log \{\omega_{B'} \pi^{B'} (\pi_{0'} \pi_{1'})^{-1}\}.$$

For a holomorphic curve, $\tilde{\omega}^{A'} \pi_{A'} = \omega^{A'} \pi_{A'} = -Q^{A'B'} \pi_{A'} \pi_{B'}$ for some $Q^{A'B'} = Q^{B'A'}$. Put $Q^{A'B'} = \xi_{(A'} \eta_{B')}$. Then the holomorphic curves are given by

$$\omega^{A'} = -\lambda \ \pi^{A'} \log \{\xi_{B'} \pi^{B'} (\pi_{o'})^{-1}\} - Q^{A'B'} \pi_{B'} + t \ \pi^{A'},$$
$$\tilde{\omega}^{A'} = \lambda \ \pi^{A'} \log \{\eta_{B'} \pi^{B'} (\pi_{1'})^{-1}\} - Q^{A'B'} \pi_{B'} + t \ \pi^{A'},$$

provided $\pi_{o'} \neq 0$ and $\xi_{B'} \pi^{B'} \neq 0$ in the U_1 patch and $\pi_{1'} \neq 0$ and $\eta_{B'}\pi^{B'} \neq 0$ in the U_2 patch. Note that we apparently have a five-parameter system of curves. However $(\xi^{A'}, \eta^{A'}, t)$ and $(\theta\xi^{A'}, \theta^{-1} \eta^{A'}, t + \lambda \log \theta)$ give the same curve, for any θ, so we must identify $(\xi^{A'}, \eta^{A'}, t)$ with $(\theta\xi^{A'}, \theta^{-1}\eta^{A'}, t + \lambda \log \theta)$ for all θ. Then if $\lambda \neq 0$ (non-flat case) we may gauge t away completely. We assume $\lambda \neq 0$, so we may take for the holomorphic curves

$$\omega^{A'} = -\lambda \ \pi^{A'} \log \{\xi_{B'} \pi^{B'} (\pi_{o'})^{-1}\} - Q^{A'B} \pi_{B'} ,$$
$$\tilde{\omega}^{A'} = \lambda \ \pi^{A'} \log \{\eta_{B'} \pi^{B'} (\pi_{1'})^{-1}\} - Q^{A'B'} \pi_{B'} .$$

Then the metric turns out to be given by

$$ds^2 = -4(\lambda - \xi_{B'} \eta^{B'}) \, d\xi_{A'} \, d\eta^{A'} + \left(\frac{2\lambda - \xi_{B'}\eta^{B'}}{\lambda - \xi_{C'} \eta^{C'}}\right)(\xi_{A'} d\eta^{A'} - \eta^{A'} \, d\xi_{A'})^2.$$

Clearly we may extend our coordinate patch so that all $\xi_{A'} \in C^2 - \{0\}$, $\eta_{A'} \in C^2 - \{0\}$ such that $\xi_{A'} \eta^{A'} \neq 0$, $\xi_{A'} \eta^{A'} - \lambda \neq 0$ are allowed. We do not examine the "horizon" $\xi_{A'} \eta^{A'} = \lambda$ here.

There is a four-parameter group of Killing symmetries given by

$$\xi_{A'} \mapsto t_{A'}{}^{B'} \xi_{B'} , \quad \eta_{A'} \mapsto \tilde{t}_{A'}{}^{B'} \eta_{B'}, \text{ where } t_{A'}{}^{B'} \tilde{t}^{A'C'} = \varepsilon^{B'C'},$$

the group being GL(2,C). The corresponding Lie algebra of Killing vectors is given by

$$\xi_{B'}\, m_{A'}{}^{B'} \frac{\partial}{\partial \xi_{A'}} + \eta_{B'}\, m^{B'}{}_{A'} \frac{\partial}{\partial \eta_{A'}}$$

for arbitrary $m_{A'}{}^{B'}$. The particular case $\xi_{A'} \mapsto \theta \xi_{A'}$, $\eta_{A'} \mapsto \theta^{-1} \eta_{A'}$ corresponds to "time" translation and is the transformation of the space-time determined by the global function Q on the twistor space via the Poisson bracket structure for twistor space $(\frac{1}{2}\varepsilon^{AB} \frac{\partial}{\partial \omega^A} \otimes \frac{\partial}{\partial \omega^B})$; i Q gives rise to the Hamiltonian vector field $P^\alpha \frac{\partial}{\partial Z^\alpha}$ with integral curves $Z^\alpha \mapsto Z^\alpha + i\, \lambda\, P^\alpha$ leading to the time translation of the space-time. Every "quadric" in the four-parameter set $\{\alpha Q + S^{A'B'}\, \pi_{A'}\, \pi_{B'}\}$ such that $S^{A'B'} \neq 0$ corresponds to an orbit of the time translation Killing vector, so each such quadric is ruled by a one-parameter system of holomorphic curves representing the points of the orbit of the Killing vector. When $S^{A'B'} = 0$, the holomorphic curves disappear (the singularity). Each quadric also possesses a one-parameter family of (non-compact) holomorphic curves with a natural linear structure: if Z^α is a point on such a curve, then the others are given by $Z^\alpha + \beta\, P^\alpha$ as β varies. They are the integral curves of the Hamiltonian vector field $P^\alpha \frac{\partial}{\partial Z^\alpha}$. They appear to represent families of points of the conformal infinity of the space-time. The space-time is of algebraic type (2,2).

§3.5. TWISTOR CONSTRUCTION FOR LEFT-HANDED GAUGE FIELDS *by R.S. Ward*

This note is concerned with generalizing the twisted photon construction of §3.2 to non-Abelian gauge theories. We begin with an outline of gauge theory (for more details, see the references listed in subsection (e) of §3.1). Let G be a Lie group and $\{L_j\}$ a matrix representation of its Lie algebra; suppose that the L_j are $n \times n$ matrices. A *gauge potential* $\Phi_a = \Phi_a^j L_j$ is an $n \times n$ matrix of 1-forms which is a linear combination of the L_j's. The corresponding *gauge field* F_{ab} is defined by

$$F_{ab} = 2\nabla_{[a}\Phi_{b]} + [\Phi_a, \Phi_b],$$

and is said to be *left-handed* iff $F^*_{ab} \equiv \frac{1}{2}e_{abcd}F^{cd} = -iF_{ab}$. Such a field automatically satisfies the Yang-Mills equations

$$\nabla^a F_{ab} + [\Phi^a, F_{ab}] = 0.$$

The gauge field may be pictured geometrically as a *connection* on an n-dimensional complex vector bundle V over complex Minkowski space-time CM. The connection tells one how to propagate a vector $\psi \in V$ along a curve γ in CM: the propagation law is

$$v^a(\nabla_a + \Phi_a)\ \psi = 0, \tag{1}$$

where v^a is tangent to γ and where ψ is regarded as a column n-vector, so that Φ_a acts on ψ by matrix multiplication.

If one propagates ψ round a closed path in CM, it does not, in general, return to its original value: in other words, the propagation law (1) is not integrable. But suppose we restrict attention to closed curves which

(a) can be continuously shrunk to a point without crossing singularities of Φ_a;

(b) lie in totally null 2-planes in CM (of the type having tangent vectors of the form $\beta^A \pi^{A'}$, with $\pi^{A'}$ fixed and β^A arbitrary).

Then it is not hard to show that (1) is integrable, provided the gauge field is left-handed. In fact, this integrability characterizes left-handed gauge fields.

Let us assume that Φ_a is holomorphic in the future tube CM^+. The space of totally null 2-planes in CM^+ is just PT^+. Consider the space K of pairs (Z,ψ), where Z is a totally null 2-plane in CM^+, and where ψ is a section of V over Z, satisfying (1). If the gauge field is left-handed, then there is an n-complex-dimensional family of such sections. It follows that K has the structure of an n-dimensional vector bundle over PT^+. So, starting with a left-handed gauge field in CM^+, we have built a vector bundle K over PT^+ (it is possible to think of K as a deformation of a space of n twistors all of which are proportional to one another). The crucial point is that K contains, in its complex structure, all the information about the gauge field. Given K, one can reconstruct Φ_a; one way of doing this is as follows.

Let us suppose that PT^+ is covered by two patches U and $\hat{U}$ and that K is determined by the transition matrix f(Z). So f(Z) is an $n \times n$ matrix of twistor functions homogeneous of degree zero and holomorphic on $U \cap \hat{U}$. Write $F(x^a,\pi_{A'}) = f(ix^{AA'}\pi_{A'}, \pi_{A'})$. Then for fixed x, $F(x^a, \pi_{A'})$ is holomorphic on $W \cap \hat{W}$, where W and $\hat{W}$ are two patches covering the π-sphere. Now "split" F as follows:

$$F(x^a, \pi_{A'}) = \hat{H}(x^a, \pi_{A'})\, H(x^a, \pi_{A'})^{-1} , \tag{2}$$

where H and $\hat{H}$ are homogeneous of degree zero in π, and holomorphic in W and $\hat{W}$ respectively. It is always possible to find such a splitting, provided that the bundle K, restricted to the line in PT^+ corresponding to x^a, is analytically trivial. It is now a simple matter to derive the gauge field: the matrix $H^{-1} \pi^{A'} \nabla_{AA'} H$ turns out to have the form $\pi^{A'} \Phi_{AA'}(x^a)$. This defines $\Phi_a(x)$ and one can check that it is indeed a left-handed solution of the Yang-Mills equations. The splitting (2) is not unique: the choice of a particular splitting corresponds exactly to a choice of gauge for Φ_a.

Remark. The encoding of the gauge field into the vector bundle structure expresses the way in which the gauge field interacts with other fields. For example (roughly speaking) a holomorphic cross-section of K gives rise (via contour integration) to a multiplet of zero-rest-mass fields on CM^+, which are coupled to the gauge field in the correct way.

§3.6. THE GOOD CUT EQUATION FOR MAXWELL FIELDS *by E.T. Newman*

This article uses the notation of *H*-space theory, details of which may be found in the *H*-space references mentioned in §3.1.

Consider a function on $\mathscr{I}^+_C$, $A(u,\zeta,\tilde{\zeta})$ of spin-weight 1, which is holomorphic on a thickened region about real $\mathscr{I}^+$, i.e. for real u and for $\tilde{\zeta} = \bar{\zeta}$. Now let $u = L \equiv x^a L_a(\zeta,\tilde{\zeta})$ with $L_a = (1 + \zeta\tilde{\zeta})^{-1} (1 + \zeta\tilde{\zeta}, \zeta + \tilde{\zeta}, -i(\zeta - \tilde{\zeta}), -1 + \zeta\tilde{\zeta})$ and consider the differential equation

$$\eth F = A(L, \zeta,\tilde{\zeta}) \tag{1}$$

for the spin-weight 0 function $F(x^a, \zeta,\tilde{\zeta})$. If we restrict ourselves to the regular solution on the S^2 defined by $\tilde{\zeta} = \bar{\zeta}$, then $F(x^a,\zeta,\tilde{\zeta})$ generates a self-dual source-free Maxwell field by the following method. The potential γ_a is defined by

$$\gamma_a = F_{,a} + \eth h\, L_a - h\, \eth L_a \,, \tag{2}$$

where $h = L^b \eth F_{,b}$. It can be shown that $\gamma_a(x)$ is independent of ζ and $\tilde{\zeta}$ and furthermore defines by $F_{ab} = 2\, \gamma_{[b,a]}$ a self-dual Maxwell field.

This procedure, which is essentially due to G.A.J. Sparling, can be generalized to arbitrary spin massless fields and to arbitrary self-dual Yang-Mills-type gauge theories.

A question immediately arises as to the mathematical nature of the "functions" $A(L,\zeta,\tilde{\zeta})$ and $F(x^a,\zeta,\tilde{\zeta})$. [We will actually be more interested in $G = G(x^a,\zeta,\tilde{\zeta}) \equiv e^F$.] Should they be thought of as forms or as elements of cohomology groups? What is their relationship to the line bundle approach of §3.2?

One can in a certain sense understand $G = e^F$ by considering it first as a surface in a three-dimensional space with coordinates $(G,\zeta,\tilde{\zeta})$ which is regular on the $\tilde{\zeta} = \bar{\zeta}$ diagonal. If ζ is restricted to the two values $\zeta = \zeta_0$ and $\zeta = \zeta_1$ then the surface is restricted to the two lines $G = G_0(\tilde{\zeta}) \equiv G(x^a, \zeta_0, \tilde{\zeta})$ and $G = G_1(\tilde{\zeta}) \equiv G(x^a, \zeta_1, \tilde{\zeta})$. We now define for those values of $\tilde{\zeta}$ where both G_0 and G_1 are defined, the mapping from one line to the other

$$G_0\, G_1^{-1} = e^{F_0 - F_1} = \mathcal{G}(x^a, \tilde{\zeta}),$$

which is to be thought of as an element of $H^1(CP^1, \mathcal{O}^*)$ constructed over a line in twistor space. Actually one can show, by integrating (1) along ζ from ζ_0 to ζ_1, that $F_0 - F_1$ is a twistor function. The decomposition of $\mathcal{G}$ into $e^{F_0}\, e^{-F_1}$ is the Sparling splitting of the twistor function.

From this it follows that $A(L, \zeta, \tilde{\zeta})$ is essentially to be considered as the Dolbeault version of the $H^1(CP^1, \mathcal{O})$, i.e. it is a $\bar{\partial}$-closed (0,1)-form. Since the bundle is trivial, $H^1 = 0$ and the one-form is $\bar{\partial}$-exact. This is actually the content of equation (1), where F is essentially (modulo factors of $1 + \zeta\tilde{\zeta}$) the function such that $\bar{\partial}F = A$.

§3.7. THE GOOGLY PHOTON* *by R. Penrose*

Current twistor dogma presents the following picture of things:

(i) The wave function of a particle or system or particles is to be given by a "twistor function" $\psi(X^\alpha,\ldots,Z^\alpha)$, which is holomorphic in all arguments $X^\alpha,\ldots,Z^\alpha$ --possibly with the modification that some (or all) of these should be dual twistors $W_\alpha,\ldots$ instead (or, conceivably, Hughston-type multi-twistors $X^{\alpha\beta\gamma},\ldots$, etc.).

(ii) More correctly, ψ is (presumably) an element of a *sheaf cohomology group*, say $\psi \in H^k(Q,\mathcal{O})$, for some suitable open set (or closure of an open set) $Q \subset T \times T \times \ldots \times T$ say (T = twistor space), $\mathcal{O}$ being the sheaf of (germs of) holomorphic functions on Q (see Chapter 2).

(iii) When a particle interacts it does so by *deforming* the structure of twistor space, or of $Q \subset T \times \ldots \times T$, whereby ψ now plays an *active* role in determining the precise deformation in question, thence affecting the behavior of other twistor functions defined on Q.

The above is vague in various respects. But something like (iii) is strongly suggested by two examples: the *non-linear graviton* (see §3.1) and the *twisted photon* (§3.2). (In fact, for gravitons, (iii) is strictly true only in the weak-field limit--unless some kind of non-linear sheaf theory can be evoked.) In the standard procedures for performing deformations (infinitesimal ones at least) it is only H^1's that are involved. Sparling has suggested that *new* procedures for performing deformations may be needed in order to be able to use H^k's $(k > 1)$ in an active way. For the moment, I

---

* 'Googly' is a cricketing term, meaning a positive helicity cricket ball bowled with an apparently negative helicity bowling action.

prefer the more modest idea that *single* particles are to be described by H^1's, these being the things that interact *directly* with other particles. H^2's, H^3's, etc. would play roles in describing many-particle states, with elements of suitable $H^k(Q,\mathcal{O})$ groups defining k-particle states. This could supply a possible answer to a question posed to me by Feynman: how do you know whether $\psi(X^\alpha,Y^\alpha,Z^\alpha)$ describes a hadron, or three massless particles, or a lepton and a massless particle? Elements of $H^1(Q,\mathcal{O})$, $H^3(Q,\mathcal{O})$, and $H^2(Q,\mathcal{O})$ are really quite different kinds of animals, after all! But the detailed implementation of this idea has proved elusive--not the least problem being to understand "twistorially" why the spin-statistics relation should hold for many-particle states.

Perhaps the most primitive problem, in trying to push forward with (iii), lies in the fact that a twisted photon is only *half* a photon, namely the left-handed half. Thus, the twisted photon construction gives a deformation of $Q = T^+$ starting from a twistor function $f(Z^\alpha)$, homogeneous of degree zero [$f \in H^1(PT^+,\mathcal{O}(0))$], to give a *left*-handed photon. Of course a right-handed photon can be produced by using $\tilde{f}(W_\alpha)$, of degree zero. But such would be to defeat the purpose of the economy of the twistor description (i). Simply changing the helicity quantum number (or any other quantum number) should not involve us in changing the *space* Q over which the twistor function is defined. Thus maintaining the general programme (i) - (iii) seems to lead us to the view that some form of deformation of T^+ *must* be possible, which effectively encodes the information provided by an element $f \in H^1(T^+,\mathcal{O})$, when f is homogeneous of degree -4 [i.e., in effect, $f \in H^1(PT^+,\mathcal{O}(-4))$]. My suggestion for this, which relates closely to an earlier proposal due to Sparling and Ward (cf. §2.4), is as follows.

Consider, first, the standard Ward twisted photon. This is obtained by deforming the bundle T^+ where the base space remains the undeformed PT^+ and the fibre $C - \{0\}$. Furthermore, the Euler operator $Z^\alpha \frac{\partial}{\partial Z^\alpha} =: \Upsilon_Z$ remains globally defined. However, with non-trivial twisting, the forms $DZ: = \frac{1}{24} \varepsilon_{\alpha\beta\gamma\delta}\, dZ^\alpha \wedge dZ^\beta \wedge dZ^\gamma \wedge dZ^\delta$ and $\mathcal{D}Z: = \frac{1}{6} \varepsilon_{\alpha\beta\gamma\delta}\, Z^\alpha dZ^\beta \wedge dZ^\gamma \wedge dZ^\delta$ are *not* well defined; indeed, globally defined analogues of these forms do not exist at all on a non-trivially twisted Ward photon. There is a good reason for this. In the case of flat T, there is a canonical way of representing a twistor Z^α (up to the fourfold ambiguity $\pm Z^\alpha$, $\pm iZ^\alpha$) in terms of PT: namely as the projective twistor Z in PT, together with a *3-form at* Z in PT. This 3-form lifts into T along the fibre over Z; where it agrees with $\mathcal{D}Z$ defines us the point Z^α (or iZ^α, or $-Z^\alpha$, or $-iZ^\alpha$). Thus, if $\mathcal{D}Z$ is canonically known in the bundle, then the complete bundle structure is determined uniquely in terms of PT; therefore no twisting in the "photon" can occur. (This construction is the analogue of how one defines a spinor κ^A (up to sign) in terms of the celestial sphere.) Furthermore, if a $\mathcal{D}Z$ exists globally in a bundle over PT^+, then it must be unique to an overall constant factor.

Suppose, generally, we have a bundle (or holomorphic fibration) *locally*, which is just a complex 4-space over a complex 3-space with fibre a complex 1-space. Then all we know locally in the 4-space is a direction field δ_Z (1-foliation). To know $\mathcal{D}Z$ in the 4-space would be to know rather more structure than δ_Z. Being a 3-form, $\mathcal{D}Z$ is orthogonal to (i.e., annihilates) precisely one complex direction and so serves to *define* δ_Z. (I shall always assume $\mathcal{D}Z$ to be restricted to be orthogonal to δ_Z, so as to qualify to be a "$\mathcal{D}Z$.") Now $\mathcal{D}Z$ also defines a volume 4-form DZ by $d\mathcal{D}Z = 4DZ$. Furthermore $\mathcal{D}Z$ and DZ together define the Euler operator Υ_Z, roughly speaking by "$\Upsilon_Z = \mathcal{D}Z \div DZ$," or more precisely by $\mathcal{D}Z = \Upsilon_Z \lrcorner\, DZ$. Conversely, this relation

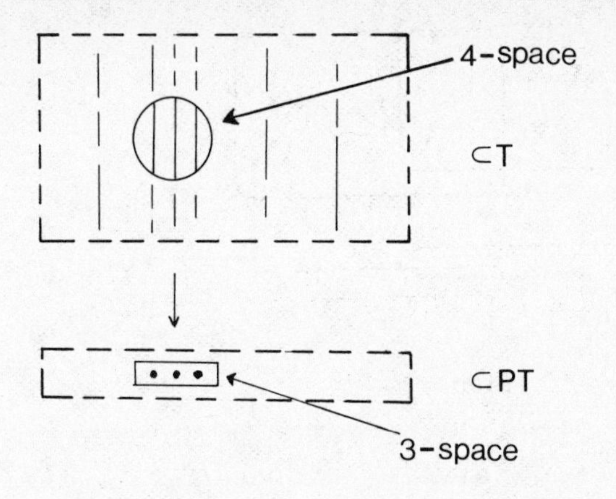

Figure A

shows us that $\mathcal{D}Z$ is determined by the *pair* (T_Z, DZ). To know one or the other of T_Z, DZ is, by itself, not sufficient to determine $\mathcal{D}Z$, but the two together are equivalent to $\mathcal{D}Z$. Furthermore, assuming that T_Z is to point along δ_Z (where δ_Z is given), there is precisely as much information in T_Z locally as there is in DZ. Each provides us with a kind of local scaling, but it is a "homogeneity degree 4" scaling for DZ and a "homogeneity degree 0" scaling for T_Z. In a sense DZ and T_Z seem to be sorts of duals to one another.

We can regard the twisted photon (§3.2) as arising when we retain only the T_Z scaling and throw out DZ. Let us try to do the "dual" thing and retain DZ while throwing out T_Z. I shall proceed in a fairly explicit way, assuming that two "coordinate" patches are given, where a standard twistor description is given in each patch, with X^α on the left and Z^α on the right. I am assuming there is no monkey-business in the base space, so $X^\alpha \propto Z^\alpha$ may be assumed on the overlap region. (The fibres in each half are given when X^α or Z^α is held constant up to proportionality.) The hypothesis is now

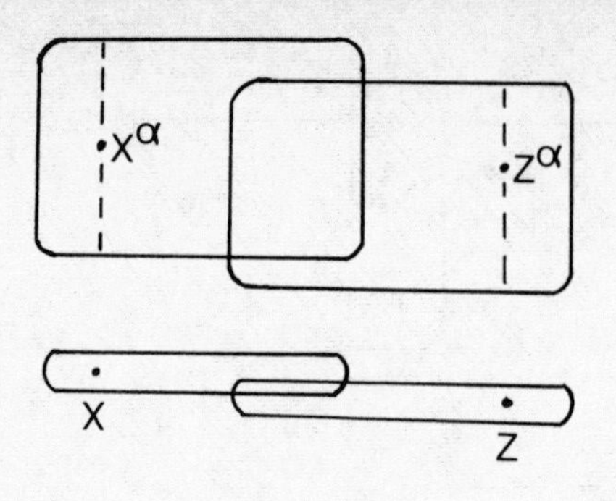

Figure B

$DX = DZ$ on the overlap (instead of Ward's $\tau_X = \tau_Z$), i.e., $d\mathcal{D}X = d\mathcal{D}Z$, i.e., $d\{\mathcal{D}X-\mathcal{D}Z\} = 0$. But $\delta_X = \delta_Z$, so $\mathcal{D}X \propto \mathcal{D}Z$. Put $\mathcal{D}X = \{1 + f(Z^{\alpha})\}\mathcal{D}Z$; then $d\{f(Z^{\alpha})\mathcal{D}Z\} = 0$, which holds iff $f(Z^{\alpha})$ is *homogeneous of degree -4*. The transition relation is then

$$X^{\alpha} = \{1 + f_{-4}(Z^{\alpha})\}^{\frac{1}{4}} Z^{\alpha}. \tag{1}$$

This appears to be a particular case of a Sparling-Ward construction given earlier for massless fields of *any* helicity.

We run into trouble owing to branch points arising from the fourth root unless we exclude an extended region about the origin of each fibre. But things are okay near ∞ on the fibres. Thus, I envisage the fibres as being like C with some bounded (probably connected) region removed (i.e., biholomorphic to $C - \{z : |z| < 1\}$). Note that τ_Z is not preserved by the patching, but related by $\tau_X = (1 + f(Z^{\alpha}))\tau_Z$ (which does strange things to the notion of a homogeneous function).

The intention is to regard $f(Z^{\alpha})$ as a twistor function for a right-handed photon. But, in accordance with (ii), such an $f(Z^{\alpha})$ is really describing an element of a sheaf cohomology group. Thus we need to check that (1) is appropriately cohomological. Suppose we have a covering of the base space

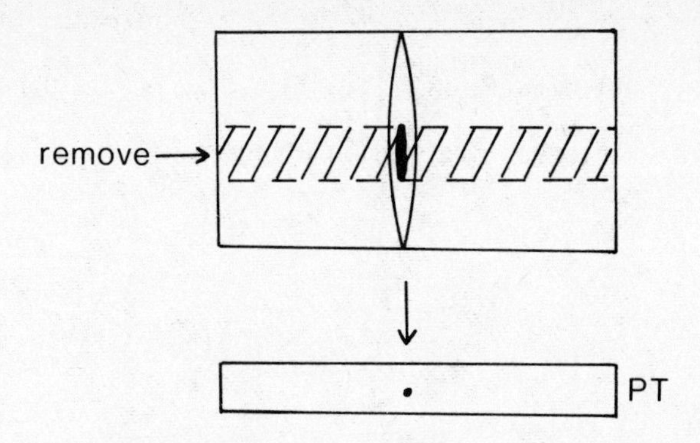

Figure C

(say PT^+) by a number of open sets $\mathcal{U}_i$. Then on each $\mathcal{U}_i \cap \mathcal{U}_j$ we need $f_{ij} = -f_{ji}$ such that on each $\mathcal{U}_i \cap \mathcal{U}_j \cap \mathcal{U}_k$, $f_{ik} = f_{ij} + f_{jk}$ (cocycle condition). If we use standard twistor "coordinates" Z^α for the entire base space, then the f_{ij} will simply be homogeneous functions of Z^α, of degree -4. Thus, $f_{ij}(Z^\alpha) = -f_{ji}(Z^\alpha)$, $f_{ik}(Z^\alpha) = f_{ij}(Z^\alpha) + f_{jk}(Z^\alpha)$. The cocycle defining the right-handed photon can be taken to be just this collection of homogeneous functions. We piece together our "bundle" by taking "coordinates" $\underset{i}{Z^\alpha}$ for the portion lying above $\mathcal{U}_i$, where on each overlap (say, the portion of the bundle lying above $\mathcal{U}_i \cap \mathcal{U}_j$) we have $\underset{i}{Z^\alpha} = \{1 + f_{ij}(\underset{j}{Z^\beta})\}^{\frac{1}{4}}\underset{j}{Z^\alpha}$. We must first check the consistency of this with $f_{ij} = -f_{ji}$. Avoiding messy indices we have (1) : $X^\alpha = \{1 + f(Z^\beta)\}^{\frac{1}{4}}Z^\alpha$ and $Z^\alpha = \{1 + g(X^\beta)\}^{\frac{1}{4}}X^\alpha$, and we need to check that $g = -f$. Now $Z^\alpha = \{1 + g(X^\beta)\}^{\frac{1}{4}}\{1 + f(Z^\beta)\}^{\frac{1}{4}}Z^\alpha$, so we require $(1 + g(X^\beta))(1 + f(Z^\beta)) = 1$; that is

$$
\begin{aligned}
1 &= (1 + g(\{1 + f(Z^\beta)\}^{\frac{1}{4}}Z^\sigma))\ (1 + f(Z^\rho)) \\
&= (1 + \frac{g(Z^\sigma)}{1 + f(Z^\beta)})(1 + f(Z^\rho))
\end{aligned}
$$

$$= (1 + f(Z^\rho) + g(Z^\rho))$$

(because of the -4 homogeneity of g) so $g(Z^\rho) = -f(Z^\rho)$ as required. Next we check the compatibility of (1) with $f_{ik} = f_{ij} + f_{jk}$. Again avoiding indices we have, on the triple intersection region about $\mathcal{U}_i \cap \mathcal{U}_j \cap \mathcal{U}_k$, relations

$$Y^\alpha = \{1 + p(X^\rho)\}^{\frac{1}{4}}X^\alpha;\ X^\rho = \{1 + f(Z^\sigma)\}^{\frac{1}{4}}Z^\rho;\ Y^\alpha = \{1 + q(Z^\sigma)\}^{\frac{1}{4}}Z^\alpha .$$

For consistency we require

$$\{1 + p(\{1 + f(Z^\sigma)\}^{\frac{1}{4}}Z^\rho\})\{1 + f(Z^\beta)\} = 1 + q(Z^\sigma)$$

$$\text{i.e.}\quad \{1 + \frac{p(Z^\rho)}{1 + f(Z^\rho)}\}\{1 + f(Z^\rho)\} = 1 + q(Z^\rho)$$

i.e. $q(Z^\rho) = p(Z^\rho) + f(Z^\rho)$ as required. We need also to show that if the cocycle is a coboundary, i.e., if $f_{ij} = f_j - f_i$ for each i,j, then the bundle is the same as that for flat twistor space (f_i being holomorphic throughout $\mathcal{U}_i$). For this we set $Z^\alpha = \{1 + f_i(Z^\beta_i)\}^{\frac{1}{4}}Z^\alpha_i$ and find that the Z^α, so constructed, is defined *globally* over the whole bundle. The compatibility over each $\mathcal{U}_i \cap \mathcal{U}_j$ follows by a calculation which is basically the same as the ones given above. Finally we need the fact that *if* the bundle is the same as that for flat twistor space (where here and above "the same" must be suitably interpreted in terms of "analytically extendible to"--owing to the gaping holes in the fibres), then the $\{f_{ij}\}$-cocycle is a coboundary. To prove this, we simply reverse the above argument. The flatness implies the existence of a global Z^α for the bundle. The local Z^α_i for each patch (above $\mathcal{U}_i$) must be related to Z^α by a formula $Z^\alpha = \{1 + f_i(Z^\beta_i)\}^{\frac{1}{4}}Z^\alpha_i$ from which follows (by the reverse of the above calculation)

$f_{ij} = f_j - f_i$ above $\mathcal{U}_i \cap \mathcal{U}_j$.

The argument just given effectively shows that although (1) appears to be a highly non-linear relation, the system of bundles so constructed has nevertheless a *linear* structure (given by simply adding the sheaf cohomology group elements). This may be contrasted with Sparling's method of patching together non-linear gravitons:

$$X^{\alpha} = \exp \{I^{\rho\sigma} \frac{\partial f(Z^{\tau})}{\partial Z^{\rho}} \frac{\partial}{\partial Z^{\sigma}}\} Z^{\alpha}.$$

For this, the cocycle condition *fails*. This is a manifestation of the very non-linearity of the non-linear graviton. On the other hand, the "4" in the above construction is *not* essential for linearity and could be replaced by other powers; i.e., f of homogeneity -n, with $X^{\alpha} = \{1 + f(Z^{\beta})\}^{1/n} Z^{\alpha}$. When n = 1, "0", -1, -2, -3,... this is implicit in the Sparling-Ward construction. However, when $n < 0$ the global structure of the fibres would have to be something different (because $(1 + z^{-n})^{1/n} z \to 1$ as $|z| \to \infty$ if $n < 0$, whereas $(1 + z^{-n})^{1/n} z \to z$ as $|z| \to \infty$ if $n > 0$). In any case, the motivation from "preservation of DZ" exists only when n = 4.

A direct construction of the right-handed Maxwell field $\tilde{\phi}_{A'B'}$ (satisfying $\nabla^{AA'} \tilde{\phi}_{A'B'} = 0$) from the bundle structure has not yet emerged. But the most serious problem is that of fitting the right-handed and left-handed ways of deforming T^+ together into one bundle...? One possibility for doing this is a method suggested by Hughston in §3.8. However it is unclear, as yet, how to extract left- and right-handed Maxwell fields which do not interact with one another. If this problem can be resolved, then among other things, it *might* suggest an analogous approach to an ambidextrous graviton!

Remarks concerning the ambidextrous photon

The information of a *left*-handed Maxwell field can be stored in a holomorphic line bundle T over a suitable portion R of PT, the Euler operator Υ being well-defined on T and pointing along the fibres of T. Likewise, the information of a *right*-handed Maxwell field can be stored in a 1-manifold-fibration T over R, but where now it is the holomorphic volume 4-form Δ(=DZ) which is well-defined on T instead of Υ.

In §3.8, Hughston suggests putting the right-handed construction on top of the left-handed one, but it is not obvious that this preserves any particular structure on T. In fact it does, namely the operator I shall call DIV. This is a map from vector fields V, pointing along the fibration, to scalars, satisfying $DIV(\lambda V) = \lambda DIV(V) + V(\lambda)$, where λ is a scalar field. It differs from an ordinary div operation only in that it need not act on any vector field which does not point along the fibration. Clearly, if Δ exists, then DIV exists, defined by $DIV(V) = div_{\Delta} V$, where div_{Δ} is the ordinary divergence based on Δ (i.e., use Δ to convert V into a 3-form $V \lrcorner \Delta$, take the exterior derivative of $V \lrcorner \Delta$, and then divide out by Δ to obtain a scalar). Also, if Υ exists, then DIV exists, defined by $\Upsilon \times DIV(V) = [\Upsilon,V] + 4V$.

Let us now suppose that the 4-dimensional space T, which is a fibration over R (all holomorphic), possesses just the structure DIV. Cover R by open sets $\{U_i\}$ and lift these to give a covering $\{U_i\}$ of T. Define a Δ_i and an Υ_i in each U_i which are compatible, in the sense above, with DIV. These compatibility requirements can be expressed as

$$DIV(\Upsilon_i) = 4 \quad \text{and} \quad \pounds_{\Upsilon_i} \Delta_i = 4\Delta_i$$

(the latter being equivalent to $div_{\Delta_i} \Upsilon_i = 4$). Now define $f_{ij} = \log(\Delta_j/\Delta_i)$.

We have $\pounds_{T_i} f_{ij} = 0$ and $f_{ij} - f_{ik} + f_{jk} = 0$ on $\mathcal{U}_i \cap \mathcal{U}_j \cap \mathcal{U}_k$, so f_{ij} defines a 1-cocycle on R. This is the left-handed (homogeneous of degree 0) "part" of the photon. Whenever this cohomology class vanishes, Δ exists globally. Suppose, next, that Δ *is* global. Define $F_{ij} = (T_j - T_i) \lrcorner \Delta$. We find $\pounds_{T_i} F_{ij} = 0$ and $F_{ij} - F_{ik} + F_{jk} = 0$ on $\mathcal{U}_i \cap \mathcal{U}_j \cap \mathcal{U}_k$, so F_{ij} defines a 1-cocycle 3-form on R. This is the right-handed (back-handed) (homogeneous of degree -4) "part" of the photon.

In fact, this is exactly what we had before, although phrased somewhat differently, so that we don't even notice the $(\ldots)^{\frac{1}{4}}$ unpleasantness! (Incidentially, another way of disguising this is to write the $\hat{Z} = (1 + 4g)^{\frac{1}{4}} Z$ as $\hat{Z} = e^{gT} Z$, where $g(Z)$ is homogeneous of degree -4. The power series do check!)

The trouble, of course, is the same as before: if there *is* a left-handed part (Δ *not* global) then there seems to be no way of extracting an $H^1(R, \mathcal{O}(-4))$ element to measure the non-globalness of T (without bringing in more structure, that is). This is the old problem--the right-handed part appears to be *charged* with respect to the left-handed part. To decouple the two parts, it seems that strict conformal invariance must be broken (e.g. $F_{ij} = (T_j - T_i) \lrcorner \Delta_i$ or $F_{ij} = (T_j - T_i) \lrcorner (\Delta_j - \Delta_i)$ won't do because the cocycle condition fails). Thus we must bring in I (the "infinity twistor"). This works in principle, but is, as yet, inelegant.

§3.8. A GENERALIZATION OF THE GOOGLY PHOTON *by L.P. Hughston*

One of the most intriguing ideas of twistor theory is that it should eventually be possible to see how to incorporate the quantum states of elementary particles "intrinsically" into the geometry of twistor space. Roughly speaking, the idea runs as follows. Initially, a quantum state is thought of as being represented in terms of certain classes of *holomorphic functions* defined on some (background) twistor geometry. But then these functions are *reinterpreted* in such a way that they appear in an *active* guise: they are incorporated into the structure of a set of *holomorphic coordinate transition functions* which are used to define the complex analytic structure of a new "curved" twistor space. The new curved twistor space now carries built directly into its structure the information of the original quantum state.

While the general methodology for such a scheme remains yet to be defined, nevertheless one of the requirements which has tentatively been adopted in this connection is that the entire construction should be "cohomologically natural": by this what is meant is that the deforming of the twistor space should depend only on the *sheaf cohomology class* of the twistor function that represents the state of the particle. Thus particle states are, in effect, interpreted as being elements of twistor sheaf cohomology groups[1], and these in turn induce finite deformations of twistor space itself.

The right-handed photon construction[2] can be given a natural cohomological description in a setting of considerable generality. The background space on which the photon is defined (and which the photon subsequently "twists up") can be essentially any curved twistor space (or generalized twistor space) on which the *Euler homogeneity operator* can be globally defined. Suppose the twistor space is covered by coordinate patches U_i, with coordinates Z_i^α, and transition functions H_{ij}^α defined by $Z_i^\alpha = H_{ij}^\alpha(Z_j^\alpha)$ in[3] U_{ij}. The condition for the existence of a global Euler operator is that H_{ij}^α should be homogeneous of degree one. To see this note that in U_{ij} one has $\partial/\partial Z_j^\alpha = (\partial H_{ij}^\beta/\partial Z_j^\alpha)\partial/\partial Z_i^\beta$ from which it follows, transvecting this expression with Z_j^α, that $Z_j^\alpha\partial/\partial Z_j^\alpha = Z_j^\alpha(\partial H_{ij}^\beta/\partial Z_j^\alpha)\partial/\partial Z_i^\beta$. The Euler operator Υ_j in the patch U_j is defined by $Z_j^\alpha\partial/\partial Z_j^\alpha$. Requiring that Υ_j agrees with Υ_i in U_{ij} now gives the desired result $Z_j^\alpha\ \partial H_{ij}^\beta/\partial Z_j^\alpha = H_{ij}^\beta$.

The transition functions $H_{ij}(Z_j)$ are also, of course, required to satisfy the usual *compatibility conditions* $H_{ij} \circ H_{jk} \circ Z_k = H_{ik} \circ Z_k$ on the triple overlap region U_{ijk}, as well as the *inverse relations* $H_{ij} \circ H_{ji} \circ Z_i = Z_i$, on U_{ij}[4]. Denoting this background space M now consider an element q in the cohomology group $H^1(M,\mathcal{O}(-4))$, where $\mathcal{O}(-4)$ is the sheaf of germs of holomorphic functions homogeneous of degree minus four. The element q defines a photon state *relative to the background provided by the space M*. A *representative* for q will consist of a certain collection of holomorphic functions $q_{ij}(Z_j)$ defined on the intersection regions U_{ij}. These functions must satisfy the *cocycle* condition. If Z_j is the coordinate of the point p in the patch U_j then the cocycle condition is that $q_{ij}(Z_j) + q_{jk}(Z_k) + q_{ki}(Z_i) = 0$ for each $p \in U_{ijk}$[5]. The functions q_{ij} must also satisfy the *skew-symmetry* condition, which is $q_{ij}(Z_j) + q_{ji}(Z_i) = 0$. (Note that if $q_{ij}(Z_j)$ is modified by the addition of a term $-q_i(Z_i) + q_j(Z_j)$, where $q_i(Z_i)$ is a holomorphic function definable over the whole of U_i, then both the cocycle condition and the skew-symmetry condition are left invariant. Such a term is a *coboundary* term. Two cocycles which differ by a coboundary term define the same element in the cohomology group.)

Now a new space M(q) ["M, deformed by q"] will be defined by the transition relations $Z_i = [1 + q_{ij}(Z_j)]^{\frac{1}{4}}\, H_{ij}(Z_j)$. If this patching is denoted by $Z_i = \tilde{H}_{ij}(Z_j)$, then it must be demonstrated that *(a)* $\tilde{H}_{ij}$ *correctly satisfies the compatibility conditions and the inverse relations,* and *(b)* $\tilde{H}_{ij}$ *is cohomologically natural,* i.e. if q_{ij} is modified by the addition of coboundary terms, the complex analytic structure of M(q) is left untainted (so that the manifold M(q) depends only on the sheaf cohomology class to which q_{ij} belongs).

Proof of (a). We must prove the compatibility conditions $\tilde{H}_{ij} \circ \tilde{H}_{jk} \circ Z_k = \tilde{H}_{ik} \circ Z_k$, and the inverse relations $\tilde{H}_{ij} \circ \tilde{H}_{ji} \circ Z_i = Z_i$. Note that providing the compatibility conditions hold true the inverse relations are equivalent to $\tilde{H}_{ii}(Z_i) = Z_i$. Since $H_{ii}(Z_i) = Z_i$, and $q_{ii} = 0$ (by the skew condition), it follows that the inverse relations will certainly hold for $\tilde{H}_{ij}$ once the compatibility conditions have been ensured. Now if we are given $Z_i = [1 + q_{ij}(Z_j)]^{\frac{1}{4}} H_{ij}(Z_j)$ together with $Z_j = [1+ q_{jk}(Z_k)]^{\frac{1}{4}} H_{jk}(Z_k)$ we wish to substitute so as to obtain the composition $Z_i = [1 + q_{ik}(Z_k)]^{\frac{1}{4}} H_{ik}(Z_k)$:

$$\begin{aligned} Z_i &= [1 + q_{ij}(Z_j)]^{\frac{1}{4}} H_{ij}(Z_j) \\ &= [1 + \{1 + q_{jk}(Z_k)\}^{-1} q_{ij} \circ H_{jk} \circ Z_k]^{\frac{1}{4}} [1 + q_{jk}(Z_k)]^{\frac{1}{4}} H_{ij} \circ H_{jk} \circ Z_k \\ &= [1 + q_{jk}(Z_k) + q_{ij}(Z_j)]^{\frac{1}{4}} H_{ik}(Z_k) \\ &= [1 + q_{ik}(Z_k)]^{\frac{1}{4}} H_{ik}(Z_k) \ . \end{aligned}$$

Note that the last step of the proof employs the cocycle condition on $q_{ij}(Z_j)$. □

Proof of (b). Suppose, for simplicity, that q_{ij} is cohomologically trivial, i.e. a pure coboundary, of the form $q_{ij} = -q_i(Z_i) + q_j(Z_j)$. We must show, under these circumstances, that M(q) is equivalent to M on a complex manifold. Thus there must exist a set of holomorphic functions ϕ_i on U_i such that $Z_i = \phi_i^{-1} \circ H_{ij} \circ \phi_j(Z_j)$, where $\phi_i^{-1} \circ \phi_i = 1$. One obtains:

$$Z_i = [1 - q_i(Z_i) + q_j(Z_j)]^{\frac{1}{4}} H_{ij}(Z_j)$$

$$= [1 - \{1 + q_j(Z_j)\}^{-1} q_i(Z_i)]^{\frac{1}{4}} [1 + q_j(Z_j)]^{\frac{1}{4}} H_{ij}(Z_j)$$

$$= [1 - q_i \circ H_{ij} \circ \phi_j(Z_j)]^{\frac{1}{4}} H_{ij} \circ \phi_j(Z_j),$$

where the function ϕ_j has been defined by $\phi_j(Z_j) = [1 + q_{ij}(Z_j)]^{\frac{1}{4}} Z_j$. If we define similarly $\phi_j^{-1} = [1 - q_j(Z_j)]^{\frac{1}{4}} Z_j$, it can be verified with no difficulty that $\phi_j^{-1} \circ \phi_j = 1$. Referring back now to the last step in the calculation above, the desired result $Z_i = \phi_i^{-1} \circ H_{ij} \circ \phi_j(Z_j)$ is obtained. Thus it has been shown that if q_{ij} is cohomologically trivial, then $M(q) = M$.

More generally, one can show, using essentially the same sort of argument, that if q_{ij} is replaced with $q_{ij} - q_i + q_j$ then for the new transition functions one obtains the result $Z_i = \phi_i^{-1} \circ \tilde{H}_{ij} \circ \phi_j(Z_j)$, with ϕ_j defined as before. These transition relations give the same complex manifold as $Z_i = \tilde{H}_{ij}(Z_j)$, and thus it follows, as desired, that $M(q)$ is completely independent of the choice of representative cocycle for the cohomology element q. □

Remarks. The algebra involved in the proofs of (a) and (b) is for all practical purposes the same as that involved in the flat space construction for the right-handed photon, just modified a bit for the accommodation of the more general structure of the background space.

The non-linear graviton[6], the left-twisted photon[7], and the combined non-linear graviton - left-twisted photon are all examples of spaces for which the coordinate transition functions H_{ij} are homogeneous of degree one. These spaces accordingly all qualify as examples of background spaces acting on which the construction described above can be initiated.

Notes

1. For the basic ideas of twistor sheaf cohomology, see Chapter 2 in this volume. Also see Jozsa 1976, Hughston 1979, Lerner and Sommers 1979, and numerous other references.
2. See §3.7.
3. The notation here is: $U_{ij} = U_i \cap U_j$, $U_{ijk} = U_i \cap U_j \cap U_k$, etc..
4. Note that from here on twistor indices are suppressed, and only cohomology indices are retained. Note that at no stage is it really necessary to assign values to the cohomology indices; their role is essentially combinatorial. It would be interesting in this connection for someone to try to devise some sort of an "abstract index calculus" for cohomology.
5. The notation can be simplified here if we say by convention that the index on the coordinate appearing in the argument of a cochain should always be the same as the *last* index on the cochain itself (suggested by N.M.J.W.). With this convention q_{ij} stands for $q_{ij}(Z_j)$, and similarly the cocycle condition reduces to $q_{ij} + q_{jk} + q_{ki} = 0$, for $p \in U_{ijk}$. Another useful convention is to introduce a 'restriction' map ρ_i which restricts the domain of whatever it's applied to to its intersection with U_i. Thus the cocycle condition can be written $\rho_k q_{ij} + \rho_i q_{jk} + \rho_j q_{ki} = 0$, or simply, $\rho_{[k} q_{ij]} = 0$. Similarly q_{ij} is a coboundary if there exists a q_j such that $q_{ij} = \rho_{[i} q_{j]}$, and because of $\rho_{[i}\rho_{j]} = 0$ it follows obviously that coboundaries are always cocycles, and so forth. For other purposes it is useful to refer a cohomological relation

entirely to one patch. For such purposes the coordinate transition functions must be used. Referred to the patch U_j, for example, to cocycle condition above is expressed $q_{ij}(Z_j) + q_{jk} \circ H_{kj}(Z_j) + q_{ki} \circ H_{ij}(Z_j) = 0$.

6. See Penrose (1976).
7. See §3.2 and §3.3.

REFERENCES

Hughston, L.P. 1979 *Twistors and Particles*. Springer Lecture Notes in Physics, Volume 97.

Jozsa, R.O. 1976 M.Sc. Thesis, University of Oxford.

Lerner, D., and P. Sommers 1979 *Complex Manifold Techniques in Theoretical Physics*. Pitman: London.

Penrose, R. 1976 Gen. Rel. Grav. 7, 31-52.

§3.9. A GOOGLY GRAVITON? *by R. Penrose*

The so-called "non-linear graviton construction" (cf. §3.1c) enables the local structure of any anti-self-dual complex solution of Einstein's vacuum equations to be encoded into the global structure of a deformed twistor space. The weak field limit of this construction yields the $H^1(\ldots, \mathcal{O}(2))$ "twistor function" description of linearized gravity (helicity -2). For weak fields we also have the $H^1(\ldots, \mathcal{O}(-6))$ "twistor function" description in the *self*-dual case (helicity +2). Is there a non-linear version of this, providing a deformed-twistor-space method of describing *self*-dual complex Einstein vacuums? (Of course, one could apply the original non-linear graviton construction to *dual* twistor space, but that is not what is meant here.) If so, then there would be the hope that by combining both types of deformation, one might be able to encode the local (?) structure of a *general* solution of the vacuum equations (i.e. just $R_{ab} = 0$)!

But there would seem to be an insurmountable problem right at the start. The key to the non-linear graviton construction lies in the interpretation of a point in the complex "space-time" M as a compact holomorphic curve in P$\mathcal{T}$ ($\mathcal{T}$ being a suitable deformed portion of flat twistor space T), with null separation in M being defined as intersection of the corresponding holomorphic curves in P$\mathcal{T}$. This leads at once to the fact that M is conformally anti-self-dual, because fixing a point in P$\mathcal{T}$ we get a 2-parameter family of curves through it, providing a corresponding α-surface in M — 3 parameters worth of α-surfaces in all. Altering the definition of "null separation" seems unpromising and to redefine these as β-surfaces would be cheating. However, an alteration in the definition of a "space-time point" to obtain the opposite helicity is suggested by the following.

Consider the standard exact sequence (Penrose & MacCallum 1972) which expresses the Poincaré invariant structure of *flat* twistor space T, where S is spin-space, * means dual, $\sim$ means complex conjugation, $\leftarrow\cdots\rightarrow$ means isomorphic via complex conjugation, and where the maps i and p are defined by $i : \omega^A \mapsto (\omega^A, 0)$, $p : (\omega^A, \pi_{A'}) \mapsto \pi_{A'}$

$$0 \longrightarrow S \xrightarrow{i} T \xrightarrow{p} \tilde{S}^* \longrightarrow 0$$

and *dually*

$$0 \longleftarrow S^* \longleftarrow T^* \longleftarrow \tilde{S} \longleftarrow 0.$$

Now, the non-linear graviton construction for anti-self-dual vacuums depends upon the existence of a deformed version of the map $p : T \to \tilde{S}^*$, this being the given holomorphic fibration. The points of M are *cross-sections* of this fibration (to which the Euler vector field Υ is tangent). In the flat case, we can think of these as linear maps q back from $\tilde{S}^*$ to T such that the composition $p \circ q$ is the identity of $\tilde{S}^*$ and which provide one way of "splitting" the sequence, i.e. of expressing T as a product, compatibly with the sequence

$$0 \longrightarrow S \underset{j}{\overset{i}{\rightleftarrows}} T \underset{q}{\overset{p}{\rightleftarrows}} \tilde{S}^* \longrightarrow 0.$$

Quite equivalent, in this flat case, is to "split" the sequence in the dual way with a map j from T to S such that $j \circ i$ is the identity on S. We may envisage a "deformed" version of this situation, however, in which T loses its linear space structure and is replaced by some suitable complex manifold $\mathcal{T}$ which, in some suitable sense, has structure providing an exact sequence

$$0 \longrightarrow S \xrightarrow{i} \mathcal{T} \xrightarrow{p} \tilde{S}^* \longrightarrow 0.$$

We know, from the original non-linear graviton, that the definition of "space-time point" analogous to the q-maps will provide us with an anti-self-dual M. Furthermore, because complex conjugation and duality are so intimately related in twistor theory, we may anticipate that an appropriate analogue of the j-maps will provide us with a space $\tilde{M}$ that is a (general) self-dual complex vacuum.

To see what this is likely to entail, consider the flat case again and, for simplicity of description, we examine the projective space PT. (Fig. A)

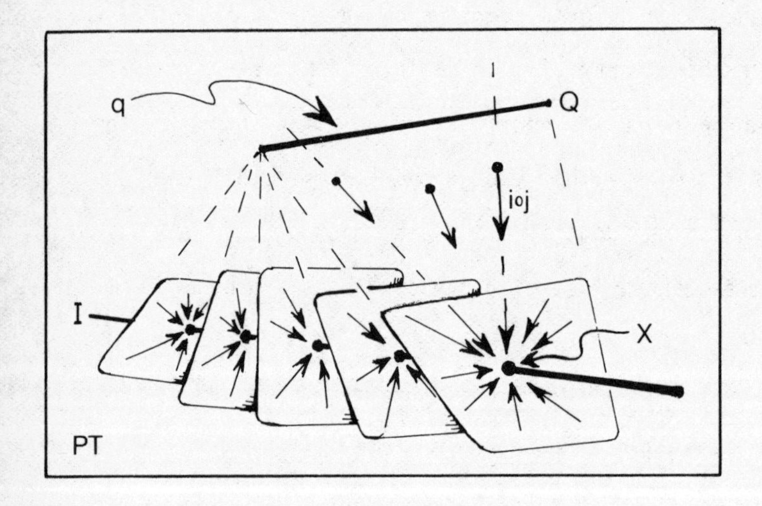

Figure A

The various maps q correspond to the various lines in PT not meeting the line I. Now I corresponds to the image of i, so each map j (or, more correctly, $i \circ j$) may be thought of as a projection from PT down onto I. The inverse image of the zero of S $(j^{-1}(0))$ is, however, not represented in this projection. This is the line Q in PT that arises in the above "q-map". Every other point of PT - Q takes part in the "$i \circ j$" projection. It is mapped to where the plane joining it to Q meets the line I. Thus, the inverse image of any point X of I under this map is a *plane* through Q and through the point X (excluding the line Q itself). This family of planes, each containing a different point X on I and none containing I itself, provides another way of representing the line Q, and hence another way of representing a space-time point. It is essentially the *dual* description to that of the point-locus Q since here we have a "plane-locus", namely a point-locus in PT*. But when PT is deformed to the space *PT*, we do not expect planes to survive as global loci. Indeed, it is this fact that enables the "dual" construction to differ from the original non-linear graviton one. The "planes" need only exist in the neighbourhood of I and do not need to extend far enough out that they actually have a holomorphic curve "Q" in common. At each point X of I there will be a two-parameter family of possible "planes" through X. The totality of these "planes" constitutes a complex 3-manifold which is a fibration over I - giving the space *PT** - and the "dual points", giving the space $\tilde{M}$, would be the cross-sections of this fibration *PT**. For this "dual" construction to work, therefore, we require that the space *PT** that is to arise in this way be similar in structure to the usual space "*PT*" of the original non-linear graviton. One important feature of this structure, however, is that the fibres are *not* flat spaces. If they were, then we should have, in effect, a vector bundle over CP^1 and all information would be lost, the cross-sections providing a manifold with only a flat structure. But if our deformed space *PT* is too "well-behaved" near I, then this is exactly what will happen. Our "planes", being well-defined near I, would be expected to have good tangent planes at their intersections with I, and to be determined by these tangent planes. Regularity at I would imply that the required PT^* space is simply the canonical *flat* one and all our supposed information concerning self-dual curvature would be lost. Thus, for this "dual" construction to work, it is necessary that the neighbourhood of I in PT be in some sense "singular" or "peculiar".

In order to investigate what kind of peculiarity to expect, consider a space-time W that is about as well-behaved as one could possibly ask for. W is to be real, (+ - - -), analytic, vacuum, asymptotically simple (Penrose 1965), with a regular point i^+ (future timelike infinity) and a point i^0 (spacelike infinity) that is as regular as i^+ when viewed intrinsically from within $\mathscr{I}^+$ (though i^0 would be singular for W). It seems that many such space-times *ought* to exist, having the full functional freedom of two real functions of three real variables. They describe imploding-exploding free gravitational waves (and, of course, Minkowski space is a special case). The curved twistor space T that we are trying to obtain is to be the *asymptotic twistor space of W*. Then the original non-linear graviton construction will yield Newman's $\tilde{H}$-space as our M. The idea is that, in some appropriate sense, the above "dual" construction ought to yield Newman's H-space as our $\tilde{M}$. In order for this to work, the information of the structure of *dual* twistor space must also be contained in T and by the preceding discussion, it would be expected to reside in the peculiar way that the line I sits in its immediate neighbourhood in T.

Recall that the points of PT-I are represented on $C\mathscr{I}^+$ by "twistor lines", that is, by null geodesics, other than generators of $C\mathscr{I}^+$, that lie in the β-planes on $C\mathscr{I}^+$ (Penrose & MacCallum 1972, Penrose 1975). (Fig. B). Each point of I would be represented by an entire α-plane on $\mathscr{I}^+$. For each fixed β-plane on $C\mathscr{I}^+$, the twistor lines lying in it form a 2-parameter

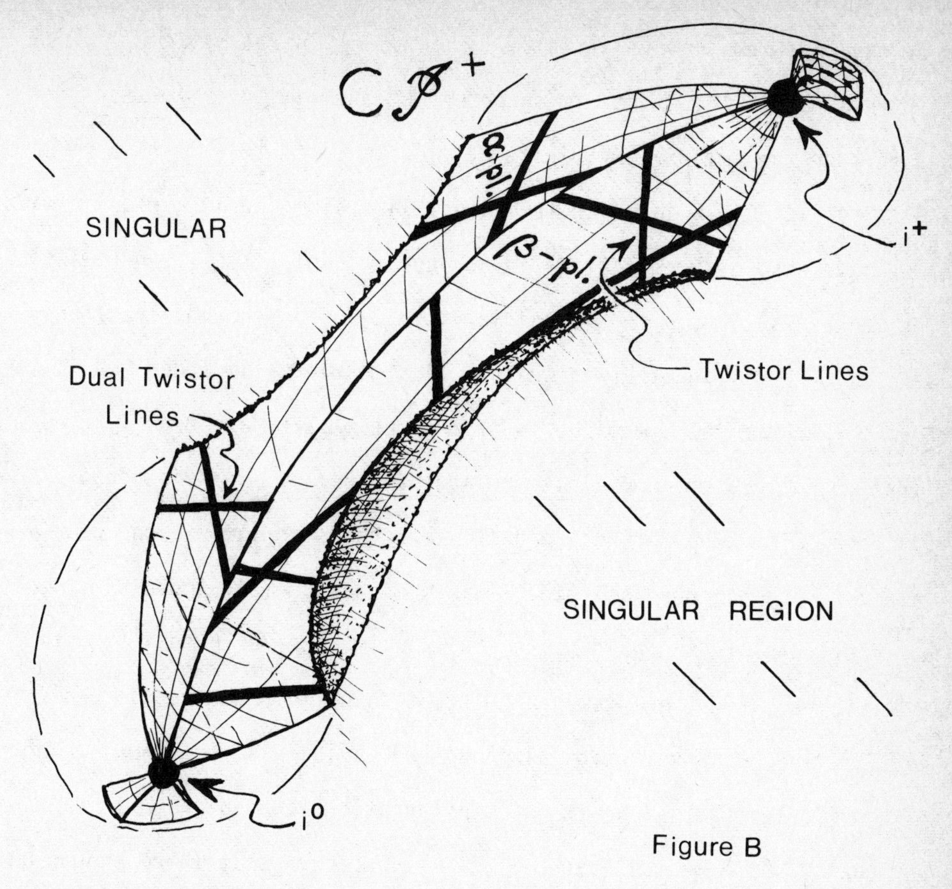

Figure B

family representing the points of a special 2-surface in PT through I. Now consider a sequence of points, on such a special 2-surface, which approach a point X of I. This is represented by a sequence of twistor lines, in the corresponding fixed β-plane, which turn around and approach a generator of $C\mathscr{I}^+$, this generator being the intersection of the α-plane ξ on $C\mathscr{I}^+$, representing X, with the given β-plane. If we vary the special 2-surface through I, the β-plane on $C\mathscr{I}^+$ correspondingly varies, so we get different generators of ξ as the limit of the sequence of twistor lines. It is thus by no means obvious that the points of I need be regular points of PT.

We have assumed, however, that i^+ is a regular point of $C\mathscr{I}^+$. It is not unreasonable, therefore, to suppose that the family of twistor lines that pass through the immediate vicinity of i^+ will in fact, give rise to a region of PT - call it PT_+ - in which the line I lies imbedded in a regular way. But we have also assumed that i^0 is a regular point of $C\mathscr{I}^+$. The twistor lines that pass through the immediate vicinity of i^0 should, therefore, also provide a region of PT in which I lies regularly imbedded - call this one PT_-. However there will be twistor lines passing arbitrarily close to i^0 that do not reach the vicinity i^+ and *vice versa*. This is because, roughly speaking, they are disconnected by the "singular region" of $C\mathscr{I}^+$. On the other hand, there are other twistor lines, namely those close to *real* generators of $C\mathscr{I}^+$, which pass arbitrarily close both to i^+ and to i^0. As such lines vary, there comes a point at which they become disconnected. (Fig. B). Thus we have a typical situation giving rise to a *non-Hausdorff* manifold PT, having regions in which bifurcation takes place. In fact, in order that PT be a manifold at all, it is necessary that there should be *two copies* of the line I, labelled I_+ and I_0, where I_+ sits regularly imbedded in PT_+ and I_0 regularly imbedded in PT_0. The open manifolds PT_+ and PT_0 can each be Hausdorff, but they are glued together (because of the existence of twistor lines passing close *both* to i^+ and to i^0) in a way that results in a non-Hausdorff combined manifold.

We may ask whether this non-Hausdorff feature is *essential* or not, i.e. whether the gluing of PT_+ to PT_0 cannot or can be extended until the resulting manifold is Hausdorff. The answer is that it *is* essential (at least in the general case under consideration). To see this, we recall a statement made earlier to the effect that a *regular* imbedding of I in PT would lead to a (conformally) *flat* structure for $\tilde{M}$. As applied to PT_+, this would give us the complex Minkowski space $\tilde{M}_+$ that arises from the "good cuts" of $C\mathcal{J}^+$ arbitrarily near i^+; as applied to PT_0 this gives another complex Minkowski space $\tilde{M}_0$ arising from "good cuts" arbitrarily near i^0 (cf. Newman & Penrose 1966). When W is non-flat we expect that the presence of outgoing gravitational radiation in W will imply that the good cuts near i^+ and those near i^0 will be related by a non-trivial *supertranslation* (Sachs 1962) at least in the general case, but perhaps always. The complex Minkowski spaces $\tilde{M}_+$ and $\tilde{M}_0$, which describe, in effect, the first neighbourhoods of I_+ in PT_+ and of I_0 in PT_0, respectively, thus differ from one another in an essential way and cannot be identified. It follows that I_+ cannot be identified with I_0 while retaining a smooth complex manifold structure, so PT seems to be an *essentially* non-Hausdorff manifold.

The presence of the above supertranslation tells us that the first neighbourhoods of I_+ and I_0 differ. The structure of this difference is described in terms of a holomorphic function of two variables (e.g. of the usual ζ and $\tilde{\zeta}$ coordinates for $C\mathcal{J}^+$). This function is obtainable, by integration, from the required self-dual radiation field Ψ_4^0 (Newman & Penrose 1962) which will determine the structure of $\tilde{M}$, but represents only *part* of the information contained in Ψ_4^0 - necessarily, because Ψ_4^0 is a function of *three* variables instead of two. But the non-Hausdorff feature of PT involves an incompatibility not only between the first neighbourhood of I_+ and I_0 but between their second and higher neighbourhood as well. Taking all this into account, it seems possible that the detailed non-Hausdorff nature of PT does indeed contain sufficient information that Ψ_4^0, and thus $\tilde{M}$, can, in principle, be determined.

Two immediate problems for this programme suggest themselves. In the first place, we may examine "infinitesimally non-Hausdorff spaces" (i.e.

where the map describing the gluing of T_+ to T_0 is given by an infinitesimal displacement of a portion Q of flat twistor space T), to see whether these correspond to cohomology groups $H^1(PQ, \mathcal{O}(-6))$, with $PN \subset PQ \subset PT$, in any way. In the second, we must find a way to characterize the required "j-maps" so that they indeed provide the sought-for space $\tilde{M}$. This involves finding a suitable regularity condition for these j-maps that contains the information of the location of the *dual* twistor lines (lying on α-planes on $C\mathscr{I}^+$). The idea is that this information ought already to be contained in the non-Hausdorff nature of PT (or perhaps of T; the j-maps are presumably to be defined on the space $T_+ \cap T_0$). More generally, a thorough study of non-Hausdorff (but well-behaved) complex manifolds seems called for. A further problem is to find a *third* definition of "space-time point" (perhaps some kind of "average" of j and q) to yield W itself - a generic solution of Einstein's vacuum equations!

REFERENCES

Newman, E.T. & Penrose, R. 1962 J. Math. Phys. 3, 566.

Newman, E.T. & Penrose, R. 1966 J. Math. Phys. 7, 863-870.

Penrose, R. 1965 Proc. Roy. Soc. A284, 159.

Penrose, R. 1975 *Twistor Theory, its Aims and Achievements*. In: *Quantum Gravity*, eds. C.J. Isham, R. Penrose & D.W. Sciama. Oxford: University Press.

Penrose, R. & MacCallum, M.A.H. 1972 Phys. Reports 6C, 241-315.

Sachs, R.K. 1962 Phys. Rev. 128, 2851.

4 Twistors and elementary particles

§4.1 INTRODUCTION *by L.P. Hughston*

It can be argued that perhaps the most singularly speculative aspect of twistor theory is embodied in the so-called *twistor particle hypothesis*. For it is here that twistor theory impinges directly on the particle physicists' sacrosanct territory, and in doing so casts aside much of the dogma of quantum chromodynamics (a theory firmly rooted into the framework of a spacetime description, and plagued with difficulties arising out of mathematical vagueness) with the intention of supplying in its place a new theory of hadrons based on the principles of algebraic geometry and complex analysis.

The basic ideas of the twistor particle hypothesis were worked out and put into what is their more or less present form during the latter part of 1974. The subject begins with the consideration of the *n-twistor internal symmetry groups* (cf. §§4.7 & 4.8, and references therein). Any massive system with spin can be thought of, classically, as being composed of two or more twistors. If these twistors are labelled Z^{α}_{i} with $i = 1,\dots,n$, then the total momentum and angular momentum of the system are represented by the *kinematic twistor* $A^{\alpha\beta}$ as follows:

$$A^{\alpha\beta} = 2Z^{(\alpha}_{i}I^{\beta)\gamma}\bar{Z}^{i}_{\gamma} \; . \qquad (1)$$

For a fixed value of $A^{\alpha\beta}$ (i.e. for a definite fixed momentum and angular momentum) it is possible to chose Z^{α}_{i} in many different ways such that equation (1) holds. Thus we have some "internal" degrees of freedom. More precisely, the n-twistor internal symmetry group is defined to be the set of linear transformations of Z_i^α and $\bar{Z}_\alpha^i$, preserving the conjugacy relations of Z_i^α and $\bar{Z}_\alpha^i$, such that $A^{\alpha\beta}$ is left invariant. It turns out that transformations satisfying these conditions must necessarily be of the form

$$Z_i^\alpha \to U_i^j (Z_j^\alpha + I^{\alpha\beta} \Lambda_{jk} \bar{Z}_\beta^k), \tag{2}$$

where U_i^j is unitary and Λ_{ij} is skew-symmetric. See §§4.7 and 4.8 for a detailed proof of this fact. For n = 3 the transformations (2) evidently contain SU(3) as a subgroup. One of the ingredients of the twistor particle hypothesis is that the flavour SU(3) of particle physics has its origin ultimately in association with the internal symmetries of massive 3-twistor systems.

Quantum mechanical observables are introduced by applying the standard twistor quantization rule

$$\bar{Z}_\alpha^i \to \hat{Z}_\alpha^i := -\partial/\partial Z_i^\alpha \ . \tag{3}$$

Thus, by applying (3) to (1) we obtain the following expression for the operators for momentum and angular momentum:

$$\hat{A}^{\alpha\beta} = 2Z_i^{(\alpha} I^{\beta)\gamma} \hat{Z}_\gamma^i \ . \tag{4}$$

For other observables we hypothesize that it is appropriate to consider *holomorphic differential operators with polynomial coefficients*, i.e. expressions of the form

$$R + R_{\alpha j}^{i\beta} \hat{Z}_i^\alpha Z_\beta^j + R_{\alpha\beta k\ell}^{ij\gamma\delta} Z_i^\alpha Z_j^\beta \hat{Z}_\gamma^k \hat{Z}_\delta^\ell + \ldots \tag{5}$$

where R, $R_{\alpha j}^{i\beta}$, etc., are constants. In order that observables of the form (5) should exhibit real eigenvalues, it is necessary that all of the R's be Hermitian, i.e. satisfy $R = \bar{R}$, $R_{\alpha j}^{i\beta} = \bar{R}_{j\alpha}^{\beta i}$, and so forth. The kinematic operator $\hat{A}^{\alpha\beta}$ defined in equation (4) is, of course, not Hermitian as it stands: in order to form the operator appropriate to some *specific* component of momentum or angular momentum one takes an expression of the form

$$P_{\alpha\beta}\hat{A}^{\alpha\beta} + \bar{P}^{\alpha\beta}\hat{A}_{\alpha\beta} \quad , \tag{6}$$

where $\hat{A}_{\alpha\beta}$ is the Hermitian conjugate of $\hat{A}^{\alpha\beta}$, and where $P_{\alpha\beta}$ is some fixed twistor which picks out the component of interest.

By an *internal observable* we mean a Hermitian operator of the form (5) which has the further property that it *commutes with the kinematic operator (4)*. Thus an internal observable has the property that its measurement can be made compatibly with a measurement of any component of the momentum or angular momentum of a state.

According to the twistor particle hypothesis, *all* the observables of physics can ultimately be expressed in the form (5), for suitable values of n. In the case n = 3, for example, one expects with appropriate choices of the R's to be able to build all the observables necessary for the description of elementary hadrons. This turns out indeed to be the case, as we shall now proceed to describe. For further details see, e.g., Popovich 1978; and Hughston 1979a, chapter 6.

Let us label our three twistors Z^{α}_{i} according to the scheme

$$Z^{\alpha}_{i} = (U^{\alpha}, D^{\alpha}, S^{\alpha}) \quad , \tag{7}$$

and for the three operators $\hat{Z}^{i}_{\alpha}$ let us put

$$\hat{Z}^{i}_{\alpha} = (\hat{U}_{\alpha}, \hat{D}_{\alpha}, \hat{S}_{\alpha}) \quad . \tag{8}$$

Then for the *baryon number* $\hat{B}$, the *electric charge* $\hat{Q}$, and the *hypercharge* $\hat{Y}$ we have the following expressions:

$$\begin{cases} \hat{B} = -\frac{1}{3}(U^\alpha \hat{U}_\alpha + D^\alpha \hat{D}_\alpha + S^\alpha \hat{S}_\alpha) + 2 & (9) \\ \hat{Q} = -\frac{1}{3}(2U^\alpha \hat{U}_\alpha - D^\alpha \hat{D}_\alpha - S^\alpha \hat{S}_\alpha) & (10) \\ \hat{Y} = -\frac{1}{3}(U^\alpha \hat{U}_\alpha + D^\alpha \hat{D}_\alpha - 2S^\alpha \hat{S}_\alpha). & (11) \end{cases}$$

The *total isospin operator* is given by the formula

$$\hat{I}^2 = (\hat{I}_1)^2 + (\hat{I}_2)^2 + (\hat{I}_3)^2 \ , \tag{12}$$

where $\hat{I}_1$, $\hat{I}_2$, and $\hat{I}_3$ are the twistor analogues of the Pauli matrices:

$$\begin{cases} \hat{I}_1 = -\frac{1}{2}(U^\alpha \hat{D}_\alpha + D^\alpha \hat{U}_\alpha) & (13) \\ \hat{I}_2 = -\frac{i}{2}(U^\alpha \hat{D}_\alpha - D^\alpha \hat{U}_\alpha) & (14) \\ \hat{I}_3 = -\frac{1}{2}(U^\alpha \hat{U}_\alpha - D^\alpha \hat{D}_\alpha) \ . & (15) \end{cases}$$

Note, incidentally, that we have the well-known relation

$$\hat{Q} = \hat{I}_3 + \hat{Y}/2 \ , \tag{16}$$

due to Gell-Mann and Nishijima.

Given the kinematic operator $\hat{A}^{\alpha\beta}$ it is straightforward to derive twistor expressions for the operators for *mass* and *spin*. The mass-squared operator, for example, is given by

$$M^2 = I_{\alpha\beta} I^{\gamma\delta} Z^\alpha_i Z^\beta_j \hat{Z}^i_\gamma \hat{Z}^j_\delta \ , \tag{17}$$

and thus corresponds to an expression of the form (5) with the only non-vanishing $R^{:::}_{:::}$ given by

$$R^{ij\gamma\delta}_{\alpha\beta k\ell} = I_{\alpha\beta} I^{\gamma\delta} \delta^i_k \delta^j_\ell \ . \tag{18}$$

The remaining two basic observables for elementary (i.e., one-particle state) hadrons are the *SU(3) Casimir operators* $\hat{C}_2$ and $\hat{C}_3$. These can be conveniently defined by the formulae

$$\begin{cases} \hat{C}_2 = \tilde{A}^{[i}_i \tilde{A}^{j]}_j , & (19) \\ \hat{C}_3 = \tilde{A}^{[i}_i \tilde{A}^{j}_j \tilde{A}^{k]}_k , & (20) \end{cases}$$

where $\tilde{A}^i_j$ is given by

$$\tilde{A}^i_j = Z^{\alpha}_j \hat{Z}^i_{\alpha} - \frac{1}{3} \delta^i_j Z^{\alpha}_k \hat{Z}^k_{\alpha} . \tag{21}$$

The precise definitions chosen for $\hat{C}_2$ and $\hat{C}_3$ tend to vary slightly from author to author, so the reader should be warned not to apply our formulae without taking due care.

As an aside, it is perhaps worth dwelling a moment longer on $\hat{C}_2$ and $\hat{C}_3$. The eigenvalues of these operators characterize the SU(3) representation to which the state upon which the operators act belongs. SU(3) representations can be classified most conveniently by a pair of numbers λ and μ, related to the eigenvalues C_2 and C_3 by the formulae

$$\begin{cases} C_2 = \frac{1}{3}(\lambda^2 + \mu^2 + \lambda\mu + 3\lambda + 3\mu) , & (22) \\ C_3 = \frac{1}{27}(\lambda-\mu)(2\lambda + \mu + 3)(\lambda + 2\mu + 3) . & (23) \end{cases}$$

If we write

$$\begin{cases} \rho = \frac{-(2\lambda+\mu)}{3} -1 & (24) \\ \sigma = \frac{\lambda+2\mu}{3} +1 & (25) \\ \tau = \frac{\lambda-\mu}{3} , & (26) \end{cases}$$

then a short calculation shows

$$\begin{cases} 2C_2 = \rho^2 + \sigma^2 + \tau^2 - 2 & (27) \\ -3C_3 = \rho^3 + \sigma^3 + \tau^3 \ , & (28) \end{cases}$$

along with $\rho + \sigma + \tau = 0$. One can solve explicitly for λ and μ in terms of C_2 and C_3 as follows:

$$\lambda = -2\rho - \sigma - 1 \quad , \quad \mu = \rho + 2\sigma - 1 \tag{29}$$

$$\rho = \frac{1}{3}(\alpha+\beta) \quad , \quad \sigma = \frac{1}{3}(\omega^2\alpha + \omega\beta) \tag{30}$$

$$\{\alpha^3, \beta^3\} = \frac{3}{2}(-9C_3 \pm [81C_3^2 - 12(C_2 + 1)^3]^{1/2} \tag{31}$$

where ω is the cube-root of unity. (These formulae were worked out by A. Popovich and Z. Perjés.)

In an SU(3) representation it is always possible, by a judicious application of epsilon tensors, to arrange for an expression which is totally symmetric and tracefree over all the SU(3) indices. Then λ is the number of upstairs indices, and μ is the number of downstairs indices. Thus, as characterized by the pair $\{\lambda,\mu\}$ a quark state is $\{1,0\}$, an anti-quark $\{0,1\}$, an octet $\{1,1\}$, a decimet $\{3,0\}$, a 27-plet $\{2,2\}$, and so forth. This notation is used by Perjés and Sparling in §4.2. One important conclusion that they draw is a set of inequalities characterizing 3-twistor hadron states, given in their equation (12). In terms of ρ, σ, and τ, as defined above, these inequalities are:

$$J + \frac{1}{2}B + \rho + 1 \leq 0 \tag{32}$$

$$J - \frac{1}{2}B - \sigma - 1 \leq 0 \tag{33}$$

$$J + \frac{1}{2}B + \tau \geq 0 \tag{34}$$

$$J - \frac{1}{2}B - \tau \geq 0 \quad , \tag{35}$$

where J is the spin and B is the baryon number. Combining (32) and (33) we obtain the fundamental inequality

$$J \leq \frac{\lambda+\mu}{2}, \qquad (36)$$

which characterizes 3-twistor systems. This inequality shows us that higher spin hadron resonances cannot, apparently, be described as 3-twistor particles. One way of possibly getting around this crux is hinted at in §4.2: namely, the consideration of negative J values (since the inequalities 32 - 35 above are derived, it seems, on the assumption that J is positive); this idea has not yet been developed very thoroughly (cf., however, Penrose, Sparling & Tsou 1978). Another approach, which in my own mind is a bit more plausible, is to regard most of the higher mass hadron resonances as *compound systems*, of a nature which will be outlined shortly. This is also the point of view taken in §4.3 by M. Sheppard. [A third way around the perplexities of (36) is simply to take the inequality at face-value and propose that higher-spin resonances belong, in fact, to very large SU(3) multiplets. This idea is not unreasonable, and deserves further exploration: its success would ultimately hinge on some argument that would suppress all the superfluous unwanted (?) states that arise on account of the extravagantly-sized SU(3) multiplets under consideration. At the moment I am not terribly optimistic over the prospects of this particular scheme.]

In order to proceed further in this discussion we need to introduce some contour integral formulae. Such formulae, as applied to massive fields, are first described in Penrose & MacCallum 1972, and are further explored in Penrose 1975a, Hodges 1975, and in numerous other references; the formulae specifically relevant to the twistor particle hypothesis first appear in Penrose 1975b, and are studied further in Penrose 1977, Popovich

1978, Hughston 1979a, and elsewhere. An n-twistor particle state is represented by a holomorphic function $f(Z^{\alpha}_{i})$. This twistor function is to be thought of, heuristically, as a representative cocycle for an element of a suitable cohomology group. At the moment it is not known what, in fact, this cohomology group is. One approach which seems rather promising (cf. Hughston 1979a,b) is the consideration of the group $H^{2n-1}(X,\mathcal{O}(r))$, where X is the domain $Z^{\alpha}_{i}\bar{Z}^{i}_{\alpha} > 0$. At any rate, one can proceed in a formal fashion, without a detailed knowledge of the relevant cohomology, as follows. Let us write

$$Z^{\alpha}_{j} = (\omega^{A}_{j}, \pi_{A'j}) \tag{37}$$

for the spinor parts of Z^{α}_{j}, and write

$$\hat{\pi}^{Aj} = -\partial/\partial\omega_{Aj} \quad . \tag{38}$$

For any twistor function $f(Z^{\alpha}_{j})$ we write

$$\rho_{x}\, f(Z^{\alpha}_{j}) = f(ix^{AA'}\pi_{A'j}, \pi_{A'j}) \tag{39}$$

for $f(Z^{\alpha}_{j})$ restricted to the spacetime point $x^{AA'}$. If the function $f(Z^{\alpha}_{j})$ corresponds to a particle state with a definite set of quantum numbers, then the field associated with that state is some component of a multiplet $\phi^{\vdots\vdots}_{\vdots\vdots}(x)$ given by the contour integral formula

$$\phi^{\vdots\vdots}_{\vdots\vdots}(x) = \oint \rho_{x}\, \hat{\pi}\ldots\ \pi\ldots\ f(Z)\Delta\pi \ . \tag{40}$$

Here $\Delta\pi$ is the appropriate projective differential form on the P^{2n-1} of the $\pi_{A'i}$ space. Note that the restriction to the space-time point $x^{AA'}$ is performed only after $\hat{\pi}$ has operated as many times as necessary on f(Z). The field $\phi^{\vdots\vdots}_{\vdots\vdots}(x)$ has both group indices and spinor indices. This is why we refer to it as a "multiplet". If $f(Z)$ is in a definite eigenstate of all the various internal observables appropriate for the description of a state, then the multiplet $\phi^{\cdots}_{\cdots}(x)$ will only have one non-vanishing component, corresponding to the particle with precisely the quantum numbers exhibited by the twistor function $f(Z)$.

For example, if $f(Z)$ has the quantum numbers appropriate for membership in the basic baryon 56-plet supermultiplet, then the relevant contour integral formula is

$$B^{ijk}_{ABC}(x) = \oint \rho_x \hat{\pi}^i_A \hat{\pi}^j_B \hat{\pi}^k_C f(Z) \Delta\pi \ , \tag{41}$$

where, of course, $n = 3$. At this point it is important to draw a distinction between what we are doing here and the standard quark model with colour and flavour. In particular, we treat the low-lying baryons (i.e. members of the basic baryon octet and decimet, comprising the 56-plet in equation 41 above) as *single particle* states, rather than as three-particle bound states as does the quark model. [At this point the seasoned particle physicist will commence to elaborate eloquently on various aspects of the parton model, deep inelastic scattering, the electron-positron annihilation R value, and a host of other odds and ends favouring, in one way or another, evidence for the three-quark internal structure of the proton. Internal structure I do not for one moment deny — but I would prefer to think that these experiments are in effect *probing the internal twistor geometry* of hadrons, which can be expected in any case to be rich and intricate. In fact, I share a view which has been put forward by Penrose from time to time: namely, that ordinary space-time structure actually *breaks down* at length scales appropriate to elementary particle sizes, i.e. $\sim 10^{-13}$ cm. This view, it should be remarked, is quite distinctly at odds with the opinion that appears to be widely prevalent among particle physicists in general, and indeed among many relativists as well — their opinion being that space-time structure breaks down (at least in some sense) only at around 10^{-33} cm. Of course the precise manner in which the normal notions of space-time structure become inadequate at 10^{-13} cm is not at present known. It is perhaps the case that space-time is a *secondary* notion that emerges only after systems of sufficient complexity are under consideration (analogous in this respect to the macroscopic variables of statistical mechanics, for example). This viewpoint receives some support from *spin network theory* (cf. §§5.13 - 5.16), and in any case has, I feel a certain aesthetic appeal to it — i.e. it would make perfectly good sense and it is quite plausible that the world really is organized in a manner consistent with this way of thinking. In summary, we expect the customary description of space-time to cease being adequate at the 10^{-13} cm level: geometries of a more elaborate character (e.g. products of twistor spaces, and deformations of portions thereof) must be invoked for an accurate description of all the relevant phenomena.] The most strikingly convincing evidence for the internal quark structure of baryons is, phenomenologically, the general pattern of the spectrum of excited baryon states actually seen: but this is precisely the sort of evidence that can be accounted for in alternative ways by means of twistor theory, and thus the quark model has nothing really to say for itself on this score.

Since baryons are treated, in effect, as single particle states (although they may exhibit some internal twistor structure of a sort which is incapable of manifesting itself directly in space-time terms) there is no need to introduce *colour* degrees of freedom — the Fermi statistics problem that one encounters in the quark model simply does not arise. Although baryons are not three-quark bound states, it is useful to consider mesons nevertheless, as *quark-antiquark pairs* — in this respect our ideas about mesons are much closer to the "establishment" view (in certain respects) than are our ideas about baryons. A single quark state can be thought of as a twistor function f(Z) for which the relevant contour integral formula is

$$Q_A^i(x) = \oint \rho_x \hat{\pi}_A^i \, f(Z)\Delta\pi \; . \tag{42}$$

Thus f(Z) must have B = 1/3, J = 1/2, $\{\lambda,\mu\} = \{1,0\}$, and likewise must exhibit appropriate electric charge, hypercharge, and isospin eigenvalues. Whether free quark states occur in nature or not must simply be regarded as an open question — I can see no particularly compelling argument against the possibility of free quarks. Perhaps there is some mechanism involved in the structure of deformed twistor spaces which singles out integral charges — such a mechanism is indeed hinted at in §3.3, but it should be evident that the argument would have to be elevated to a considerably higher level of sophistication before it would be applicable to the case at hand. For an antiquark the relevant contour integral formula is

$$Q_i^{A'}(x) = \oint \rho_x \, \pi_i^{A'} \, f(Z)\Delta\pi \; , \tag{43}$$

where now f(Z) describes a state with B = -1/3, J = 1/2, $\{\lambda,\mu\} = \{0,1\}$, etc.. In order to describe a bound state of a quark and an antiquark it is necessary to consider a function of six twistors $f(Z_1, Z_2)$. The first three twistors relate to the first particle, and the second three twistors relate to the second particle. On account of Fermi statistics it will not be apparent which of the two particles is the quark, and which is the antiquark. For the spinor parts of Z_1^α and Z_2^α let us write

$$\underset{1}{Z}^{\alpha} = (\alpha_i^A, \alpha_{A'i}) \ , \quad \underset{2}{Z}^{\alpha} = (\beta_i^A, \beta_{A'i}) \ , \tag{44}$$

and likewise for the associated spinor operators let us put

$$\hat{\alpha}^{Ai} = -\partial/\partial\alpha_{Ai}; \quad , \quad \hat{\beta}^{Ai} = -\partial/\partial\beta_{Ai} \quad . \tag{45}$$

Then for a quark-antiquark pair (in an S-state) the appropriate contour integral expression is

$$\phi_j^{iAB}(x) = \oint \rho_x[\hat{\alpha}^{iA}\ \beta_j^B - \ \hat{\beta}^{iA}\ \alpha_j^B] f(\underset{1}{Z},\underset{2}{Z})\Delta\pi \ , \tag{46}$$

the minus sign being inserted for Fermi statistics. Let us try to clarify a bit what is meant by saying that (46) is the "appropriate contour integral expression" for a meson state. What is meant is that $\hat{\alpha}^{iA}\ \beta_j^B$ and $\hat{\beta}^{iA}\ \alpha_j^B$ are the only two expressions which, when integrated with $f(\underset{1}{Z},\underset{2}{Z})$, give a non-vanishing answer — moreover, the result in each case must be identical, apart from a sign change. The supermultiplet ϕ_j^{iAB} contains the spin 0 pseudoscalar meson nonet, and the spin 1 vector meson nonet.

In order to obtain higher spin meson resonances it is necessary to introduce orbital angular momentum between the quark and the antiquark. This can be achieved with an operator $\hat{L}_{AB'}$ defined by

$$\hat{L}_{AB} = (\overset{1}{P}_{AA'} - \overset{2}{P}_{AA'})P_B^{A'} \ , \tag{47}$$

where $\overset{1}{P}_{AA'}$ and $\overset{2}{P}_{AA'}$ are the momentum operators for the two systems $\underset{1}{Z}$ and $\underset{2}{Z}$, and $P_{AA'}$ is the *total* momentum operator given by $\overset{1}{P}_{AA'} + \overset{2}{P}_{AA'}$. The definition of $\hat{L}_{AB}$ must be altered suitably if particles with distinct masses are under consideration. This can be realized by introducing parameters γ and δ such that

$$\hat{L}_{AB} = (\gamma \overset{1}{P}_{AA'} - \delta \overset{2}{P}_{AA'}) P_B^{A'} \tag{48}$$

is *automatically symmetric*. With the operator $\hat{L}_{AB}$ at hand the relevant contour integral formula for a state with orbital angular momentum L is given by

$$\phi_j^{iABCD\ldots}(x) = \oint \rho_x \, [\hat{\alpha}^{iA} \beta_j^B \pm \hat{\beta}^{iA} \alpha_j^B] \hat{L}^{CD} \ldots f(\overset{1}{Z}, \overset{2}{Z}) \Delta\pi \tag{49}$$

where $\hat{L}_{AB}$ appears L times. The higher meson resonance multiplets that arise out of such a scheme are described by M. Sheppard in §4.3, and the agreement with phenomenology is quite reasonable, at least insofar as qualitative aspects of the spectrum are concerned. The choice of sign in (49), incidentally, is given by $-(-1)^L$, and is governed by the requirements of Fermi statistics.

In recent years physicists have contemplated the possibilities of observing *diquonium* states — that is to say, states composed of two quarks and two antiquarks. Some of the proposed states can be modelled quite neatly in twistor terms, and these are described in §4.4. An important feature which arises in this connection is the appearance of *diquarks* as fundamental dynamical entities. Physicists often toy around with diquarks in various hadron models, but there usually what they have in mind is some sort of tightly bound system of two quarks — that is, diquarks are generally regarded as *two-particle states*. But according to our scheme, there should exist *one-particle* diquark states (i.e. elementary hadrons with B = 2/3, and J = 0 or 1). For diquarks the relevant contour integral formula is

$$D_{AB}^{ij}(x) = \oint \rho_x \, \hat{\pi}_A^i \, \hat{\pi}_B^j \, f(Z_i^\alpha) \Delta\pi \, . \tag{50}$$

Our diquarks can either have spin 0 or spin 1. The spin 0 diquarks belong to a {0,1} representation of SU(3), whereas the spin 1 diquarks belong to a {2,0} representation. As described in §4.4, we consider a diquonium to be a bound state of a diquark and an anti-diquark.

Diquarks also figure in to the structure of baryon resonances. For the low-lying baryons we use 3-twistor states, for which the relevant contour integral formula is given in equation (41). This description appears to be quite satisfactory in many respects: for example, one can account for certain aspects of the magnetic moments of these states (§4.5), and also one can calculate matrix elements for semi-leptonic processes (§4.6). However, for the higher baryon resonances it is necessary to adopt the viewpoint that these states are *quark-diquark systems* that is to say, we regard these particles as two-particle states, composed of a quark and a diquark. This model is described in §4.3, by M. Sheppard; further details can be found in Hughston 1979a.

It is interesting to contemplate the consequence of reactions of the form BB → DDD, wherein two baryons, colliding with sufficient energy, produce a set of three diquarks. It is not inconceivable that reactions of this sort are of relevance to the later stages of stellar evolution, although at the moment I cannot envisage precisely what role they would play. The point is, however, that since diquarks are *bosons* (and, according to the picture presented here, are *not* in any sense composed of two quarks) the reaction BB → DDD could, under suitable conditions, provide a mechanism for the elimination of Fermi pressures in a dense system composed of baryons. This matter will be explored in greater depth elsewhere.

REFERENCES

Hodges, A. 1975 *The Description of Mass in the Theory of Twistors.* Ph.D. Thesis, Birkbeck College, London.

Hughston, L.P. 1979a *Twistors and Particles.* Springer Lecture Notes in Physics, Volume 97.

Hughston, L.P. 1979b *Some New Contour Integral Formulae.* In: *Complex Manifold Techniques in Theoretical Physics,* eds. D. Lerner & P. Sommers. Pitman: London.

Penrose, R. 1975a *Twistor Theory, Its Aims and Achievements.* In: *Quantum Gravity: An Oxford Symposium,* eds. C.J. Isham, R. Penrose & D.W. Sciama. Clarendon Press: Oxford.

Penrose, R. 1975b *Twistors and Particles.* In: *Quantum Theory and the Structure of Time and Space,* eds. L. Castell, M. Drieschner, & C.F. von Weiszacker. Verlag: München.

Penrose, R. 1977 Reports Math. Phys. 12, 65.

Penrose, R. & MacCallum, M.A.H. 1972. *Twistor Theory: An Approach to the Quantization of Fields and Spacetime.* Phys. Reports Vol. 6C, No.4.

Penrose, R., Sparling, G.A.J., & Tsou, S.T. 1978 J. Phys. A: Math. Gen. Vol. 11, No. 9, L231.

Popovich, A. 1978 *Twistor Classification of Elementary Particles.* M.Sc. Thesis, Oxford University.

§4.2. THE TWISTOR STRUCTURE OF HADRONS *by Z. Perjés and G.A.J. Sparling*

Attempts to formulate a relativistic theory of hadronic internal symmetries have been numerous but never thoroughly successful in the past. In addition, many painful problems are presented already by the SU(3) quark picture (cf. Lipkin 1973). And since the early successes of the theory, marked, for instance, by the Gell-Mann-Okubo formulas, very little has been achieved in producing figures out of theories.

Here we are outlining an intrinsically relativistic quantum theory of hadronic structure. Our basic assumption is that hadrons are composed of three twistors. A *single* twistor is mathematically a spinor of the group SU(2,2) and it possesses the physical properties of a classical zero-mass particle. There exist some theoretical arguments in favour of a *two-twistor* structure of leptons. Systems of three twistors have been shown (Perjés 1975) to admit a group of internal symmetries isomorphic to ISU(3), the semi-direct product group of SU(3) rotations with complex translations in the complex linear space C^3 (cf. §§4.7 & 4.8).

The Lie algebra of the ISU(3) group is given by the commutation relations

$$\left.\begin{aligned} &[A^i_j, A^k_l] = \delta^i_l A^k_j - \delta^k_j A^i_l \\ &[d^i, A^j_k] = \delta^i_k d^j - \tfrac{1}{3}\delta^j_k d^i \\ &[d^i, d^k] = 0, \quad [d^i, \bar{d}_k] = 0 \end{aligned}\right\} \qquad (1)$$

$$+ \text{ Hermitian conjugates} \quad ,$$

where the infinitesimal generators A^i_j of the su(3) subalgebra satisfy

$$A^i_j = \bar{A}^i_j, \qquad A^i_i = 0 \qquad (2)$$

and the indices range through 1, 2 and 3 with summation convention understood.

The map

$$C: \quad A^i_j \to - A^j_i, \qquad d^i \to \bar{d}_i, \qquad \bar{d}_i \to d^i \tag{3}$$

is an automorphism of the Lie algebra. An essential implication of twistor theory is (cf. Penrose 1975) that the two Casimirians of the group,

$$m^2 = 2\, d^i\, \bar{d}_i \tag{4}$$

and

$$\vec{J}^2 = \frac{1}{4} B^2 - \frac{3}{2} B + \frac{1}{2} A^k_\ell\, A^\ell_k - 2m^{-2} d^i \bar{d}_k A^k_\ell A^\ell_i \tag{5}$$

will be identified with the operators of rest-mass squared and spin squared, respectively. The operator B ('baryon number operator') can be defined by

$$m^2\, B = 2\, d^i\, \bar{d}_k\, A^k_i. \tag{6}$$

We do not consider here singular representations with $m=0$.

Hadrons are eigenstates of B, the standard hypercharge operator Y, and isospin projection operator I_3, since they are to be represented by homogeneous twistor functions (cf. Perjés 1977). Further we require the isospin I and the Casimirians of the SU(3) subgroup, giving a pair of observables λ and μ (whose eigenvalues are integers) as diagonal operators. States are labeled by the complete set of operators

$$|\alpha\rangle = |J,m;\, B,\, (\lambda,\mu),\, Y,I,I_3\, \rangle. \tag{7}$$

The dependence on the SU(3) projection quantum numbers Y, I and I_3 of the matrix elements of the SU(3) tensor operators d^i can be factored out by the use of the Wigner-Eckart theorem;

$$\langle\beta|d^i|\alpha\rangle = \langle\alpha;\, (1,0)^i\, |\beta\rangle\, \langle\beta|\,\, |d|\,\, |\alpha\,\rangle \tag{8}$$

where $<\alpha; (1,0)^i|\beta>$ are SU(3) 'quark coupling' Clebsch-Gordan coefficients. The reduced matrix elements arising from the Clebsch-Gordan series of the product $(1,0) \otimes (\lambda,\mu)$ are given as

$$U^2 = \frac{\dim(\lambda+1),\mu)}{\dim(\lambda,\mu)} |<B - \tfrac{2}{3}, (\lambda+1,\mu)||d||B, (\lambda,\mu)>|^2 \Delta^{-1}$$

$$D^2 = \frac{\dim(\lambda,\mu-1)}{\dim(\lambda,\mu)} |<B - \tfrac{2}{3}, (\lambda,\mu-1)||d||B, (\lambda,\mu)>|^2 \Delta^{-1} \tag{9}$$

$$S^2 = \frac{\dim(\lambda-1,\mu+1)}{\dim(\lambda,\mu)} |<B - \tfrac{2}{3}, (\lambda-1, \mu+1)||d||B, (\lambda,\mu) > |^2 \Delta^{-1}$$

where we have:

$$\dim(\lambda,\mu) = \frac{1}{2} (\lambda+1)(\mu+1)(\lambda+\mu+2),$$

$$\Delta = m^2/2 \qquad . \tag{10}$$

Using the SU(3) recoupling techniques developed by Hecht (1975), or, alternatively, highest-state methods (work by G.A.J. Sparling), we obtain:

$$U^2 = \frac{(-\frac{1}{2}B-J+1+x)(-\frac{1}{2}B+J+2+x)}{(\lambda+1)(\lambda+\mu+2)} \quad , \qquad x = \frac{2\lambda+\mu}{3}$$

$$D^2 = \frac{(-\frac{1}{2}B-J-1-y)(-\frac{1}{2}B+J-y)}{(\mu+1)(\lambda+\mu+2)} \quad , \qquad y = \frac{\lambda+2\mu}{3} \tag{11}$$

$$S^2 = \frac{(\frac{1}{2}B+J+z)(-\frac{1}{2}B+J+1-z)}{(\lambda+1)(\mu+1)} \quad , \qquad z = \frac{\lambda-\mu}{3} \qquad .$$

The quantities U^2, D^2 and S^2 are non-negative for unitary representations. These representations will be labeled by the quantum number J since their structure is independent of m. In the $\{B,(\lambda,\mu)\}$ space, the boundary planes of the representations are determined by the zeroes of U, D and S and of the corresponding conjugate matrix elements. We find that J and $\frac{3}{2}B$ are simultaneously either integers or half-integers.

Each representation consists of an infinite lattice of SU(3) multiplets in the $\{B,(\lambda,\mu)\}$ space. For J = 0 their points lie on a quadrant of a plane through the C-conjugation symmetry axis $\{0,(\lambda,\lambda)\}$ and the anti-diquark triplet $\{-\frac{2}{3}, (1,0)\}$. We identify the points $\{0,(0,0)\}$ as the singlet and $\{0,(1,1)\}$ as the octet of pseudoscalar mesons. For $J > 0$, the representations are characterized by the inequalities

$$
\begin{aligned}
&J + \frac{1}{2}B - x \leq 0 \\
&J - \frac{1}{2}B - y \leq 0 \\
&J + \frac{1}{2}B + z \geq 0 \\
&J - \frac{1}{2}B - z \geq 0 \quad .
\end{aligned}
\tag{12}
$$

and form an infinite wedge. Again we find for $J = \frac{1}{2}$ the baryon octets $\{\pm 1,(1,1)\}$, for J = 1 the vector meson octet $\{0,(1,1)\}$, and for $J = \frac{3}{2}$ the baryon decimets $\{1,(3,0)\}$ and $\{-1,(0,3)\}$ lying on the boundary.

For unitary ISU(3) representations, the inequality $\lambda+\mu \geq 2J$ holds. Therefore, we conjucture that the higher resonances should be associated with *non-unitary representations* (and non-local states).

It is our next purpose to show that the many Gell-Mann-Okubo mass splitting formulas for various SU(3) multiplets merge into a single equation in the present theory, by the simple assumption that the mass perturbing octet operator is a function of the elements of the Lie algebra. From the translation operators we can construct a Hermitian tensor operator $\Delta^i_k = d^i \overline{d}_k$, of positive charge (C) parity as required by CPT invariance. The theorems of Okubo (1962) imply that there are five algebraically independent nonet operators constructed out of A^i_k and Δ^i_k. Remarkably, there is a unique combination of them with positive charge parity. Thus our mass splitting operator is obtained in the form

$$H^i_k = a\Delta^i_k + b[2\overline{f}_k f^i - 3(\overline{d}_k f^i + \overline{f}_k d^i) + A^i_k\Delta + \delta^i_k B\Delta] , \tag{13}$$

where

$$f^i = d^k A^i_k , \quad \Delta = m^2/2. \tag{14}$$

For some inexplicable reason often physicists believe that mass splitting formulas ought to be linear in baryon masses and quadratic in meson masses. However, our mass operator (5) has a universal structure, and consequently, we take, for the observed mass-squared

$$M^2 = c^2 + \langle\alpha|H^3_3|\alpha\rangle . \tag{15}$$

M^2 has the familiar dependence on Y and I,

$$M^2 = c^2 + m_o^2 + m_1^2 Y + m_2^2 [I(I+1)-(\tfrac{1}{2} Y)^2]. \tag{16}$$

The coefficients m_0^2, m_1^2 and m_2^2 are algebraic functions of the reduced matrix elements (11) and the parameters a and b. Here a,b and c^2 may possibly depend on J. Using the phenomenological values quoted in the Particle Data Tables we find that $b = (- 0.048 \pm 0.01)a$.

If one plots the values of c^2 as a function of the parameter a for the spin 0 octet, the spin 1 octet, the spin $\frac{1}{2}$ octet, and the spin $\frac{3}{2}$ decimet, a striking pattern emerges: *the dependence of* c^2 *upon* a *is well approximated by a linear relation*. We may speculate on the nature of a(J), which represents intrinsic dynamical information, and conjecture that a(J) defines a pair of "reduced" Regge trajectories, one for fermions and another for bosons.

We wish to thank Prof. R. Penrose for providing stimulating ideas. One of us (Z.P.)acknowledges suggestions by Prof. Tibor Nagy and by members of the Theory Department of the Central Research Institute for Physics, Budapest, and Z.P. also wishes to acknowledge helpful correspondence with Prof. K.T. Hecht. G.A.J.S. owes thanks to the members of the Oxford relativity group, particularly L. Hughston. G.A.J.S. also wishes to thank the Central Research Institute for Physics, Budapest, for its generous hospitality.

REFERENCES

Hecht, K.N. 1965 Nucl. Phys. 62, 1.

Lipkin, H.J. 1973 Physics Reports 8C, 173.

Okubo, S. 1962 Prog. Theor. Phys. 27, 949.

Penrose, R. 1975 *Twistors and Particles*. In: *Quantum Theory and the Structure of Time and Space*, eds. L. Castell, M. Drieschner & C.F. von Weizsäcker. Verlag: München.

Perjés, Z. 1975 Phys. Rev. D. 11, 2031.

Perjés, Z. 1977 Reports Math. Phys. 12, 193.

§4.3 THE SIX-TWISTOR DESCRIPTION OF HADRON RESONANCES *by M. Sheppard*

There exists an immense amount of evidence for resonant hadron states which exhibit the same SU(3) quantum numbers as the low lying states (the baryon $J^P = \frac{1}{2}^+$ octet and $J^P = 3/2^+$ decimet, and the two meson nonets, with $J^{PC} = 0^{-+}, 1^{--}$) but having higher masses and/or spins than these basic states. There is evidence also for resonant baryon states with SU(6) quantum numbers differing from those of the observed low lying states. We shall describe the way in which these various resonant states are accommodated in the six-twistor model for hadrons. The notation used here is described in §4.1. (Cf. also Hughston 1979, pp. 72-91.)

The Meson Resonances

There are two nonets of low lying mesons with $J^{PC} = 0^{-+}$ and $J^{PC} = 1^{--}$. The observed 0^{-+} states are given by linear combinations of the following fields:

$$\phi^i_j(x) = \oint_x (\hat{\alpha}^i_A \, \beta^{A'}_j - \hat{\beta}^i_A \, \alpha^{A'}_j) \, P^A_{A'} \, f(Z)\Delta\pi \quad , \tag{1}$$

while the observed 1^{--} states are given by linear combinations of the fields

$$\phi^{iAB}_j(x) = \oint_x [\hat{\alpha}^{i(A}\beta_{A'j} - \hat{\beta}^{i(A}\alpha_{A'j}]P^{B)A'} \, f(Z)\Delta\pi. \tag{2}$$

It should be observed that the identification of the 0^{-+} and 1^{--} mesons as six-twistor states, as proposed here, is different from the choice made in §4.2 by G.A.J.S. and Z.P.

We can generate further meson resonances by including the orbital angular momentum operator L_{AB} in the expressions (1) and (2). Before proceeding with this program we notice that both (1) and (2) are antisymmetric with respect to the exchange $\alpha \leftrightarrow \beta$. We shall require in what follows that the spinor coefficients for the resonant meson states are always *antisymmetric* with respect to $\alpha \leftrightarrow \beta$, so that we must combine even numbers of L_{AB} with antisymmetric coefficients of the form $(\hat{\alpha}^{iA}\,\beta_{A'j} - \hat{\beta}^{iA}\,\alpha_{A'j})$ and odd numbers of L_{AB} with symmetric coefficients of the form $(\hat{\alpha}^{iA}\,\beta_{A'j} + \hat{\beta}^{iA}\,\alpha_{A'j})$. This single rule serves to fix the correct charge conjugation quantum numbers for the observed meson resonances. (Charge conjugation is dealt with in detail in Hughston 1979 §7.3.)

<u>The L = 1 mesons</u>

There are four possible L = 1 meson nonets. We shall just consider the spinor coefficients for these in unreduced form (the reduction merely involves taking the trace over the group indices). Since L = 1 we combine a single L_{AB} with the symmetric (w.r.t. $\alpha \leftrightarrow \beta$) spinor coefficient structure:

The 2^{++} nonet

These mesons are associated with a spinor coefficient of the form

$$[\hat{\alpha}^{i(A}\,\beta_{jA'} + \hat{\beta}^{i(A}\alpha_{A'j}]L^{BC}p^{D)A'} \quad .$$

This nonet is well established and incorporates the following states (all data taken from the 1978 particle tables):

States	Isospin	Mass MeV	Decays
$A_2(1310)$	1	1312 ±5	$\rho\pi$ $\eta\pi$ $\omega\pi\pi$
$\kappa^*(1430)$	½	1434 ±5	$K\pi$ $K^*\pi$ $K^*\pi\pi$
f'(1515)	0	1516 ±10	$K\bar{K}$
f (1270)	0	1271 ±5	$\pi\pi$ $2\pi^+$ $2\pi^-$

<u>Table 1. 2^{++} Mesons</u>

The 1^{++} nonet

There exists a fairly well established nonet of 1^{++} mesons, encompassing the following states:

State	Isospin	Mass MeV	Decays
$A_1(1100)$	1	$\sim$ 1100	$\rho\pi$
Q_A	½	1200-1400	
E (1420)	0	1416 ± 10	$K\bar{K}\pi$
D (1285)	0	1282 ± 5	$K\bar{K}\pi$

Table 2. 1^{++} Mesons

The E(1420) is the most recently established state and its J^P value waits to be confirmed. The 1^{++} states are generated by spinor coefficients of the form:

$$[\hat{\alpha}^{iA}\,\beta_{jA'} + \hat{\beta}^{iA}\,\alpha_{jA'}]P^{A'(B}L^{C)}{}_{A} \quad .$$

The 1^{+-} mesons

We can construct a spinor coefficient of the form:

$$[\hat{\alpha}^{iA}\,\beta_{jA'} + \hat{\beta}^{iA}\,\alpha_{jA'}]P^{A'}{}_{A}L^{BC} \quad ,$$

which yields a nonet of 1^{+-} states. Experimentally, the only candidates for this nonet found so far are the I=1 B(1235) state (seen as a resonance in $\pi\pi$ final states) and the I=½ Q_B(1200-1400).

The 0^{++} mesons

With one L^{AB} there remains a single further spinor coefficient we can construct, which has the form:

$$[\hat{\alpha}^{iA}\ \beta_{jA'} + \hat{\beta}^{iA}\ \alpha_{jA'}]P^{A'B}L_{AB}\ , \tag{3}$$

giving J^{PC} values of 0^{++}.

The experimental evidence relating to 0^{++} mesons is still rather confused. There is evidence for an I=½, S=1, 0^{++} resonance with mass $\sim$ 1510 MeV in $K\pi$ final states, and there are indications in $\pi^- p \to K^0 K^- p$ of the existence of an I=1 resonance with mass $\sim$ 1300 MeV to accompany the already well established $\delta(980)$. As for the I=0 states, analysis of reactions such as $\pi^- \rho \to \pi^+ \pi^- \eta$ would seem to imply the existence of three I=0, 0^{++} resonances with the following masses:

$$M_\varepsilon \sim 800 \text{ MeV}$$
$$M_{\varepsilon'} \sim 1550 \text{ MeV}$$
$$M_{S^*} \sim 1005 \text{ Mev}$$

Thus, altogether, there appears to be more than a nonet's worth of 0^{++} mesons! The relevant states are listed as follows:

State	Isospin
κ (1510)	½
δ (980)	1
δ'(1300)	1
ε (800)	0
S* (998)	0
ε'(1540)	0

Table 3. 0^{++} Mesons

One possible way of accounting for all these states would be to accommodate some of the states (for example the S*,ε,δ and κ) in the nonet with spinor coefficients (3) and invoke exotic spinor coefficients of the form

$$\alpha_{iA'}\,\alpha_{jB'}\,\hat{\beta}^{pA}\,\hat{\beta}^{qB} + \beta_{iA'}\,\beta_{jB'}\,\hat{\alpha}^{pA}\,\hat{\alpha}^{qB}$$

to account for the remaining states. Such coefficients would generate an SU(3) 27-plet. These exotic (diquonium) states are dealt with in §4.4. It is interesting to note that while the coloured quark model predicts two distinct varieties of diquonium (referred to as M and T diquonium) the twistor model would give only one (essentially T type) such variety.

To summarize, we can account for the observed mesons with masses of up to about 1550 MeV as L = 1 excitations of the L = 0 nonets by accommodating them in the following multiplets (although the 0^{++} assignments are a little tentative):

$(J^P)C_n$	{SU(3),L}	I = 1	I = 0	I = 1/2
$(0^-)^+$	$\{\underline{9}^1,0^+\}$	π	η,η'	K
$(1^-)^-$	$\{\underline{9}^3,0^+\}$	$\rho(770)$	$\omega(783),\phi(1020)$	K*(892)
$(1^+)^-$	$\{\underline{9}^1,1^-\}$	B(1235)		Q_B
$(0^+)^+$	$\{\underline{9}^3,1^-\}$	$\delta(980)$	S*(980),$\varepsilon(1300)$	$\kappa(1400)$
$(1^+)^+$	$\{\underline{9}^3,1^-\}$	$A_1(1100)$	D(1285),E(1420)	Q_A
$(2^+)^+$	$\{\underline{9}^3,1^-\}$	$A_2(1310)$	f(1270),f'(1515)	K*(1430)

Table 4. L = 0,1 Meson Multiplets

The L = 2 mesons

By adding in two L_{AB} operators into the spinor coefficients, taking now the antisymmetric (w.r.t. $\alpha \leftrightarrow \beta$) form of the spinor coefficient structure, we can generate further resonances. At this level the experimental evidence is already getting a little thin and we shall not consider any higher excitations than L = 2.

The well established 3^{--} states $\omega(1670,I=0)$, $g(1680,I=1)$ and the 3^{-} $\kappa^*(1780,I=\frac{1}{2})$ will be adequately accommodated by the nonet with spinor coefficients:

$$[\hat{\alpha}^{i(A}\,\beta_{jA'} - \hat{\beta}^{i(A}\,\alpha_{jA'}]P^{A'B}{}_{L}{}^{CD}{}_{L}{}^{EF)} \quad .$$

The spinor coefficients

$$[\hat{\alpha}^{iA}\,\beta_{jA'} - \hat{\beta}^{iA}\,\alpha_{jA'}]P_{A}^{A'}{}_{L}{}^{(CD}{}_{L}{}^{EF)}$$

would provide accommodation for a nonet of 2^{-+} mesons containing the I=1 resonance $A_3(1640)$ and there is evidence for some L=2, $J^{PC} = 1^{--}$ states which would be accounted for by spinor coefficient structures of the form:

$$[\hat{\alpha}^{i(A}\,\beta_{jA'} - \hat{\beta}^{i(A}\,\alpha_{jA'}]P^{B)A'}{}_{L}{}_{(AB}{}^{L}{}_{CD)} \quad .$$

The Baryon Resonances

Resonances with baryon number 1 are seen in the various $p\pi$ and pK channels and it is possible to describe these states by holomorphic functions of six twistors $f(Z_1^{\alpha}, Z_2^{\alpha})$ associated with spinor coefficient structures of the form

$$\hat{\alpha}_i^A\,\hat{\alpha}_j^B\,\hat{\beta}_k^C \quad \pm \quad \hat{\beta}_i^A\,\hat{\beta}_j^B\,\hat{\alpha}_k^C \quad .$$

We shall characterize the various resonances by the notation [SU(6) multiplet, L^P] where L is the number of occurences of L_{AB} in the appropriate spinor coefficient, and P is the parity. Thus the low-lying baryons are given as [$\underline{56}$, 0^+]. Since these quantum numbers are unaffected by the symmetry with respect to the exchange $\alpha \leftrightarrow \beta$ we shall work with coefficients of the form $\hat{\alpha}\hat{\alpha}\hat{\beta}$ only. These coefficients will in some respects yield equivalent fields to those associated with $\hat{\beta}\hat{\beta}\hat{\alpha}$ coefficients and it may be that some further principle fixes the $\alpha \leftrightarrow \beta$ symmetry (e.g. Bose statistics).

Baryons with L = 1

The L = 1 baryons are to be associated with spinor coefficients of the generic form

$$\hat{\alpha}_i^A \hat{\alpha}_j^B \hat{\beta}_k^C L^{DE}$$

where we associate P = -1 with the inclusion of the L^{DE}. We may segregate the possible states with this coefficient structure into a [$\underline{56}$, 1^-] and a [$\underline{70}$, 1^-] as follows:

The [$\underline{70}$,1^-] Supermultiplet

The [$\underline{70}$, 1^-] may be decomposed in terms of the dimension of its SU(3) representation, the intrinsic spin multiplicity μ, and the total spin J: [$^{\mu}dim$, J^P]. We obtain the following states:

$$\hat{\alpha}^A_{(i} \hat{\beta}_{Aj} \hat{\alpha}_{k)}{}^{(B} L^{CD)} \leftrightarrow [{}^2\underline{10}, 3/2^-]$$

$$\hat{\alpha}^A_{(i} \hat{\beta}_{Aj} \hat{\alpha}^B_{k)} L^C_B \leftrightarrow [{}^2\underline{10}, 1/2^-]$$

$$\varepsilon^{ijk} \hat{\alpha}_j^{(A} \hat{\beta}_k^B \hat{\alpha}_m^C L^{DE)} \leftrightarrow [{}^4\underline{8}, 5/2^-]$$

$$\varepsilon^{ijk} \hat{\beta}_j^A L_A{}^{(B} \hat{\alpha}_k^C \hat{\alpha}_m^{D)} \leftrightarrow [{}^4\underline{8}, 3/2^-]$$

$$\varepsilon^{ijk}\, \hat{\beta}^{A}_{j}\, \hat{\alpha}^{(B}{}_{k}\, L_{AB}\, \hat{\alpha}^{\ C)}_{m} \quad \leftrightarrow \quad [{}^{4}\underline{8},\ 1/2^{-}]$$

$$\varepsilon^{ijk}\, \hat{\alpha}^{(A}_{j}\, \hat{\alpha}_{Am}\, \hat{\beta}^{B}_{k}\, L^{CD)} \quad \leftrightarrow \quad [{}^{2}\underline{9},\ 3/2^{-}]$$

$$\varepsilon^{ijk}\, \hat{\alpha}^{A}_{m}\, \hat{\alpha}_{(Aj}\, \hat{\beta}_{B)k}\, L^{BC} \quad \leftrightarrow \quad [{}^{2}\underline{9},\ 1/2^{-}]$$

The nonets can be further reduced, each into an octet and a singlet. Representative members of all of these multiplets have been observed and we shall list in Table 5 the observed N, Δ, Λ content for each multiplet, taking all data from the 1978 particle tables and using the normal particle table notation.

J^P SU(3)	NΔ	Λ	Λ
$1/2^-$ $\underline{8}$, $\underline{1}$	$N(1535)S_{11}$	$\Lambda(1405)S_{01}$	$\Lambda(1670)\ S_{01}$
$1/2^-$ $\underline{8}$	$N(1700)S_{11}$		
$3/2^-$ $\underline{8}$, $\underline{1}$	$N(1520)D_{13}$	$\Lambda(1690)D_{03}$	$\Lambda(1520)\ D_{03}$
$3/2^-$ $\underline{8}$	$N(1700)D_{13}$		
$5/2^-$ $\underline{8}$	$N(1670)D_{15}$		
$1/2^-$ $\underline{10}$	$\Delta(1650)S_{31}$		
$3/2^-$ $\underline{10}$	$\Delta(1670)D_{35}$		

Table 5. The L = 1 Baryon Resonances

The [$\underline{56}$, 1^-] *Supermultiplet*

The [$\underline{56}$, 1^-] has the following decomposition:

$$[{}^{4}\underline{10},\ 5/2^{-}] \quad [{}^{4}\underline{10},\ 3/2^{-}] \quad [{}^{4}\underline{10},\ 1/2^{-}]$$

$$[{}^{2}\underline{8}\ ,\ 3/2^{-}] \quad [{}^{2}\underline{8}\ ,\ 1/2^{-}]$$

The evidence for the [$\underline{56}$, 1^-] is rather tentative; however, there do exist the following candidates for this multiplet;

N(2040 $3/2^-$

Δ(1850) $1/2^-$

Δ(1960) $5/2^-$.

Baryon Resonances with L > 1

Clearly, higher spin resonances may be produced by introducing more L_{AB} operators into the spinor coefficients. Thus, for example, two L_{AB} operators produce a [$\underline{56}$,2^+] and a [70, 2^+] with spinor coefficients of the generic form

$$\hat{\alpha}_i^{\ A} \ \hat{\alpha}_j^{\ B} \ \hat{\beta}_K^C \ L^{DE} \ L^{FG}$$

which may be reduced as in the L=1 case. The [$\underline{56}$, 2^+] (which appears as a Regge recurrence of the [$\underline{56}$, 0^+]) seems to be well represented with six established N, Δ states along with a Λ state and two Σ states as possible candidates. This supermultiplet appears in Table 6:

J^P	SU(3)	NΔ	Λ
$3/2^+$	$\underline{8}$, $\underline{1}$	N(1810) P_{13}	
$5/2^+$	$\underline{8}$, $\underline{1}$	N(1688) F_{15}	Λ(1815)F_{65}
$1/2^+$	$\underline{10}$	Δ(1910) P_{31}	
$3/2^+$	$\underline{10}$	Δ(1690) P_{33}	
$5/2^+$	$\underline{10}$	Δ(1890) F_{35}	
$7/2^+$	$\underline{10}$	Δ(1950) F_{37}	

Table 6. The [56, 2^+] Supermultiplet

The evidence for the [$\underline{70}$, 2^+] on the other hand is meagre, the possible candidates being:

N(1970) $7/2^+$

Λ(1900) $3/2^+$

Λ(2100) $5/2^+$.

The [$\underline{70}$,0^+] Supermultiplet

The six-twistor model also allows for the construction of a (presumably low mass) [$\underline{70}$, 0^+] unless some further principle is invoked to forbid it. The appropriate spinor coefficients take the form:

$$\varepsilon^{ijk}\,\hat{\alpha}_m{}^A\,\hat{\alpha}_j{}^B\,\hat{\beta}_k{}^C \quad \leftrightarrow \quad {}^4\underline{8},\ {}^2\underline{8}$$

$$\hat{\alpha}^A{}_{(i}\,\hat{\alpha}_j{}^B\,\hat{\beta}_{k)}{}^B \quad \leftrightarrow \quad {}^2\underline{10}$$

$$\varepsilon^{ijk}\,\hat{\alpha}_i{}^{(A}\,\hat{\alpha}_{jA}\,\hat{\beta}_k{}^{B)} \quad \leftrightarrow \quad {}^2\underline{1} \quad .$$

The evidence for such states is extremely tentative but we shall list possible candidates along with the PDG star ratings:

J^P	SU(3)	NΔ	Λ	Λ
$1/2^+$	$\underline{8}$, $\underline{1}$	N(1780) ***	Λ(1800) **	Λ(1600) **
$3/2^+$	$\underline{8}$	N(1540) *	Λ(1860)***	
$1/2^+$	$\underline{10}$	N(1550) *		

Table 7

There are also the following Σ states:

$\Sigma(1660)$ P_{11} ***

$\Sigma(1840)$ P_{13} *

$\Sigma(1770)$ P_{11} *

$\Sigma(1880)$ P_{11} * .

The Strangeness +1 Baryons (Z*)

Finally we shall mention the exotic S=1 states. Partial wave analysis of K^+p channels gives some indication of positive strangeness resonances and in particular there is the possibility of a P_{13} $3/2^+$ (I=1) resonance at $\sim$ 1800 MeV. There is also the possibility of a P_{01} $1/2^-$ (I=0) resonance at $\sim$ 1780 MeV in the K^+n and K^0p channels. If such states exist they might be accounted for by exotic spinor coefficients of the following form:

$$\hat{\alpha}_i^A \hat{\alpha}_j^B \hat{\alpha}_k^C \hat{\alpha}_\ell^D \beta_{A'}^m .$$

A more extensive account of the six-twistor hadron model is in preparation in collaboration with A. Popovich and L.P. Hughston.

§4.4. THE TWISTOR $QQ\overline{QQ}$ MESON SPECTRUM *by L.P. Hughston and Tsou S.T.*

The purpose of this note is to point out a possibly important prediction of twistor theory with regards to the existence of certain types of excited meson states. The results can be summarized as follows: Insofar as the ordinary $Q\overline{Q}$ type mesons are concerned, the spectrum of states that we would anticipate on the grounds of twistor theory agrees, qualitatively, with what would be expected on the basis of the standard quark model with colour. However, a difference of considerable significance arises when one goes on to consider $QQ\overline{QQ}$ type mesons. Now it is not known with any certainty at the moment whether $QQ\overline{QQ}$ states ("diquoniums") do in fact exist in nature. However, if it turns out that these states *can* be formed (and there is - as we shall discuss shortly - a certain amount of evidence available that supports this possibility) then on the basis of the quark theory with colour one would expect two very distinct families of $QQ\overline{QQ}$ states to arise - called "M-diquoniums" and "T-diquoniums"; whereas within the framework of twistor theory it turns out, as far as we can see, that only *one* of these families should exist: namely, the T-diquonium family. Therefore, assuming that further data will indeed be forthcoming verifying the existence of $QQ\overline{QQ}$ states, we should have at our disposal soon a very direct test of some of the principles which have gone into the development of twistor theoretic hadron models.

The $Q\overline{Q}$-Spectrum

Within the quark model it is assumed that the $Q\overline{Q}$ mesons have the following generic structure:

$$Q^{iA}_{\rho} \; \overline{Q}^{A'\rho}_{j}$$

Here $i = 1,2,3$ is the flavour index, $\rho = 1,2,3$ is the colour index, and $A = 1,2$ is the spin index. The state Q_ρ^{iA} represents a coloured quark, and the state $\overline{Q}_j^{A'\rho}$ represents a coloured antiquark. The colour indices are contracted so that the resultant meson state is in a colour singlet. By reducing the spin and flavour indices in various ways, a variety of irreducible multiplets can be obtained; further multiplets are produced when the quarks are put into orbital motion relative to each other. Phenomenologically speaking, the resulting picture is, in many ways, very satisfactory.

As mentioned above, the same spectrum emerges, qualitatively, from twistorial considerations. For the $Q\overline{Q}$ mesons a representation can be built up based on the properties of holomorphic functions of six twistors (we exclude charmed mesons, etc., from the present discussion). The relevant contour integral formula is

$$\phi_j^{AA'i}(x,y) = \oint \rho_x^1 \rho_y^2 (\hat{\alpha}^{Ai} \beta_j^{A'} \pm \hat{\beta}^{Ai} \alpha_j^{A'}) f(\underset{1}{Z}{}_i^\alpha, \underset{2}{Z}{}_i^\alpha) \Delta\pi , \tag{1}$$

where we have used the following notation:

$$\underset{1}{Z}{}_i^\alpha = (\alpha_i^A, \alpha_{A'i}), \ \underset{2}{Z}{}_i^\alpha = (\beta_i^A, \beta_{A'i}), \ \hat{\alpha}^{Ai} = -\partial/\partial\alpha_{Ai} , \ \hat{\beta}^{Ai} = -\partial/\partial\beta_{Ai}.$$

The sign appearing in formula (1) is determined by the number of units of orbital angular momentum possessed by the state and the requirements of Fermi statistics.

The $QQ\overline{Q}\overline{Q}$ Spectrum

Within the coloured quark model the spectrum of a $QQ\overline{Q}\overline{Q}$ state is governed by the fact that Fermi statistics must be imposed on the two quarks within the QQ pair and on the two anti-quarks within the $\overline{Q}\overline{Q}$ pair. Thus we have two distinct kinds of "diquarks": namely, an M-type, given by

$$M^{ABij}_{\rho\sigma} = \underset{1}{Q}^{Ai}_{(\rho} \underset{2}{Q}^{Bj}_{\sigma)} - \underset{1}{Q}^{Bj}_{(\rho} \underset{2}{Q}^{Ai}_{\sigma)} ,$$

and a T-type, given by

$$T^{ABij}_{\rho\sigma} = \underset{1}{Q}^{Ai}_{[\rho} \underset{2}{Q}^{Bj}_{\sigma]} + \underset{1}{Q}^{Bj}_{[\rho} \underset{2}{Q}^{Ai}_{\sigma]} .$$

In both cases the demands of Fermi statistics are satisfied, since the states are antisymmetric under interchange of $\underset{1}{Q}$ and $\underset{2}{Q}$. Accordingly, as is well known, there are *two distinct types* of colour singlet $QQ\overline{QQ}$ states: the so-called M-diquoniums, given by

$$M^{ABij}_{\rho\sigma} \; \overline{M}^{\rho\sigma A'B'}_{k\ell} , \tag{2}$$

and the so-called T-diquoniums, given by

$$T^{ABij}_{\rho\sigma} \; \overline{T}^{\rho\sigma A'B'}_{k\ell} . \tag{3}$$

It is straightforward to verify that the multiplets associated with (2) and (3) exhibit a wide range of differences.

Certain types of diquonium states can be represented as functions of six twistors. In this case the relevant contour integral formula is given as follows:

$$\phi^{ABA'B'ij}_{k\ell}(x,y) = \oint \rho_x \, \rho_y \, (\hat{\alpha}^{Ai} \hat{\alpha}^{Bj} \beta^{A'}_{k} \beta^{B'}_{\ell} \pm \hat{\beta}^{Ai} \hat{\beta}^{Bj} \alpha^{A'}_{k} \alpha^{B'}_{\ell}) f(\underset{1}{Z}^{\alpha}_{i}, \underset{2}{Z}^{\alpha}_{i}) \Delta\pi \tag{4}$$

It is not difficult to verify that the set of multiplets associated with the field ϕ above is precisely of the T-diquonium type (i.e. symmetric under the interchange of the pairs Ai and Bj, etc.). In fact, it is impossible to produce M-diquonium states within the six-twistor framework. (To produce states with M-diquonium quantum numbers, functions of at least *twelve* twistors are required!) Thus, on the basis of twistorial considerations, we would at moderately low energies expect a range of T-diquonium states to be observable, but no M-diquonium states.

The Experimental Picture

What does experiment say? First, it will be useful to introduce some terminology. Experimentalists working in meson spectroscopy have found recently states which cannot readily be explained by the old-fashioned ideas of $Q\overline{Q}$ states. Many of these new states are narrow and have the peculiar property of preferring to decay into a baryon-antibaryon pair. Let us call these experimentally seen states "baryoniums". Diquoniums, on the other hand, are certain mathematical constructions defined by equations (2) and (3) in the quark model, and by equation (4) in twistor theory. The diquonium construction can explain most of the properties of baryoniums better than any other models, such as radial excitations of $Q\overline{Q}$ states, baryon-antibaryon bound states in a nuclear potential, etc.. However, the final confirmation that baryoniums are indeed diquoniums has yet to come.

Here we shall not discuss why it is thought that baryoniums are diquoniums; instead, we shall present a brief survey of baryoniums seen to date. In the 1978 Particle Data Group tables one finds three baryonium states listed, labelled S, T, and U, at masses of 1.936, 2.18, and 2.35 GeV. These are the most well established states. There is possibly a fourth resonance, tentatively called V, at 2.5 GeV [Montanet 1978]. All these states are seen in proton-antiproton annihilation, which is expected to be a fertile ground for baryonium physics in general.

Another well-studied channel is e^+e^-, whose spectrum yields several states not readily accounted for. From DCI at Orsay two relatively narrow states are reported at 1.66 and 1.77 GeV [Protopopescu 1978], and from ADONE at Frascati comes evidence for two narrow states at 2.13 GeV and 1.83 GeV. These states, however, are in need of further confirmation.

The states reported so far are good candidates for T-diquoniums [see Chan H.-M. 1978; Tsou S.T. 1977]. Unfortunately, in most cases we do not have an unmistakable signal for distinguishing between T and M states. Most analyses base their results on some sort of specific model calculation. We do not want to limit ourselves at this stage by such details. Nonetheless from fairly general dynamical considerations one can conclude within the framework of the quark model that T-diquoniums are easily formed in $p\bar{p}$ annihilation and are fairly wide except near threshold. M-diquoniums, on the other hand, prefer to be produced and are rather hard to form, the difference between the two processes being that in production the resonance is produced with other particles. M-diquoniums are in general very narrow, one of their preferred decay modes being that of cascade.

The two fairly clear resonances at 2.020 and 2.204 GeV [Bankhieri et al 1977] could be candidates for M-diquoniums, since they are produced backwardly in πp collision, and are not found when searched for in $p\bar{p}$ formation [Montanet 1978]. Further confirmation would be needed to set these states on a firm basis. We note, in passing, that there was reported [Evangelista et al 1977] a resonance at 2.95 GeV which would have been an ideal candidate for M-diquonium because of its cascade decay - however, when it was searched for again in a later experiment with ten times the statistics its existence failed to be confirmed [S. Ozaki 1979].

Many experiments are now being performed to study baryonium physics. At present the picture seems to be that the existence of baryoniums is fairly well established and that they are very probably diquoniums. If so, many of the better resonances are T-states. There are a couple of resonances that could be candidates for M-diquoniums, but they need confirmation. In other words, the question of colour is still tantalizingly unsolved. Optimists believe that a couple of good experiments in a couple of year's time will settle the issue. Meanwhile, one can only wait.

REFERENCES

Benkhieri, P., et. al. 1977 Phys. Lett. 68B, 483.

Chan H.-M. 1978 talk given at the IV European Antiproton Symposium at Barr; to appear in the proceedings.

Evangelista, C. et. al. 1977 Phys. Lett. 72B, 139.

Montanet, L. 1978 *Proceedings of the XIIIth Recontre de Moriond, Les Arcs,* Volume 1. Ed. J. Tran Thanh Van.

Ozaki, S. 1979 *Proceedings of the 19th International Conference on High Energy Physics, Tokyo 1978,* eds. S. Homma et. al.

Protopopescu, S.D. 1978 Brookhaven Laboratories preprint BNL-23612.

Tsou Sheung Tsun, 1978 Nucl. Phys. B141, 397; also "Baryoniums in e^+e^- processes". The Mathematical Institute, Oxford, 1979 preprint.

§4.5 BARYON MAGNETIC MOMENTS *by L.P. Hughston and M. Sheppard*

The magnetic moments of several of the low-lying baryons are known with a reasonable degree of accuracy. In units of $e\hbar/2m_p c$ these are as follows:

$$
\begin{array}{llll}
p \ldots & 2.7928456 & \Sigma^{+} \ldots & 2.83 \\
 & \pm.0000011 & & \pm.25 \\
n \ldots & -1.91304211 & \Sigma^{-} \ldots & -1.48 \\
 & \pm.00000088 & & \pm.37 \\
\Lambda \ldots & -0.606 & \Xi^{-} \ldots & -1.85 \\
 & \pm.034 & & \pm.75
\end{array}
\qquad (1)
$$

It has been known for some time that it is possible, in the context of the standard quark model with spin, to account to some extent for the various ratios of these magnetic moments. It is our purpose here to present a twistorial procedure for deriving these ratios.

As has been discussed in §§4.1 and 4.2, we shall adopt the view that the low-lying baryons are to be described in terms of holomorphic functions of three twistors. Thus, for example, if $f_p(Z_i^\alpha)$ represents a proton state, then the field $\phi_p^A(x)$ associated with this state is given by the following contour integral formula:

$$\phi_p^A(x) = \oint \rho_x \, \hat{u}^A \, \hat{u}_B \, \hat{d}^B \, f(Z_i^\alpha)\Delta\pi \qquad (2)$$

The notational conventions used in the expression above and throughout this discussion have been described in Hughston 1979.

We shall require in what follows that the twistor functions $f(Z_i^\alpha)$ correspond to definite eigenstates of the operator $\hat{S}_z$ for the directional component of spin. For the purposes of this discussion we shall restrict our attention to momentum eigenstates. Assuming z^a, the direction in which the spin is to be measured, satisfies

$$z^a z_a = -1, \quad P_z z^a = 0, \tag{3}$$

it is straightforward to show that there exists a spinor dyad

$$\{o^A, \iota^A \mid o_A \iota^A = 1\}, \tag{4}$$

determined uniquely up to phase, such that P^a and z^a are given by the following expressions:

$$\begin{aligned} P^{AA'} &= m2^{-1/2}\,(o^A \bar{o}^{A'} + \iota^A \bar{\iota}^{A'}) \\ z^{AA'} &= 2^{-1/2}\,(o^A \bar{o}^{A'} - \iota^A \bar{\iota}^{A'}) \end{aligned} \tag{5}$$

If $f(Z_i^\alpha)$ represents a spin eigenstate with directional eigenvalue $s_z = 1/2$ (i.e. "spin up") then the associated field $\phi^A(x)$ will be of the form

$$\phi^A(x) = ke^{-ip.x}\, o^A, \tag{6}$$

where k is a normalization factor; and similarly if $s_z = -1/2$ then we have

$$\phi^A(x) = ke^{-ip.x}\, \iota^A. \tag{7}$$

Now if the twistor function $f(Z_i^\alpha)$ corresponds to a definite baryon state, then there exists a unique linearly independent spinor coefficient structure Σ associated with $f(Z_i^\alpha)$ giving rise to a nonvanishing contour integral. Thus in the case of a proton if $f_p(Z_i^\alpha)$ has $s_z = 1/2$ inspection of equations (2) and (6) shows that the appropriate spinor coefficient structure is

$$\Sigma = \hat{u}^1(\hat{u}^0\,\hat{d}^1 - \hat{u}^1\hat{d}^0), \tag{8}$$

where we have written $\hat{u}^1 = \iota_A \hat{u}^A$ and $\hat{u}^0 = o_A \hat{u}^A$, etc.. What (8) means is that when we take components of (2) with the basis (4) the only terms which contribute non-vanishing answers are $\hat{u}^1\hat{u}^0\hat{d}^1$ and $\hat{u}^1\hat{u}^1\hat{d}^0$. Moreover, the answer obtained from each of these terms is the same, apart from a sign.

The magnetic moment of a state can be obtained by taking the expectation value of a certain twistor operator $\hat{\mu}$ defined as follows:

$$\hat{\mu} = \mu m^{-1} \varepsilon^i_j \sigma^\beta_\alpha Z^\alpha_i \hat{Z}^j_\beta \quad , \tag{9}$$

where the constant twistor σ^β_α is given by

$$\sigma^\beta_\alpha = \begin{pmatrix} 0 & 0 \\ z^{BA'} & 0 \end{pmatrix} \quad , \tag{10}$$

with $z^{BA'}$ as in (5), and where ε^i_j is the standard SU(3) charge matrix, given by

$$\varepsilon^i_j = \begin{pmatrix} 2/3 & 0 & 0 \\ 0 & -1/3 & 0 \\ 0 & 0 & -1/3 \end{pmatrix} . \tag{11}$$

The constant μ eventually drops out, in what follows, when we take ratios.

In spinor terms $\hat{\mu}$ can be expressed somewhat more simply as:

$$\hat{\mu} = \mu m^{-1} \varepsilon^i_j z^{AA'} \hat{\pi}^j_A \pi_{iA'} \quad , \tag{12}$$

where we have put

$$\hat{\pi}^i_A = -\partial/\partial\omega^A_i \ , \quad Z^\alpha_i = (\omega^A_i, \pi_{iA'}) \ , \tag{13}$$

as usual.

If $|f_p\rangle$ represents a normalized proton eigenstate, then $\hat{\mu}|f_p\rangle$ will not represent that same eigenstate. Instead $\hat{\mu}|f_p\rangle$ will correspond to a mixture of states as follows:

$$\hat{\mu}|f_p\rangle \quad \alpha|f_p\rangle + |g\rangle \,, \tag{14}$$

where $|g\rangle$ is a superposition of fields satisfying

$$\langle g|f_p\rangle = 0 \,, \tag{15}$$

and α is a certain numerical coefficient. Accordingly, we have

$$\langle f_p|\ \hat{\mu}\ |f_p\rangle = \alpha \tag{16}$$

for the expectation value of $\hat{\mu}$ for a proton state. Similarly, for a neutron state $|f_n\rangle$ we have

$$\begin{aligned} &\hat{\mu}|f_n\rangle = \beta|f_n\rangle + |h\rangle \\ &\langle h|f_n\rangle = 0 \end{aligned} \tag{17}$$

from which it follows that for the expectation value we have

$$\langle f_n|\ \hat{\mu}\ |f_n\rangle = \beta. \tag{18}$$

Thus, the ratio of the magnetic moments is given by α/β.

Now we must demonstrate the procedure according to which the numbers α, β, etc., can be calculated. To simplify matters, let us consider the example of a state $f_q(Z_i^\alpha)$ for which the appropriate spinor coefficient structure consists of a single $\hat{\pi}_A^i$. The magnetic moment μ_q of this state is given by the relation

$$\mu_q\ \phi_A^i(x) = \oint \rho_x\ \hat{\pi}_A^i\ \hat{\mu}\ f_q(Z_i^\alpha)\Delta\pi \,, \tag{19}$$

where $\phi_A^i(x)$ is the field corresponding to $f_q(Z_i^\alpha)$, given by the formula:

$$\phi^i_A(x) = \oint \rho_x \, \hat{\pi}^i_A \, f_q(Z^\alpha_i)\Delta\pi \, . \tag{20}$$

We assume that $f_q(Z^\alpha_i)$ is in a momentum eigenstate with eigenvalue $P_{AA'}$. Using (1), equation (19) can be rewritten as

$$\mu_q \, \phi^i_A(x) = \mu m^{-1} \, \varepsilon^j_k \, z^{BB'} \oint \rho_x \, \hat{\pi}^i_A \, \hat{\pi}^k_B \, \pi_{jB'} \, f_q(Z^\alpha_i)\Delta\pi \, . \tag{21}$$

Since the spinor coefficient structure corresponding to $f_q(Z^\alpha_i)$ must involve a single $\hat{\pi}$, it follows that a *trace* must be taken in (21), within the integral, over the indices i and j. Since $\hat{\pi}^i_A \, \pi_{iB'}$ is the momentum operator $\hat{P}_{AB'}$, it follows that (21) reduces to

$$\mu_q \, \phi^i_A(x) = \mu m^{-1} \, \varepsilon^j_k \, z^{BB'} P_{B'A} \oint \rho_x \, \hat{\pi}^k_B \, f_q(Z^\alpha_i)\Delta\pi \, . \tag{22}$$

If one defines

$$\sigma^{AB} = o^{(A}\iota^{B)} \quad , \tag{23}$$

then (5) can be used in order to reduce equation (22) to

$$\mu_q \, \phi^i_A(x) = \mu\varepsilon^i_j \, \sigma^B_A \oint \hat{\pi}^j_B \, f_q(Z^\alpha_i)\Delta\pi \, , \tag{24}$$

from which we obtain

$$\mu_q \, \phi^i_A(x) = \mu\varepsilon^i_j \, \sigma^A_B \, \phi^j_B \, . \tag{25}$$

Now, for example, if we take $\hat{\pi}^i_A$ in (20) to be $\hat{u}_A$, and assume ϕ^i_A to be in a "spin up" state, it follows from (6), (11), and (23) that $\mu_q = 2\mu/3$.

It should be evident that this procedure can be generalized so as to accommodate any spinor coefficient structure. When applied in the case of the proton and the neutron, where for the spinor coefficient structures we take

$$\Sigma_p = \hat{u}^1(\hat{u}^1 \hat{d}^0 - \hat{u}^0 \hat{d}^1)$$

$$\Sigma_n = \hat{d}^1(\hat{u}^1 \hat{d}^0 - \hat{u}^0 \hat{d}^1) \quad , \tag{26}$$

one obtains

$$\mu_p = \mu \ , \quad \mu_n = -2\mu/3 \ . \tag{27}$$

This yields the well-known result

$$\mu_p/\mu_n = -3/2 \ , \tag{28}$$

which is in fairly good agreement with the values quoted in (1).

These techniques can be extended to evaluate magnetic moments and various transition rates for both baryons and mesons. A more detailed account of these matters will appear elsewhere. We are most grateful to Alex Popovich for useful discussions in connection with this material.

REFERENCE

Hughston, L.P. 1979 *Twistors and Particles*. Springer Lecture Notes on Physics, Volume 97.

§4.6 MATRIX ELEMENTS FOR BARYON SEMI-LEPTONIC PROCESSES *by A. Popovich*

One application of the conventional quark model is in deriving matrix elements, using relatively simple group theoretical techniques, for spin-½ baryon semi-leptonic processes (in the non-relativistic limit of zero momentum transfer) of the form

$$B_1 \to B_2 + \ell^+(\ell^-) + \nu_\ell(\bar{\nu}_\ell) \tag{1}$$

where B_1, B_2 are spin-½ baryons; $\ell^-(\ell^+)$, $\nu_\ell(\bar{\nu}_\ell)$ is a lepton (anti-lepton) and its associated neutrino (anti-neutrino). It appears that one may also derive such matrix elements twistorially. I shall first describe the quark model technique, and then illustrate the twistor model approach with the example of β-decay; $n \to p + e^- + \bar{\nu}_e$.

The Quark Model

It should first be emphasised that the quark model approach (cf. Dalitz 1965) is non-relativistic, interpreting spin using the properties of the group SU(2). To emphasise this, I shall denote a quark by

$$q_a^i = (u_a, d_a, s_a), \tag{2}$$

where 'i' is an SU(3)-index and 'a' is an SU(2) (non-relativistic) spin index. The corresponding antiquark is denoted by

$$\bar{q}_i^a = (\bar{u}^a, \bar{d}^a, \bar{s}^a). \tag{3}$$

Basically, the idea is to construct a current operator $\hat{J}$ whose action on a spin-½ baryon state $|B_1>$ changes constituent quarks of $|B_1>$ into other quarks, such that

$$\hat{J}\ |B_1\rangle \ = \alpha|B_2\rangle + \beta|B_3\rangle + \ldots. \tag{4}$$

where $|B_1\rangle,\ldots,$ are baryon states other than $|B_2\rangle$. The matrix element for the semi-leptonic process (1) is then just $M = \langle B_2|\ \hat{J}\ |B_1\rangle = \alpha$, assuming all states are normalized appropriately. Using M, one can then proceed to calculate the decay rate for the process (1). In the computation of M, it is assumed that $|B_1\rangle$ and $|B_2\rangle$ are in momentum eigenstates, and both have a definite third component of spin $s_z = +\frac{1}{2}$, where an axis z of quantization has been chosen. The *Cabibbo Hypothesis* (see Cabibbo 1963; cf. also Bernstein 1968, pp. 269-285) requires that $\hat{J}$ takes the form

$$\hat{J} = V + A\ , \tag{5}$$

where V, the *vector* contribution to $\hat{J}$ is given by

$$V\!: = G_V\ \{u^{\dagger a}\ d_a \cos\theta + u^{\dagger a}\ s_a \sin\theta\}\ , \tag{6}$$

and A, the *axial-vector* contribution to $\hat{J}$ is given by

$$A\!: = G_A\ \sigma_{3\,a}^{\ \,b}\ \{u^{\dagger a}\ d_b \cos\theta + u^{\dagger a}\ s_b \sin\theta\}. \tag{7}$$

In (6) and (7), the notation $q_i^{\dagger a} q_b^j$ is used to symbolize annihilation of a quark coefficient q_b^j and creation of q_a^i (or, creation of antiquark coefficient $\bar{q}_j^b$ and annihilation of $\bar{q}_i^a$); G_V and G_A are coupling constants determined experimentally; θ is the *Cabibbo angle* ($\theta \approx 0.245$ experimentally). Finally, in (7), $\sigma_{3\,a}^{\ \,b} = \begin{pmatrix}1 & 0\\ 0 & -1\end{pmatrix}$ is the third Pauli matrix. The operator $\hat{J}$ is interpreted as acting separately on each quark constituent of the incoming baryon B_1, to give a sum of three-quark terms. The latter can then be rewritten, as noted previously, in the form

$$\alpha|B_2> + \beta|B_3> + \ldots \tag{8}$$

so that

$$<B_2| \hat{J} |B_1> = \alpha . \tag{9}$$

Notice that the V contribution to (9) vanishes unless there is no change in angular momentum J between B_1 and B_2. However, the σ_3 term in A can change J by one unit, so that the A contribution is zero unless J = 0, 1, except that J = 0 → J = 0 transitions are not allowed, since σ_3 vanishes between states of zero angular momentum.

The Twistor Model

Recall that in the 3-Twistor Hadron Model (cf. Hughston 1979, Chapter 6), the $J^P = \frac{1}{2}^+$ low-lying baryon octet is represented by the space-time fields

$$\phi^i_{jA}(x) := \kappa_\phi \oint \rho_x \, \hat{\pi}^i_A \, \hat{\pi}^k_B \, \hat{\pi}^{\ell B} \, \varepsilon_{k\ell j} \, f_\phi(Z^\alpha_i) \Delta\pi \tag{10}$$

In (10), $\Delta\pi$ is an SU(3)-invariant differential form; A,B, ... are SL(2,C) (relativistic) spin indices; f_ϕ is a holomorphic function (where by function, I mean a cocycle representing a cohomology class in some suitable cohomology group) in an eigenstate of a complete set of commuting operators, the eigenvalues being the quantum numbers characterizing the baryon field ϕ^i_{jA}; ρ_x denotes evaluation of the integrand at the space-time point x; κ_ϕ is a constant chosen such that ϕ^i_{jA} is suitably normalized. The momentum operator is given by

$$\hat{p}^{AA'} := \hat{\pi}^{jA} \, \pi^{A'}_j \tag{11}$$

Thus, from (10), $p_A(x) = \phi^1_2(x)$, $n_A(x) = \phi^2_{1A}(x)$ are the space-time fields assigned to the proton and neutron respectively. For convenience, the spinor coefficient structure of a baryon B will be denoted $\hat{\Sigma}_B$. It will be necessary to work here with momentum eigenstates.

Using the result

$$i\nabla_{BB'}\,\rho_x\,f(Z_i^\alpha) = \rho_x\,\hat{P}_{BB'}\,f(Z_i^\alpha)\ , \tag{12}$$

a straightforward application of the chain rule, it follows that

$$\hat{P}_{BB'}\,f_n(Z_i^\alpha) = P_{nBB'}\,f(Z_i^\alpha)\qquad , \tag{13}$$

$$\hat{P}_{BB'}\,f_p(Z_i^\alpha) = P_{pBB'}\,f(Z_i^\alpha)\qquad , \tag{14}$$

where $P_n^{BB'}$, $P_p^{BB'}$, are the momenta of the neutron and proton respectively. The *current operator* $\hat{J}$ is now defined by

$$\hat{J} : = \tau_V + \tau_A \tag{15}$$

where

$$\tau_V : = G_V\,\pi_{iA'}\,\hat{\pi}_A^j\,P^{AA'}\,\eta_j^i \tag{16}$$

$$\tau_A : = G_A\,\pi_{iA'}\,\hat{\pi}_A^j\,Z^{AA'}\,\eta_j^i \tag{17}$$

with

$$\eta : = \begin{pmatrix} 0 & \cos\theta & \sin\theta \\ \cos\theta & 0 & 0 \\ \sin\theta & 0 & 0 \end{pmatrix} \tag{18}$$

In order to define τ_A, a spinor basis $\{o_A, \iota_A\}$ with $o_A\iota^A = 1$, is chosen, along with a z axis

$$Z^{AA'} : = \frac{1}{\sqrt{2}}\,(o^A\bar{o}^{A'} - \iota^A\bar{\iota}^{A'}) \tag{19}$$

with respect to which the momentum $P_1^{AA'}$ of the incoming state $|B_1\rangle$ is given by

$$P_1^{AA'} = \frac{m}{\sqrt{2}} (o^A \bar{o}^{A'} + \iota^A \bar{\iota}^{A'}) \qquad (20)$$

where m is the mass of $|B_1\rangle$. Then if $\sigma_{AB} := o_{(A}\iota_{B)}$, it follows that $Z^{AA'} = \frac{2}{m} P_1^{A'}{}_B \sigma^{AB}$. Thus

$$\oint \rho_x \hat{\Sigma} \; \tau_A f_{B_1}(Z_i^\alpha)\Delta\pi = \oint \rho_x \hat{\Sigma} \; G_A \; \pi_{iA'} \hat{\pi}_A^j \; Z^{AA'} \; \eta_j^i \; f_{B_1}(Z_i^\alpha)\Delta\pi \qquad (21)$$

$$= \frac{2}{m} \oint \rho_x \hat{\Sigma} \; G_A \; \pi_{iA'} \; \hat{\pi}_A^j \; \hat{P}^{A'}{}_B \sigma^{AB} \; \eta_j^i f_{B_1}(Z_i^\alpha)\Delta\pi \qquad (22)$$

whenever

$$\hat{P}^{AA'} \; f_{B_1}(Z_i^\alpha) = P_1^{AA'} \; f_{B_1}(Z_i^\alpha) \qquad . \qquad (23)$$

It is equation (22) which is actually used in the calculations with $\hat{J}$. One can easily show that τ_V and τ_A in (16) and (17) are Hermitian and Poincaré invariant.

The fields $p_A(x)$, $n_A(x)$, are now taken to be polarized with $s_z = +\frac{1}{2}$, so that with respect to the spinor basis $\{o_A, \iota_A\}$, $p_A = (p_o, o)$, $n_A = (n_o, o)$. The action of $\hat{J}$ on f_n (where now f_n, f_p,..., correspond to states n,p,... with $s_z = \frac{1}{2}$) is then given by (cf. equation 4)

$$\hat{J} f_n = \alpha f_p + \gamma f_\Lambda + \ldots \qquad (24)$$

where f_Λ,..., are holomorphic functions corresponding to states other than p. From (24) it follows that

$$|\langle f_p| \; \hat{J} \; |f_n\rangle| = |\alpha| \qquad (25)$$

The explicit calculation of $|\alpha|$ (which will not be gone into in detail here) involves considering the *dual* action of $\hat{J}$ on spinor coefficient structures $\hat{\Sigma}_p, \hat{\Sigma}_n$: by decomposing $\hat{\Sigma}_p\hat{J}$, $\hat{\Sigma}_n\hat{J}$ into SU(3)-irreducible parts, one obtains explicitly, coefficients $\beta, \tilde{\beta}$:

$$\hat{\Sigma}_n \hat{J} = \beta \hat{\Sigma}_p + \left\{ \begin{array}{l} \text{spinor coefficient structures} \\ \text{for states other than p.} \end{array} \right\} \tag{26}$$

$$\hat{\Sigma}_p \hat{J} = \tilde{\beta} \hat{\Sigma}_n + \left\{ \begin{array}{l} \text{spinor coefficient structures} \\ \text{for states other than n.} \end{array} \right\} \tag{27}$$

where $|\beta| = |\tilde{\beta}|$, essentially because $\hat{J}$ is a Hermitian operator. From (24) and (27) it follows that

$$\oint \rho_x \hat{\Sigma}_p \hat{J} f_n(Z_i^\alpha)\Delta\pi = \alpha \oint \rho_x \hat{\Sigma}_p f_p(Z_i^\alpha)\Delta\pi \tag{28}$$

$$= \tilde{\beta} \oint \rho_x \hat{\Sigma}_n f_n(Z_i^\alpha)\Delta\pi \tag{29}$$

In (29), the spinor coefficient structures on the r.h.s. of (27) other than $\hat{\Sigma}_n$ do not contribute to the contour integral, since they are incompatible with f_n. Then, from (28) and (29), and using the normalization of the general expression (10), we have

$$|\alpha| = \left| \frac{\kappa_p}{\kappa_n} \right| |\tilde{\beta}| = \left| \frac{\kappa_p}{\kappa_n} \right| |\beta| \tag{30}$$

Thus, in order to obtain $|\alpha|$, it is sufficient to decompose $\hat{\Sigma}_n \hat{J}$ as in (26), and to take the coefficient β. Considering separately the vector and axial-vector contributions to the β-decay, calculation shows that

$$|\langle f_p(s_z = +\tfrac{1}{2}) \; |\tau_V| \; f_n(s_z = +\tfrac{1}{2})\rangle| = R|G_V| \cos\theta \tag{31}$$

$$|\langle f_n(s_z = +\tfrac{1}{2}) \; |\tau_A| \; f_n(s_z = +\tfrac{1}{2})\rangle| = \frac{5}{3}|G_A| \cos\theta \tag{32}$$

where R depends only on κ_p and κ_n. Then from (31) and (32)

$$\left| \frac{\langle f_p(s_z = +\tfrac{1}{2}) \; |\tau_A| \; f_n(s_z = +\tfrac{1}{2})\rangle}{\langle f_p(s_z = +\tfrac{1}{2}) \; |\tau_V| \; f_n(s_z = +\tfrac{1}{2})\rangle} \right| = \frac{5}{3}\left| \frac{G_A}{G_V} \right| \tag{33}$$

Experimental results on β-decay indicate that the ratio $\left|\frac{G_A}{G_V}\right| \approx \frac{1}{\sqrt{2}}$.

Inserting $G_V = \sqrt{2}\, G_A$ into the expression for $\hat{J}$ (equations 5, 6, and 7), one can proceed, analogously to the above, to calculate matrix elements for other baryon semi-leptonic processes to give results in complete agreement with the predictions of the quark model. It is found that on the whole there is good agreement with experiment.

REFERENCES

Bernstein, J. 1968 *Elementary Particles and Their Currents.* Freeman.

Cabibbo, N. 1963 Phys. Rev. Letters 10, 531.

Dalitz, R.H. 1965 *Quark Models for the "Elementary Particles".* In: *High Energy Physics, Les Houches,* eds. C.M. DeWitt & M. Jacob. Gordon and Breach.

Hughston, L.P. 1979 *Twistors and Particles,* Springer Lecture Notes in Physics, Volume 97.

§4.7 A NOTE ON THE n-TWISTOR INTERNAL SYMMETRY GROUP *by R. Penrose and G.A.J. Sparling*

Consider a system of n twistors Z_i^α (i = 1,...,n) with complex conjugates $\bar{Z}_\alpha^i$. The *kinematic twistor* (i.e. the "angular momentum twistor") $A^{\alpha\beta}$ for this system is defined by

$$A^{\alpha\beta} = 2Z_i^{(\alpha}I^{\beta)\gamma}\bar{Z}_\gamma^i$$

The *n-twistor internal symmetry group* (or, simply, the *n-twistor group*) is defined to be the group of complex linear transformations of Z_i^α and $\bar{Z}_\alpha^i$ that preserve $A^{\alpha\beta}$ and the relation of complex conjugacy between Z_i^α and $\bar{Z}_\alpha^i$.

For some time it has been known that the transformations

$$T: \begin{pmatrix} Z_i^\alpha \\ \bar{Z}_\alpha^i \end{pmatrix} \longrightarrow \begin{pmatrix} U_i^j\delta_\beta^\alpha & \Lambda_{ik}\bar{U}_j^k I^{\alpha\beta} \\ \bar{\Lambda}^{ik}U_k^j I_{\alpha\beta} & \bar{U}_j^i\delta_\alpha^\beta \end{pmatrix}\begin{pmatrix} Z_j^\beta \\ \bar{Z}_\beta^j \end{pmatrix},$$

with U_i^j unitary and Λ_{ij} complex skew-symmetric, belong to the n-twistor group, and it has been suspected that they constitute the entire n-twistor group. We shall outline here a proof that, indeed, they constitute the entire identity-connected component of the n-twistor group.

The multiplication law for the group can be expressed as

$$(U_i^j,\Lambda_{k\ell}) \circ (U'^j_i, \Lambda'_{k\ell}) = (U_i^m U'^j_m, \Lambda_{k\ell} + U_k^m U_\ell^n \Lambda'_{mn}),$$

an "inhomogeneous U(n)". This group acts as

$$Z_i^\alpha \to U_i^j Z_j^\alpha + \Lambda_{ik}\bar{U}_j^k I^{\alpha\beta}\bar{Z}_\beta^j \quad ,$$

together with the corresponding complex conjugate transformation. Our argument consists of showing that the infinitesimals of the n-twistor group have the form of infinitesimal *T*'s:

$$Z_i^\alpha \to Z_i^\alpha + i\varepsilon\, H_i^j Z_j^\alpha + \varepsilon L_{ij} I^{\alpha\beta} \bar{Z}_\beta^j \quad ,$$

with H_i^j Hermitian and L_{ij} skew-symmetric. Certainly the infinitesimal elements of the n-twistor group have the form

$$Z_i^\alpha \to Z_i^\alpha + \varepsilon (R_{i\beta}^{j\alpha} Z_j^\beta + S_{ij}^{\alpha\beta} \bar{Z}_\beta^j) \quad ,$$

and preservation of the kinematical twistor leads to:

$$I^{\alpha(\beta} S_{ij}^{\gamma)\delta} + I^{\delta(\beta} S_{ji}^{\gamma)\alpha} = 0 \quad , \tag{1}$$

$$R_{k\rho}^{i(\alpha} I^{\beta)\sigma} + \delta_\rho^{(\alpha} I^{\beta)\gamma} \bar{R}_{k\gamma}^{i\sigma} = 0 \ . \tag{2}$$

By symmetrizing (1) over $\alpha\beta\gamma$ we deduce that $S_{ij}^{(\alpha\beta)} = 0$. Then contracting (1) with $I_{\alpha\rho}$ we obtain expressions which, by the use of simple lemmas, give

$$S_{ij}^{\alpha\beta} = L_{ij}\ I^{\alpha\beta}$$

for some skew-symmetric L_{ij}. By applying somewhat similar operations to (2) and using rather more involved lemmas we finally derive

$$R_{k\beta}^{j\alpha} = i\delta_\beta^\alpha\ H_k^j$$

for some Hermitian H_k^j. □

The quantized twistor operators obtained from these infinitesimal generators define the algebra which gives the classification of *n-twistor particles* (quantization: $\bar{Z}_\alpha^i \to -\partial/\partial Z_i^\alpha$). For n = 3 this algebra is the same as that of Z.P.'s inhomogeneous SU(3). Details will be described elsewhere. For further discussion (and references) see §4.8.

§4.8 A DERIVATION OF THE TWISTOR INTERNAL SYMMETRY GROUPS *by L.P. Hughston*

The twistor internal symmetry groups have a rather interesting history. These groups were being discussed extensively in Penrose's seminars at Birkbeck College, London as early as the spring of 1973. The first systematic investigation of the properties of the twistor internal symmetry groups to be published was by Perjés (1975). He examined the 2-twistor group and the 3-twistor group in great detail, and elucidated many of their properties. These groups have subsequently been the subject of a number of further discussions and investigations (e.g.: Penrose 1975, Hodges 1975, Tod 1975, Tod 1977, Penrose 1977, Perjés 1977, Popovich 1978, and Hughston 1979).

As is customary, we shall denote by Z^α_i a set of n twistor variables $(i = 1,\ldots,n)$, and we shall denote by $\bar{Z}^i_\alpha$ the corresponding complex conjugate variables. Associated with any set Z^α_i of n twistors there is a very special object known as the *kinematic twistor*, defined by

$$A^{\alpha\beta} = 2\, Z^{(\alpha}_i I^{\beta)\gamma} \bar{Z}^i_\gamma \quad , \tag{1}$$

where $I^{\alpha\beta}$ is the *infinity twistor*. The kinematic twistor describes the total momentum and angular momentum of the system Z^α_i according to the scheme,

$$A^{\alpha\beta} = \begin{pmatrix} -2i\mu^{AB} & P^A{}_{B'} \\ P_{A'}{}^B & 0 \end{pmatrix} \tag{2}$$

where P^a is the momentum, and μ^{AB} determines the angular momentum through the relation

$$M^{ab} = \mu^{AB}\varepsilon^{A'B'} + \bar{\mu}^{A'B'}\varepsilon^{AB} \tag{3}$$

The n-twistor internal symmetry group $J(n)$ is defined to be the set of all linear transformations acting on Z^α_i and $\bar{Z}^i_\alpha$ which leave the kinematic twistor invariant, and also which preserve the conjugacy relations between Z^α_i and $\bar{Z}^i_\alpha$. More explicitly, if we put

$$\begin{cases} Z^\alpha_i \longrightarrow Y^\alpha_i = R^{\alpha j}_{\beta i} Z^\beta_j + S^{\alpha\beta}_{ij} \bar{Z}^j_\beta \\ \bar{Z}^i_\alpha \longrightarrow \bar{Y}^i_\alpha = \bar{R}^{\beta i}_{\alpha j} \bar{Z}^j_\beta + \bar{S}^{ij}_{\alpha\beta} Z^\beta_j \end{cases} , \tag{4}$$

which is the most general linear transformation preserving the conjugacy relations, then $J(n)$ is defined to be the group of transformations satisfying

$$Z^{(\alpha}_i I^{\beta)\gamma} \bar{Z}^i_\gamma = Y^{(\alpha}_i I^{\beta)\gamma} \bar{Y}^i_\gamma , \tag{5}$$

for all values of Z^α_i.

Theorem 1 *The group* $J(n)$ *is constituted by the set of all transformations of the form*

$$R^{\alpha j}_{\beta i} = \delta^\alpha_\beta U^j_i \quad , \quad S^{\alpha\beta}_{ij} = I^{\alpha\beta} U^k_i \Lambda_{kj} \tag{6}$$

where U^j_i *is unitary and* Λ_{ij} *is skew-symmetric, i.e.:*

$$U^j_i \bar{U}^k_j = \delta^k_i \quad , \quad \Lambda_{ij} = -\Lambda_{ji} . \tag{7}$$

This result was first conjectured during the spring of 1973, and I do not believe that anyone at the time seriously questioned its validity. Indeed, the result is quoted and used in the 1975 papers of both Perjés and Penrose, and is employed in various other references as well. Oddly enough, rigorous justification was not forthcoming until the summer of 1977, when Penrose and Sparling (see the preceding article by R.P. and G.A.J.S.) demonstrated the theorem valid at least insofar as one considered infinitesimal transformations

- thus confirming that (6) gives the identity-connected component of $J(n)$. The general case was subsequently established by Sparling, who showed that $J(n)$ is indeed, as had tacitly been assumed all along, identity-connected. I shall present here a new derivation of the n-twistor internal symmetry groups. Furthermore, it will be demonstrated that $J(n)$ is in fact a subgroup of another yet larger group that arises naturally in twistor theory - namely, the group of linear transformations preserving the momentum and angular momentum of *zero rest-mass* n-twistor systems.

The Momentum

We shall denote the spinor parts of the twistor system Z^{α}_{i} and its complex conjugate $\bar{Z}^{i}_{\alpha}$ according to the following customary scheme:

$$Z^{\alpha}_{i} = (\omega^{A}_{i}, \pi_{A'i}) \quad , \quad \bar{Z}^{i}_{\alpha} = (\bar{\pi}^{i}_{A}, \bar{\omega}^{A'i}) \ . \tag{8}$$

With this notation the total momentum of the system is given by the expression

$$p^{A'A} = \pi^{A'}_{i} \bar{\pi}^{iA} \quad , \tag{9}$$

which is the sum of the null momenta of all the constituent zero rest-mass subsystems. It should be evident that if the momentum is to be preserved under linear transformations acting on Z^{α}_{i} and $\bar{Z}^{i}_{\alpha}$, then when $\pi^{A'}_{i}$ is transformed it must pick up no terms involving ω^{A}_{i} or $\bar{\omega}^{i}_{A}$. The most general linear transformation satisfying these conditions is given by:

$$\begin{cases} \pi^{A'}_{i} \longrightarrow R^{A'j}_{iB'}\pi^{B'}_{j} + S^{A'}_{ijB}\bar{\pi}^{jB} \\ \bar{\pi}^{iA} \longrightarrow \bar{R}^{Ai}_{jB}\bar{\pi}^{jB} + \bar{S}^{ijA}_{B'}\pi^{B'}_{j} \ . \end{cases} \tag{10}$$

When the momentum is subjected to transformation (10) it behaves as follows:

$$\pi_i^{A'}\bar{\pi}^{iA} \longrightarrow (R_i^{A'j}{}_{B'}\bar{R}^{Ai}_{kC} + S^{A'}_{ikC}\bar{S}^{ijA}_{B'})\pi_j^{B'}\bar{\pi}^{kC} \tag{11}$$

$$+ \quad (R_i^{A'j}{}_{B'}\bar{S}^{ikA}_{C'})\pi_j^{B'}\pi_k^{C'} + (\bar{R}^{Ai}_{jB}S^{A'}_{ikC})\bar{\pi}^{jB}\bar{\pi}^{kC} \ .$$

Now in order for the momentum to be preserved, only the first of the three terms on the right must survive, and the other two must vanish. Assuming that only the first term survives—and this assumption will be justified shortly—then the preservation of the momentum amounts to the following condition:

$$\pi_i^{A'}\bar{\pi}^{iA} = (R_i^{A'j}{}_{B'}\bar{R}^{Ai}_{kC} + S^{A'}_{ikC}\bar{S}^{ijA})\pi_j^{B'}\bar{\pi}^{kC} \quad , \tag{12}$$

which must hold for all values of $\pi_i^{A'}$.

Since we require expression (12) to be valid for all values of $\pi_i^{A'}$, it must hold when $\pi_i^{A'}$ happens to be of the degenerate form

$$\pi_i^{A'} = \pi^{A'}q_i \quad , \tag{13}$$

which is, in fact, just the condition that the total momentum is *null*. Substituting (13) into (12) we obtain

$$q_i\bar{q}^i\pi^{A'}\bar{\pi}^A = R_i^{A'}\bar{R}^{Ai} + S_i^{A'}\bar{S}^{Ai} \quad , \tag{14}$$

where the following helpful abbreviations have been introduced:

$$R_i^{A'} = R_i^{A'j}{}_{B'}\pi^{B'}q_j \quad , \quad \bar{R}^{Ai} = \bar{R}^{Ai}_{jB}\bar{\pi}^B\bar{q}^j \tag{15}$$

$$S_i^{A'} = S^{A'}_{ijB}\bar{\pi}^B\bar{q}^j \quad , \quad \bar{S}^{Ai} = \bar{S}^{ijA}_{B'}\pi^{B'}q_j \tag{16}$$

Note that $R_i^{A'}$ and $S_i^{A'}$ are, respectively, the complex conjugates of $\bar{R}^{Ai}$ and $\bar{S}^{Ai}$. Therefore, the right-hand side of equation (14) is an expression which consists of *a sum of future-pointing null vectors*. The only way that this sum can itself be a future-pointing null vector, such as the validity of (14) requires, is for all the constituent null vectors to be proportional—and this will be the case only if:

$$\begin{cases} R_i^{A'} = \pi^{A'} r_i \quad , \quad \bar{R}^{Ai} = \bar{\pi}^{A} \bar{r}^{i} & (17) \\ S_i^{A'} = \pi^{A'} s_i \quad , \quad \bar{S}^{Ai} = \bar{\pi}^{A} \bar{s}^{i} \ . & (18) \end{cases}$$

Note, however, that (18) is incompatible with (16) unless $S_{ijB}^{A'}$ and s_i are zero. Accordingly, $S_i^{A'}$ vanishes, and equation (14) reduces to

$$q_i \bar{q}^i \pi^{A'} \bar{\pi}^{A} = R_i^{A'} \bar{R}^{Ai} \ . \tag{19}$$

Now, substituting (17) into (15) we straightforwardly obtain

$$R_{i\ B'}^{A'j} = U_i^j \varepsilon_{B'}{}^{A'} \ , \tag{20}$$

and from equation (19) it follows immediately that the matrix U_i^j is unitary.

Thus, we have demonstrated that the most general linear transformation acting on $\pi_i^{A'}$ and $\bar{\pi}^{iA}$ that preserves the momentum $\pi_i^{A'} \bar{\pi}^{iA}$ is given simply by

$$\pi_i^{A'} \longrightarrow U_i^j \pi_j^{A'} \ , \quad \bar{\pi}^{iA} \longrightarrow \bar{U}_j^i \bar{\pi}^{jA} \ , \tag{21}$$

with U_i^j unitary.

The Angular Momentum

The angular momentum spinor appearing in equations (2) and (3) is given explicitly by

$$-\mu^{AB} = \omega_i^{(A} \bar{\pi}^{B)i} \ . \tag{22}$$

It should be clear that if μ^{AB} is to be preserved under the combined action of (21) along with linear transformations acting on ω_i^A, then ω_i^A must transform into a linear expression involving only ω_i^A and $\bar{\pi}^{Ai}$. The most general such transformation is obviously of the form

$$\omega_i^A \longrightarrow R_{iB}^{Aj}\omega_j^B + S_{ijB}^{A}\bar{\pi}^{jB} \quad . \tag{23}$$

Transformations (21) and (23) together give

$$\omega_i^{(A}\bar{\pi}^{B)i} \longrightarrow \bar{U}_k^i R_{iC}^j {}^{(A}\bar{\pi}^{B)i}\omega_j^C + \bar{\pi}^{k(A}S_{ijC}^{B)}\bar{U}_k^i\bar{\pi}^{jC} \quad . \tag{24}$$

And thus for the preservation of the angular momentum we require the following two conditions to hold for all values of ω_i^A and $\bar{\pi}^{iA}$:

$$\omega_i^{(A}\bar{\pi}^{B)i} = \bar{U}_k^i R_{iC}^{j(A}\bar{\pi}^{B)k}\omega_j^C \quad , \tag{25}$$

$$\bar{\pi}^{k(A}S^{B)}_{ijC}\bar{U}_k^i\bar{\pi}^{jC} = 0 \quad . \tag{26}$$

Imposing (13) these equations become

$$\bar{q}^i\omega_i^{(A}\bar{\pi}^{B)} = \bar{q}^k\bar{U}_k^i R_{iC}^{j(A}\bar{\pi}^{B)}\omega_j^C , \tag{27}$$

$$\bar{q}^j\bar{q}^k\bar{\pi}^{(A}S^{B)}_{ijC}\bar{U}_k^i\bar{\pi}^C = 0 \quad . \tag{28}$$

Equation (27) immediately implies

$$\omega_i^A = \bar{U}_i^j R_{jB}^{Ak}\omega_k^B \quad , \tag{29}$$

from which one readily deduces

$$R_{jB}^{Ai} = U_j^i\varepsilon_B{}^A \quad . \tag{30}$$

Equation (28), after some elementary algebra, implies

$$S_{i(j}^{AB}\,\bar{U}_{k)}^{i} = 0 \quad . \tag{31}$$

It is important—and we shall return shortly to explore this point in greater depth—to observe that (31) is the full extent of the condition imposed on S_{ijB}^{A} so long as we remain in the subspace defined by equation (13). Imposing (31) and returning to (26), a short calculation then reveals a stronger condition that must be imposed upon S_{ijB}^{A}, namely:

$$S_{i[j}^{(AB)}\bar{U}_{k]}^{i} = 0 \quad , \tag{32}$$

from which it follows that S_{ijB}^{A} must be of the form

$$S_{ijB}^{A} = U_{i}^{k}\Lambda_{kj}\varepsilon_{B}{}^{A} \quad , \tag{33}$$

with Λ_{ij} antisymmetric.

Equations (30) and (33) taken together show that the most general linear transformation acting on ω_{i}^{A} and $\bar{\pi}^{iA}$ preserving μ^{AB} is given by (21) in conjunction with

$$\omega_{i}^{A} \longrightarrow U_{i}^{j}(\omega_{j}^{A} + \Lambda_{jk}\bar{\pi}^{kA}) \quad . \tag{34}$$

Equations (21) and (34) are, taken together, equivalent to

$$Z_{i}^{\alpha} \longrightarrow U_{i}^{j}(Z_{j}^{\alpha} + I^{\alpha\beta}\Lambda_{jk}\bar{Z}_{\beta}^{k}), \tag{35}$$

and—as a glance at equations (4) and (6) will demonstrate—we have succeeded in proving *Theorem 1*.

The n-Twistor Internal Group for Massless Systems

As remarked earlier, equation (13) is just the condition that the total momentum of the n-twistor system Z_{i}^{α} should be *null*. Expressed in manifestly twistorial terms, equation (3) asserts that we have

$$Z^{\alpha}_{i}I_{\alpha\beta} = Z^{\alpha}I_{\alpha\beta}q_{i} \tag{36}$$

for an appropriate choice of Z^{α} and q_i. Equivalently, one may note that Z^{α}_{i} satisfies (36) if and only if it lies on the subvariety defined by

$$I_{\alpha\beta}Z^{\beta}{}_{[i}Z^{\gamma}_{j]}\, I_{\gamma\delta} = 0 \quad . \tag{37}$$

or, equally well, by

$$I_{\alpha\beta}Z^{\alpha}_{i}Z^{\beta}_{j} = 0 \quad . \tag{38}$$

A system Z^{α}_{i} satisfying equation (38) will be referred to as a *massless n-twistor system*. It is natural, then, to enquire about the nature of the group of transformations that preserve the kinematic twistor for massless n-twistor systems. This is a problem which—as far as I know—has hitherto been neglected. Thus, we require transformations of the form (4) which satisfy (5) not for *all* values of Z^{α}_{i}, but rather, only those values corresponding to points on the subvariety (38).

<u>Theorem 2</u>. *Linear transformations acting on* Z^{α}_{i} *and* $\bar{Z}^{i}_{\alpha}$ *that preserve* $A^{\alpha\beta}$ *for all values of* Z^{α}_{i} *satisfying equation (38) are of the form*

$$Z^{\alpha}_{i} \longrightarrow U^{j}_{i}(Z^{\alpha}_{j} + \Lambda^{\alpha\beta}_{jk}\bar{Z}^{k}_{\beta}) \quad , \tag{39}$$

where U^{j}_{i} *is unitary, and* $\Lambda^{\alpha\beta}_{jk}$ *is any tensor subject to the conditions*

$$\Lambda^{\alpha\beta}_{(ij)} = 0 \quad , \tag{40}$$

$$I_{\alpha\beta}\Lambda^{\beta\gamma}_{ij} = 0 = \Lambda^{\alpha\beta}_{ij}I_{\beta\gamma}. \tag{41}$$

The set of all such transformations forms a group.

Proof. We have already learned that the most general transformation acting on $\pi_i^{A'}$ and $\bar{\pi}^{iA}$ that preserves the momentum—even if we confine ourselves to massless systems—is given by (21). For the angular momentum, however, we can only get as far as equations (30) and (31) if massless n-twistor systems are being considered. Now equation (31) asserts that S^A_{ijB} is of the form

$$S^A_{ijB} = U^k_i \Lambda^A_{kjB} \quad , \tag{42}$$

with Λ^A_{ijB} subject to

$$\Lambda^{AB}_{(ij)} = 0 \quad . \tag{43}$$

Inserting (30) and (42) into (23) we obtain

$$\omega^A_i \longrightarrow U^j_i(\omega^A_j + \Lambda^A_{ijB}\bar{\pi}^{jB}) \quad , \tag{44}$$

which, when taken together with (21), gives (39), with (40) and (41). The verification that such transformations form a group is routine, providing that one notes that if $\underset{1}{\Lambda}{}^{\alpha\beta}_{ij}$ and $\underset{2}{\Lambda}{}^{\alpha\beta}_{ij}$ are any pair of tensors satifying equation (41), then

$$\underset{1}{\Lambda}{}^{\alpha\beta}_{ij}\,\underset{2}{\bar{\Lambda}}{}^{k\ell}_{\beta\gamma} = 0 = \underset{1}{\bar{\Lambda}}{}^{ij}_{\alpha\beta}\,\underset{2}{\Lambda}{}^{\beta\gamma}_{k\ell} \quad , \tag{45}$$

where $\bar{\Lambda}^{ij}_{\alpha\beta}$ is the complex conjugate of $\Lambda^{\alpha\beta}_{ij}$.

We shall denote the internal symmetry group for massless n-twistor systems $\mathcal{J}^o(n)$. Note that since the choice

$$\Lambda^{\alpha\beta}_{ij} = I^{\alpha\beta}\Lambda_{ij} \ , \quad \Lambda_{(ij)} = 0 \tag{46}$$

evidently satisfies equations (40) and (41), it follows that $\mathcal{J}(n)$ is a *subgroup* of $\mathcal{J}^o(n)$. A good deal more can be said about the group $\mathcal{J}^o(n)$ and its relationship to $J(n)$; these matters will be explored further elsewhere.

I would like to express my gratitude to R. Penrose, Z. Perjés, G.A.J. Sparling, and N.M.J. Woodhouse for useful discussions concerning this material.

REFERENCES

Hodges, A. 1975 *The Description of Mass in the Theory of Twistors*. Ph.D. Thesis, Birkbeck College, London.

Hughston, L.P. 1979 *Twistors and Particles*. Springer Lecture Notes in Physics, Volume 97.

Penrose, R. 1975 *Twistors and Particles*. In: *Quantum Theory and the Structure of Time and Space*, eds. L. Castell, M. Drieschner & C.F. von Weiszacker. Verlag: München.

Penrose, R. 1977 Reports Math. Phys. 12, 65.

Perjés, Z. 1975 Phys. Rev. D11, 2031.

Perjés, Z. 1977 Reports Math. Phys. 12, 193.

Popovich, A. 1978 *Twistor Classification of Elementary Particles*. M.Sc. Thesis, Oxford University.

Tod, K.P. 1975 *Massive Particles with Spin In General Relativity and Twistor Theory*. D.Phil. Thesis, Oxford University.

Tod, K.P. 1977 Reports Math. Phys. 11, 339.

5 Twistor diagrams

§5.1 INTRODUCTION *by L.P. Hughston*

Many aspects of the interpretation of analytic zero rest-mass fields as elements of sheaf cohomology groups are now fairly well understood. Although massive fields are of a more intricate nature — both with regard to their asymptotic behavior, as well as with regard to the range of quantum numbers that they must carry — there appears to be considerable scope for incorporating them into a cohomological framework also. When it comes to the question of *interactions*, there are broadly speaking two approaches that can be taken within the context of twistor theory. First, there is the theory of *curved twistor spaces* (cf. chapter 3). And second, there is the theory of *twistor diagrams*. The theory of curved twistor spaces proceeds in a spirit analogous to that of general relativity, providing a mechanism whereby certain types of one-particle states can be incorporated into the complex analytic structure of a deformed twistor space. Although the cases which have been studied explicitly thus far are perhaps rather primitive when it comes to the question of concrete physical applications, nevertheless these examples do suffice to show that the "curved twistor space technique" is a method of great power, and should lead to many new insights into the nature of gravitation as well as other interactions. The theory of twistor diagrams proceeds more in the spirit of quantum field theory, and is concerned with the problem of building up expressions for amplitudes for specific processes. Again, many of the processes are of a rather artificial nature, and are not — in any direct sense — applicable in the manner one might immediately desire. However, the *mathematical* questions raised in these seemingly artificial problems are themselves of a very intriguing nature — and, more to the point, these same questions are destined to arise again in any case in the "full" theory. It is for this sort of reason that a great deal of effort and discussion has gone, for example, into attempting to sort out the nature of the scalar product for zero rest-mass free fields, in twistor terms. Although a proper quantum field theory has not been built up in twistor terms — even at a heuristic level, for that matter — nevertheless the theory of twistor diagrams offers what is very likely a substantial step in this direction.

Rules for Twistor Diagrams

The theory of twistor diagrams was discussed at some length by Penrose in his lectures at the University of Cambridge during the spring and summer of 1970, and much of this material was presented in Penrose and MacCallum 1972. Further extensive treatments can be found in Penrose 1975, Sparling 1974, Sparling 1975, Hodges 1975, and Ryman 1975; cf. also Huggett 1976.

There is considerable scope for flexibility in the precise choice of rules for twistor diagrams; and indeed it might be said that an optimal choice has not yet emerged. What immediately follows here is a paraphrase of material found on pp. 330-337 of Penrose 1975, to which the reader is referred for further details.

Twistor diagrams consist of a number of black and white *vertices* connected by *edges*, where each edge joins a black to a white vertex (internal edge) — or else has a free and (external edge). Every edge bears a number, usually an integer. The free ends are normally labelled by letters, and they represent fixed twistors. (Sometimes the numbers and/or the fixed twistors are omitted, if their values are understood.) The vertices themselves represent variable twistors which are to be integrated over, black vertices denoting

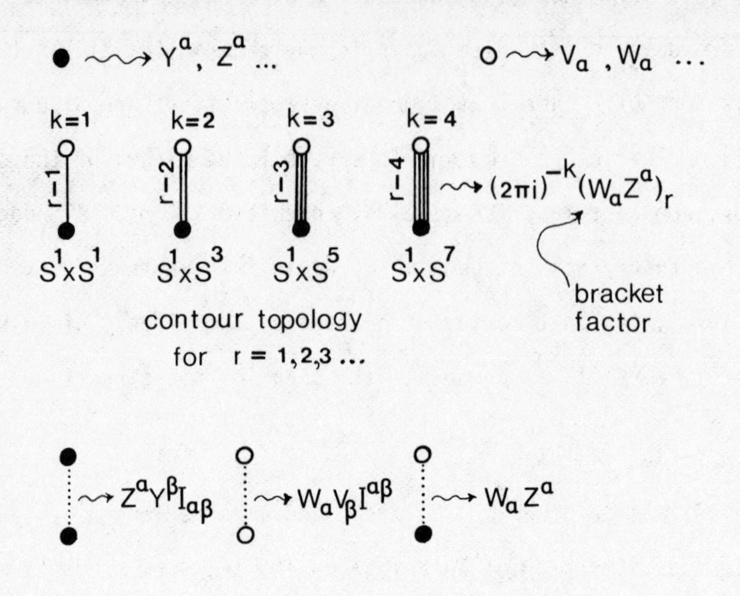

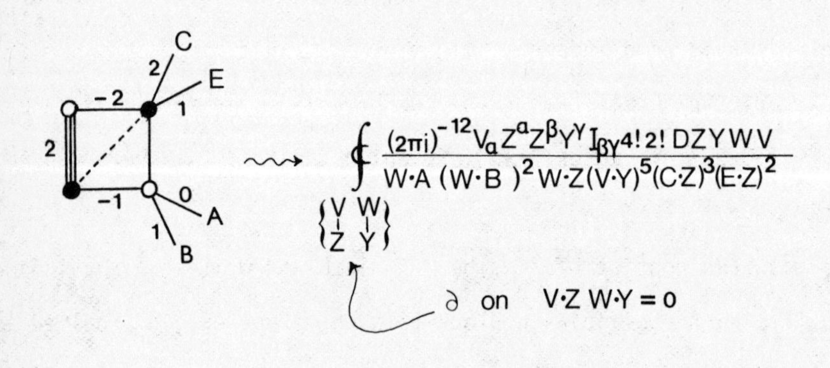

Figure A. *Rules for Twistor Diagrams*

upstairs twistors, and white vertices denoting downstairs twistors. The whole diagram defines a closed differential form. The differential form involved consists of the product of the natural holomorphic 4-forms for all the vertices, multiplied by an additional factor corresponding to each edge. To determine this factor, we add the number appearing on the edge to the multiplicity k of the edge, and call the result r. The relevant factor is then

$$(2\pi i)^{-k}\ (W_\alpha Z^\alpha)_r\ ,$$

where W_α and Z^α are the twistors represented by the ends of the edge, and $(x)_r$ is the so-called *bracket factor,* defined by:

$$(x)_r \;=\; \begin{cases} \dfrac{-\Gamma(r)}{2\pi i(-x)^r} & \text{if } r \neq 0, -1,-2,\ldots \\[2ex] \dfrac{(x)^{-r}}{(-r)!} & \text{if } r = 0, -1,-2,\ldots \end{cases} .$$

When $r \neq 0, -1,-2,\ldots$ the expression is used in an integral with a closed contour surrounding $x = 0$; however, when $r = 0, -1, -2,\ldots$ the expression is used in an integral with an open contour having boundary in the subspace $x = 0$. Finally, a factor of $2\pi i$ is included in the numerator for each free end.

It is also possible to include factors such as $Z^\alpha Y^\beta I_{\alpha\beta}$ and $W_\alpha V_\beta I^{\alpha\beta}$ in the numerator of the differential form — such factors are denoted by dotted lines joining the vertices in question. A further notational device, exploited in §§5.4, 5.6 & 5.7, is the use of *wavy lines* to indicate a subspace in which the boundary of a contour must lie. See Figure A for some examples of properly drawn diagrams.

Blowing Up. In §§5.9 & 5.10 use is made of a well-known device in algebraic geometry called "blowing up". Since blow-up techniques are also used in §§2.15 & 2.16, it seems appropriate to make a few remarks on this topic.

As an illustration, we shall indicate how to blow up a point in P^2. With this example in mind, the reader should be able to work through more complicated cases. For homogeneous coordinates on P^2 let us write Z^a, with $a = 1,2,3$. A line in P^2 is then specified by an equation of the form $L_a Z^a = 0$ for some L_a. We can denote a *pair* of lines in P^2 by L_a^A, with $A = 1,2$. The intersection of these two lines is the locus $L_a^A Z^a = 0$, which is a point (providing that L_a^A is not degenerate). Let us call this point q. To blow up q we proceed as follows. First we take $P^2 \times P^1$, with coordinates Z^a on P^2, and coordinates λ_A on P^1. Then we consider the subvariety defined by

$$\lambda_A L_a^A Z^a = 0. \tag{1}$$

This subvariety is $\tilde{P}^2$, i.e. the space P^2 with the point q blown up.

It is evident that the region $\tilde{q}$ defined by $L_a^A Z^a = 0$ in $P^2 \times P^1$ is a subset of $\tilde{P}^2$. In fact, it is not difficult to show that

$$\tilde{P}^2 - \tilde{q} \cong P^2 - q. \tag{2}$$

Thus, away from the "blown-up region", $\tilde{P}^2$ and P^2 are isomorphic — but the point q in P^2 gets blown up to $\tilde{q}$ in $\tilde{P}^2$, and $\tilde{q}$ is in fact an entire P^1.

To see (2) we note that $\tilde{P}^2 - \tilde{q}$ is the locus (1) with the additional restraint $L_a^A Z^a \neq 0$. This implies $\lambda_A \sim L_{Aa} Z^a$ (raising and lowering indices in the standard spinorial manner). Thus λ_A is fixed (up to irrelevant proportionality) by the specification of Z^a. But Z^a can lie anywhere in $P^2 - q$, and thus we have shown (2).

To see $\tilde{q} \cong P^1$ we note that (1) together with $L_a^A Z^a = 0$ implies that λ^A can take on *any* non-vanishing value.

The main point worth emphasizing here is that with a judicious application of spinor algebra it becomes possible to streamline blow-up calculations very effectively.

The Cohomology of P^n. In §§5.12 and 5.13 it is assumed that the reader is familiar with a number of basic results concerning the cohomology of complex projective spaces. These results, which are also used in various places in Chapter 2, will be reviewed briefly here.

The cohomology groups over P^1 with coefficients in the sheaf $\mathcal{O}(r)$ are very easy to describe. The only non-vanishing groups are H^0 and H^1. The group $H^0(P^1,\mathcal{O}(r))$ is the space of polynomials in λ^A, homogeneous of degree r. The group $H^1(P^1,\mathcal{O}(-2-r))$ is *dual*, as a complex vector space, to $H^0(P^1,\mathcal{O}(r))$. Now a homogeneous polynomial of degree r in λ^A is generically of the form $\phi_{AB\ldots C}\lambda^A\lambda^B\ldots\lambda^C$, where $\phi_{AB\ldots C}$ is a symmetric spinor of valence [0,r]. Thus the cohomology group $H^0(P^1,\mathcal{O}(r))$ is nothing more than the familiar complex vector space of valence [0,r] symmetric spinors; the group $H^1(P^1,\mathcal{O}(-2-r))$, correspondingly, is the space of valence [r,0] symmetric spinors. For $r < 0$, both of these groups vanish.

On P^n the set-up is very similar. Let Z^a $(a = 0,1,\ldots n)$ be homogeneous coordinates on P^n; the only nonvanishing cohomology groups are as follows:

$$\begin{cases} H^0(P^n,\mathcal{O}(r)) \cong \text{space of polynomials in } Z^a \text{ homogeneous of degree r.} \\ H^n(P^n,\mathcal{O}(-n-1-r)) \cong \text{dual of } H^0(P^n,\mathcal{O}(r)). \end{cases}$$

A polynomial homogeneous of degree r will be of the form $\phi_{ab\ldots c}Z^aZ^b\ldots Z^c$, where $\phi_{ab\ldots c}$ is a symmetric tensor of valence [0,r]; thus $H^0(P^n,\mathcal{O}(r))$ is the space of such tensors.

Similarly, $H^n(P^n, \mathcal{O}(-n-1-r)$ is the space of symmetric valence [r,0] tensors. Note again that for $r < 0$ both of these groups vanish. Note also that all the "intermediate" groups $H^1, \ldots, H^{n-1}$ vanish, for *all* r, when $r > 1$.

For various applications it is also of interest to know the cohomology groups on *products* of complex projective spaces. Let $Z^{a'}$ ($a' = 0,1\ldots m$) be homogeneous coordinates on P^m, and define $P^{n,m} := P^n \times P^m$. Let $\mathcal{O}(r,s)$ denote the sheaf of germs "twisted functions" defined over $P^{n,m}$, of twist r in Z^a and s in $Z^{a'}$. The nonvanishing cohomology groups are as follows (r unprimed indices; s primed indices):

$$H^0(P^{n,m}, \mathcal{O}(r,s)) \cong S_{a\ldots b\ a'\ldots b'}$$

$$H^n(P^{n,m}, \mathcal{O}(-n-1-r,s)) \cong S^{a\ldots b}_{a'\ldots b'}$$

$$H^m(P^{n,m}, \mathcal{O}(r,-m-1-s)) \cong S^{a'\ldots b'}_{a\ldots b}$$

$$H^{n+m}(P^{n,m}, \mathcal{O}(-n-1-r,-m-1-s)) \cong S^{a\ldots b\ a'\ldots b'} \quad ,$$

where an obvious notation is employed for the various spaces of symmetric tensors which arise. Note that these groups are nonvanishing only for $r,s > 0$. The extension of these formulae to products of larger numbers of complex projective spaces is straightforward.

Using the results cited above it becomes possible to calculate the cohomology of any algebraic projective variety. The problem consists of "resolving" the sheaf of germs of holomorphic functions on the variety in terms of sheaves defined on projective spaces (cf. Serre 1956).

For example, if the variety V is a hypersurface of degree r in P^n, then the relevant exact sequence is

$$0 \longrightarrow \mathcal{O}_{P^n}(-r+a) \xrightarrow{\alpha} \mathcal{O}_{P^n}(a) \xrightarrow{\rho_V} \mathcal{O}_V(a) \longrightarrow 0.$$

The map α above is multiplication by the polynomial whose vanishing defines V; and ρ_V is restriction down to V.

If V happens to be defined by the intersection of a *pair* of hypersurfaces f and g, of degrees r and s respectively, then the relevant exact sequence is

$$0 \longrightarrow \mathcal{O}(-r-s+a) \xrightarrow{\alpha} \mathcal{O}(-r+a) \oplus \mathcal{O}(-s+a) \xrightarrow{\beta} \mathcal{O}(a) \xrightarrow{\rho} \mathcal{O}_V(a) \longrightarrow 0.$$

Here α is multiplication by (-g,f), and β is multiplication by (f,g) followed by summation.

Once the relevant exact sequence has been established, the cohomology of $\mathcal{O}_V$ can be calculated using standard techniques — i.e. chopping up the long exact sequences into short ones, and then applying the exact cohomology sequence.

It would appear (cf. §§5.10-5.13) that many aspects of the structure of twistor diagrams can indeed be understood in terms of the cohomology of suitable projective varieties. Much work remains to be done, however, to clarify these matters — particularly when interactions are involved (e.g. in the case of the *box diagram*) and when massive fields are under consideration.

REFERENCES

Hodges, A. 1975 Ph.D. Thesis, Birkbeck College, London.

Huggett, S.A. 1976 M.Sc. Thesis, University of Oxford.

Penrose, R. 1975 *Twistor Theory, Its Aims and Achievements*. In: *Quantum Gravity*, eds. C.J. Isham, R. Penrose & D.W. Sciama. Clarendon Press.

Penrose, R. & MacCallum, M.A.H. 1972 Phys. Reports. Vol. 6C, No. 4.

Ryman, A. 1975 D. Phil. Thesis, University of Oxford.

Serre, J.P. 1956 Ann. Inst. Fourier 6, 1.

Sparling, G.A.J. 1974 Ph.D. Thesis, Birkbeck College, London.

Sparling, G.A.J. 1975 *Homology and Twistor Theory*. In: *Quantum Gravity*, eds. C.J. Isham, R. Penrose & D.W. Sciama. Clarendon Press.

§5.2 THE UNIVERSAL BRACKET FACTOR *by R. Penrose*

The standard bracket factor for twistor integrals is:

$$(x)_n = \begin{cases} \dfrac{-(n-1)!}{2\pi i(-x)^n} & \text{if } n = 1,2,3\ldots \qquad (\oint) \\ \dfrac{x^{-n}}{(-n)!} & \text{if } n = 0,-1,-2,-3\ldots \qquad (\not\int) \end{cases} .$$

The symbol $\oint$ means that the expression is to be inserted in an integral with a closed contour surrounding $x = 0$, whereas the symbol $\not\int$ means that the expression is to be integrated over an open contour with boundary in $x = 0$.

The *universal bracket factor* is

$$[x] = \sum_{n=-\infty}^{\infty} (x)_n ,$$

but this expression must be treated as *formal* at the moment (and a rigorous definition from someone would be welcome) because

$$2\pi i \sum_{n=1}^{\infty} (x)_n \overset{(\oint)}{=} \frac{1!}{x^2} + \frac{2!}{x^3} - \frac{3!}{x^4} + \ldots ,$$

which *diverges* for *all* x. However, as is described in Hardy 1973 pp.26-29, Euler was formally able to derive

$$\frac{1}{x} - \frac{1!}{x^2} + \frac{2!}{x^3} - \frac{3!}{x^4} + \ldots = e^x\{-\gamma - \log x + x - \frac{x^2}{2.2!} + \frac{x^3}{3.3!} - \ldots\},$$

which is valid with the left hand side an asymptotic series (γ = Euler's constant). Evaluating the right hand side we obtain

$$\frac{1}{x} - \frac{1!}{x^2} + \frac{2!}{x^3} - \frac{3!}{x^4} + \ldots = \int_0^\infty \frac{e^{-w}}{x+w}\,dw ,$$

which is a perfectly good function! But, unfortunately, $x = 0$ is now a branch point and not a pole, so $(\oint)$ of this expression is not strictly meaningful as it stands. Note that

$$\sum_{n=-\infty}^{0} (x)_n \overset{(\oint)}{=} 1 + x + \frac{x^2}{2!} + \frac{x^3}{3!} + \ldots = e^x = \frac{1}{2\pi i} \oint \frac{e^{-w}}{x+w}\, dw \,.$$

So (formally) we obtain

$$[x] = \{(\oint) \int_0^\infty + (\oint) \oint \} \frac{e^{-w}}{x+w}\, dw \,,$$

an observation due to G.A.J. Sparling. We have (formally) $d[x]/dx = [x]$ (with suitable contour). Thus, *irrespective of homogeneities*, we have the following formula for the "twistor transform":

$$\tilde{f}(W_\alpha) = \oint [W_\alpha Z^\alpha] f(Z^\alpha)\, d^4Z;$$

and similarly for the scalar product we have:

$$\langle f|g\rangle = \frac{1}{(2\pi i)^6} \oint f(Z^\alpha)[Z^\alpha W_\alpha] g(W_\alpha)\, d^4Z_\wedge d^4W \,.$$

The universal bracket $[Z^\alpha W_\alpha]$ has many (formal) uses, e.g. in twistor diagrams.

REFERENCE

Hardy, G.H. 1973 *Divergent Series*. Clarendon Press: Oxford.

§5.3 MASSIVE STATES AND THE UNIVERSAL BRACKET FACTOR *by A.P. Hodges*

We can define "elementary" (scalar) eigenstates of mass by the two-twistor function

$$F(X,Z) = \frac{1}{2\pi i}\int_C \frac{(m/\sqrt{2})^{2s}\,\Gamma(1-s)\Gamma(1-s)\,ds}{(\overset{\sqcap}{XZ}\underset{\sqcup}{AB})^{1+s}\,(A.X)^{1-s}\,(B.Z)^{1-s}\,\sin\pi s},$$

where C is the contour (in the s-plane) shown in Figure A.

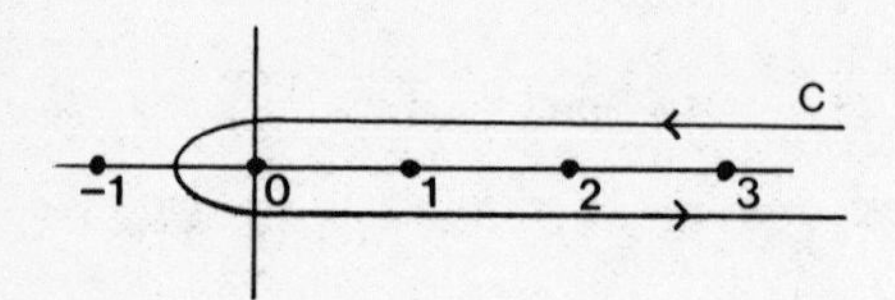

Figure A

This gives rise to a massive scalar field

$$\phi(x) \propto \frac{m\,H_1^{(1)}(m\sqrt{(x-p)^2})}{\sqrt{(x-p)^2}\;(\underset{\sqcup}{AB})},$$

where $H_n^{(1)}$ is the Hankel function of order n, and where p^a is the space-time point corresponding to the skew-symmetric twistor $A_{[\alpha}B_{\beta]}$.

If m = 0, this can easily be seen to reduce to

$$F(X,Z) = (A.X\;\overset{\sqcap}{XZ}\;B.Z\;\underset{\sqcup}{AB})^{-1},$$

which is essentially the simplest way of writing the usual scalar "elementary" massless field with a two-twistor function.

In the massive case it would be more satisfactory to have a formulation in which the value of the mass plays some role in the singularity structure. Formally, the expression

$$F(X,Z) = \frac{[A.X + B.Z]}{(\underset{\sqcup}{AB}\ \overset{\sqcap}{XZ} - \frac{1}{2}m^2)}$$

enjoys the required properties, where [x] represents the *universal bracket factor*, as defined in §5.2.

It has been observed in §5.2 that the expression

$$\sum_1^\infty \frac{-\Gamma(n)}{(-x)^n},$$

which is involved in the universal bracket factor, is an asymptotic series for the function

$$E(x) = \int_0^\infty \frac{e^{-w}}{x+w}\,dw.$$

So it is encouraging that this E function can be related to the Hankel function. In fact,

$$H_0^{(1)}(2i(uv)^{\frac{1}{2}}) = \frac{1}{2\pi i}\oint_\Gamma \frac{E(uz + v/z)dz}{z},$$

where Γ is the contour depicted in Figure B:

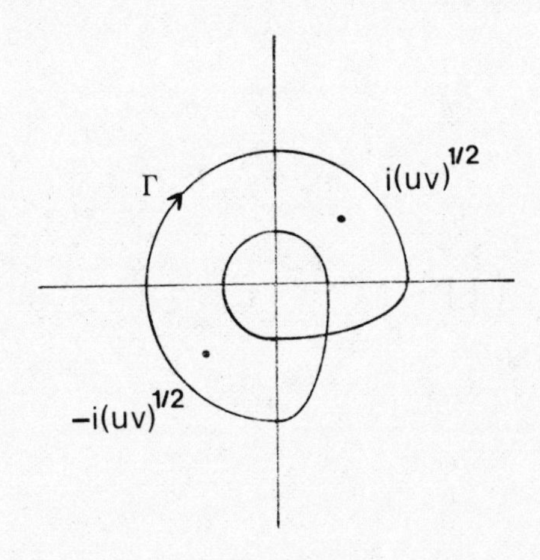

Figure B

§5.4 SOME ELEMENTARY TWISTOR INTEGRALS WITH BOUNDARY *by R. Penrose*

B
A C
D

$$\propto \oint \frac{DZ}{(A\cdot Z)(B\cdot Z)(C\cdot Z)(D\cdot Z)} \propto \frac{1}{[ABCD]}$$

Figure A

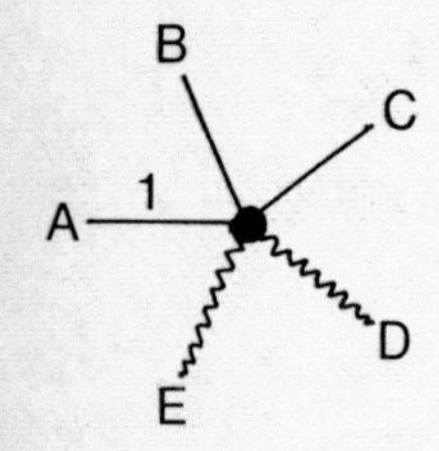

$$\propto \oint_{\{D\cdot Z,\ E\cdot Z\}} \frac{DZ}{(A\cdot Z)^2 (B\cdot Z)(C\cdot Z)} \propto \frac{[BCDE]}{[ABCD]\,[ABCE]}$$

Figure B

$$\propto \oint \frac{\log\left(\frac{D\cdot Z}{E\cdot Z}\right) DZ}{(A\cdot Z)^2 (B\cdot Z)(C\cdot Z)}$$

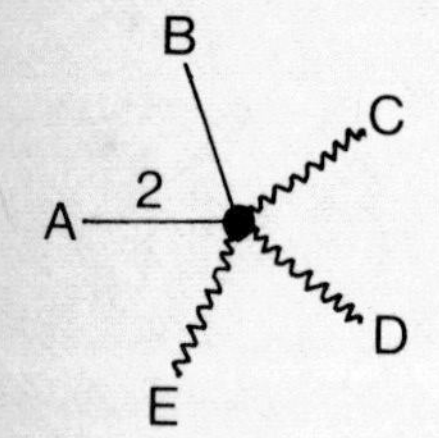

$$\propto \oint_{\{C\cdot Z,\ D\cdot Z,\ E\cdot Z\}} \frac{DZ}{(A\cdot Z)^3 (B\cdot Z)} \propto \frac{([BCDE])^2}{[ABCD]\,[ABCE]\,[ABDE]}$$

Figure C

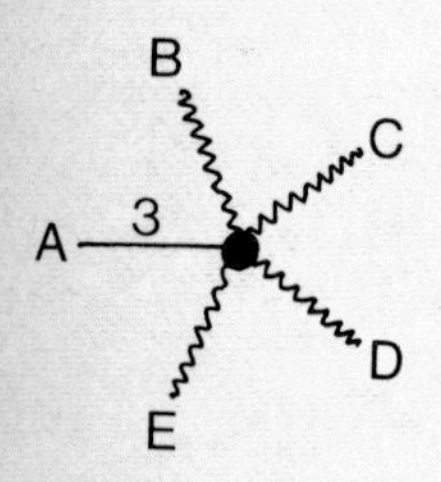

$$\propto \oint_{\{B\cdot Z,\ C\cdot Z,\ D\cdot Z,\ E\cdot Z\}} \frac{DZ}{(A\cdot Z)^4} \propto \frac{([BCDE])^3}{[ABCD]\,[ABCE]\,[ABDE]\,[ACDE]}$$

Figure D

In Figure A one finds the standard twistor vertex, as described in Penrose and MacCallum 1972. In Figures B, C and D one finds example of the evalution of some twistor integrals with boundary. For details of the notation involved here see §5.1. Note that the "4 lines at a vertex" rule is violated. G.A.J. Sparling noticed some years ago that

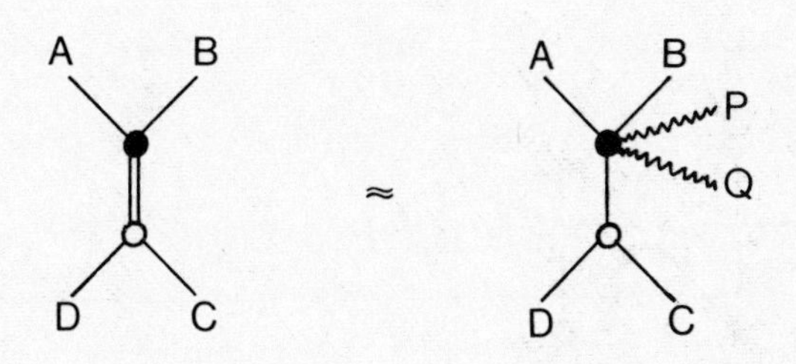

where the second diagram is evaluated by doing the filled vertex first, and the unfilled vertex second. Similarly, we have:

REFERENCES

Penrose, R. & MacCallum, M.A.H. 1972 Phys. Reports 6, 241-315.

Penrose, R. 1975 *Twistor Theory, its Aims and Achievements*. In: *Quantum Gravity*, eds. C.J. Isham, R. Penrose, & D.W. Sciama, pp. 268-407. Clarendon Press: Oxford.

§5.5 THE INTEGRATED PRODUCT OF SIX MASSLESS FIELDS *by A.P. Hodges*

It is well known that for *four* massless scalar fields, two of positive frequency (ϕ_1, ϕ_2) and two of negative frequency (ϕ_3, ϕ_4), we have:

$$\int d^4x\ \phi_1(x)\phi_2(x)\phi_3(x)\phi_4(x) \quad \propto$$

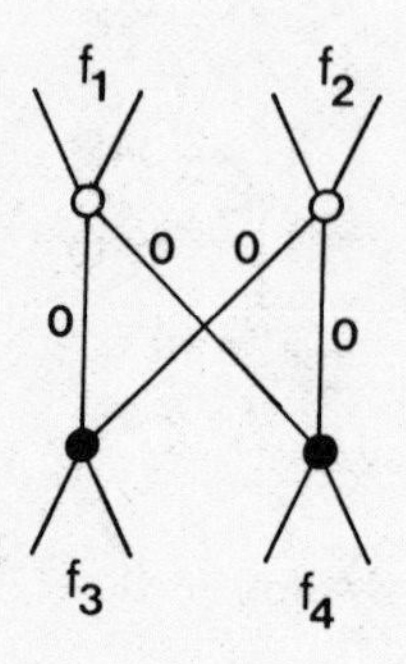

An analogous result for *six* fields (three in, three out) is:

$$\int d^4x\ \phi_1(x)\phi_2(x)\phi_3(x)\phi_4(x)\phi_5(x)\phi_6(x) \quad \propto$$

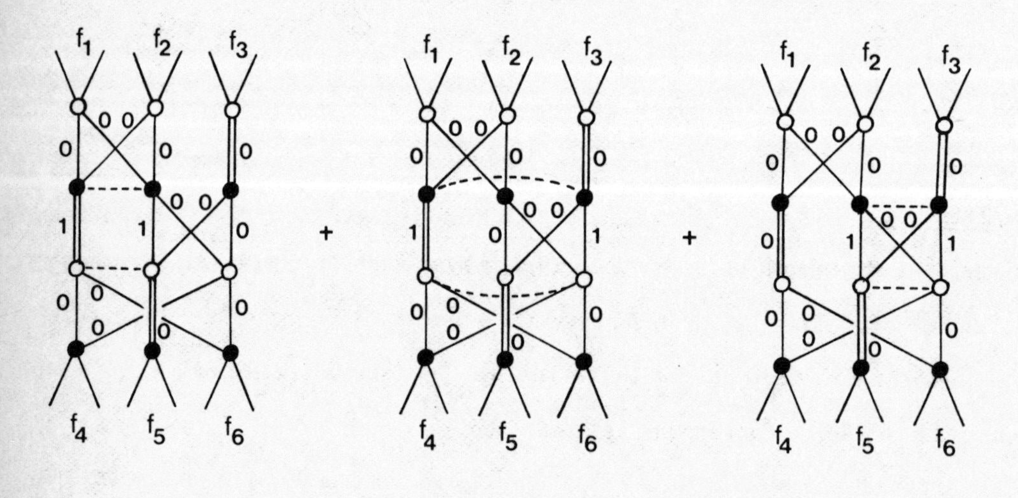

The proof, which is lengthy, has been done only for *elementary* states. Results indicated in Hodges 1975 show that it is sufficient to consider the integral

$$\int d^4x \frac{1}{[(x-p_1)^2]^2(x-p_2)^2[(x-q_1)^2]^2(x-q_2)^2} .$$

Computation, using Fourier transforms and the ϕ^4 result, shows that this is

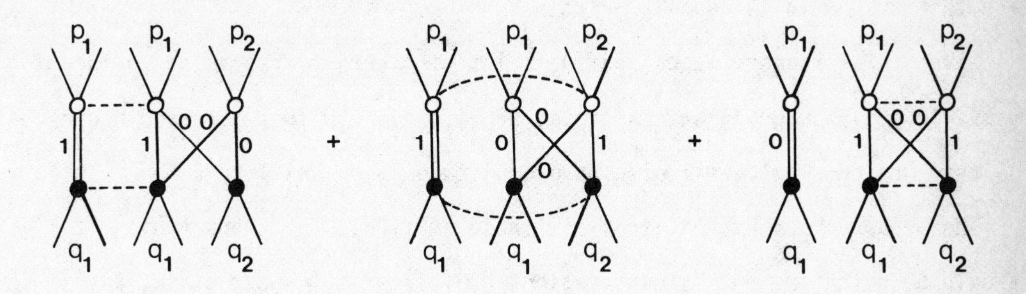

and the stated result follows from this.

The analogous problem for *eight* massless fields is much tougher, and a result has not yet been established.

REFERENCE

Hodges, A.P. 1975. *The Description of Mass in the Theory of Twistors.* Ph.D. Thesis, Birkbeck College, London.

§5.6 CROSSING AND TWISTOR DIAGRAMS *by A.P. Hodges*

Consider a process involving n external particles. In momentum space, the amplitude for it will be of form

$$F(k_1,k_2,\ldots k_n)\delta(k_1 + k_2 + \ldots + k_n)$$

where the momentum k_i of the ith particle is future-pointing if that particle is incoming, and past-pointing if it is outgoing. Actually, Lorentz invariance means that F will be a function only of the scalars $k_i^2 = m_i^2$ and $k_i.k_j$, and this is how it is usually written.

Given this process, we can define a lot of others by replacing an incoming particle by an outgoing antiparticle, or vice versa. These crossed processes, if they are physical, will also have amplitude functions like F.

The *crossing property* is that all these amplitude functions (which it should be noted are on disjoint regions of (k_i)-space) are restrictions of one single analytic function of the momenta k_i. This property is enjoyed by Feynman diagrams but also follows from the assumptions, much weaker than those of field theory, of analytic S-matrix theory.

Is there an analogue in twistor diagram theory? The trouble is that when we compute the amplitudes as functionals not of momenta but of wave-packets, we obliterate the crossing property. What happens is that we take our common function F but integrate it over a different region of k_i-space for each amplitude. Inspection of the resulting amplitudes does not reveal their common ancestry.

Example: the simplest possible, namely the massless scalar ϕ^4 integral. In momentum space the amplitudes are just $\delta(k_1 + k_2 + k_3 + k_4)$, i.e. $F(k_1,k_2,k_3,k_4) \equiv 1$. But what twistor diagrams do is to work out the amplitudes as functionals of wave-packets. Take these to be elementary, i.e. of form

$$\phi(x) = \frac{1}{(x-p)^2} \propto \int d^4k\ \delta(k^2)\ \Theta(k)e^{-ik(x-p)}$$

where $\Theta(k)$ is 1 if k is future-pointing, 0 otherwise. Then we get amplitudes as functions of p_1,p_2,p_3,p_4. Explicitly, for the processes $1,2 \to 3,4$; $1,3 \to 2,4$; $1,4 \to 2,3$, these are:

$$\int d^4k_1 \ldots d^4k_4\ \delta(k_1 + k_2 + k_3 + k_4)\ \delta(k_1{}^2)\delta(k_2{}^2)\delta(k_3{}^2)\delta(k_4{}^2)$$

$$\exp i(\sum_j k_j p_j) \quad \begin{cases} \Theta(k_1)\Theta(k_2)\Theta(-k_3)\Theta(-k_4) \\ \Theta(k_1)\Theta(k_3)\Theta(-k_2)\Theta(-k_4) \\ \Theta(k_1)\Theta(k_4)\Theta(-k_2)\Theta(-k_3) \end{cases}$$

Note that these amplitudes are defined on different regions of (p_1,p_2,p_3,p_4) space; e.g. the first requires $p_1,p_2 \in CM^+$; $p_3,p_4 \in CM^-$. The result of the computation is well-known (from twistor diagram theory!) and the amplitudes are respectively given by

$$\int_0^\infty \frac{du}{Q(u)}, \quad \int_\infty^1 \frac{du}{Q(u)}, \quad \int_1^0 \frac{du}{Q(u)} \tag{1}$$

with $Q(u) \equiv au^2 + (a+b-c)u + b$
and $a \equiv (p_1-p_4)^2(p_2-p_3)^2$, $b = (p_1-p_3)^2(p_2-p_4)^2$, $c = (p_1-p_2)^2(p_3-p_4)^2$.

These can also be written as:

$$\frac{\log\left(\frac{\kappa}{\tilde{\kappa}}\right)}{\Delta}, \quad \frac{\log\left(\frac{1-\tilde{\kappa}}{1-\kappa}\right)}{\Delta}, \quad \frac{\log\left(\frac{(1-\kappa)}{(-\kappa)}\,\frac{(-\tilde{\kappa})}{(1-\tilde{\kappa})}\right)}{\Delta}$$

where $\Delta^2 = (a+b-c)^2 - 4ab$

$$\kappa = \frac{a+b-c+\Delta}{2b}, \quad \tilde{\kappa} = \frac{a+b-c-\Delta}{2b}, \quad \text{etc.}$$

are conformal invariants, *cross-ratios* of p_1,p_2,p_3,p_4.

These functions do *not* have the crossing property themselves. In fact, they are essentially *different* functions of $p_1 \ldots p_4$. But it is clear that they enjoy a remarkable property: - they can be analytically continued to a common region in such a way that they sum to zero. Alternatively, one can think of this as a *splitting* property: the first amplitude, which is non-singular for $p_1, p_2 \in CM^+$ $p_3, p_4 \in CM^-$, can be split into (analytic continuations of) two functions, one non-singular in $p_1, p_3 \in CM^+$, $p_2, p_4 \in CM^-$; one non-singular in $p_1, p_4 \in CM^+$, $p_2, p_3 \in CM^-$.

This suggests some underlying unity. Further insight into this unity has been obtained by considering the *symmetrized* box diagram

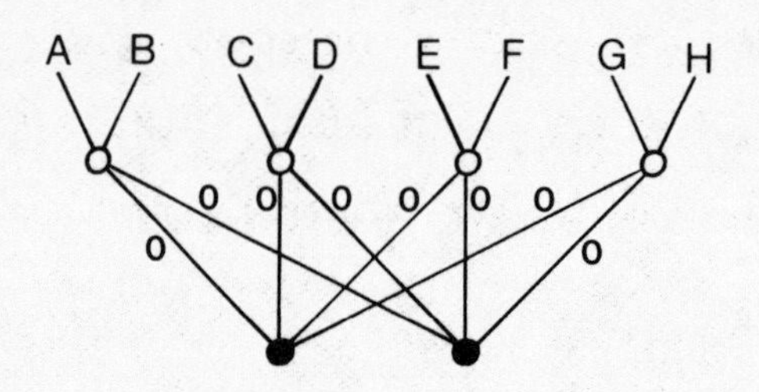

where

$$A^{[\alpha_B\beta]} \sim p_1 \quad C^{[\alpha_D\beta]} \sim p_2 \quad E^{[\alpha_F\beta]} \sim p_3 \quad G^{[\alpha_H\beta]} \sim p_4$$

It has been known for a long time that there are contours for this diagram which correspond to the three amplitudes (1). It can now be shown that there is another contour, of which those contours that give amplitudes can be considered to be the *periods*. The idea is to evaluate the integral retaining complete symmetry at every stage. We introduce boundaries, but the choice of boundary is shown to be irrelevant. We need the following result (see §5.2):

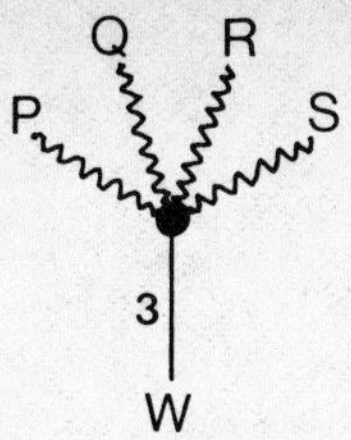

$$= \frac{(\text{PQRS})^{3}}{\text{PQRW}\ \ \text{PRSW}\ \ \text{PQSW}\ \ \text{QRSW}}$$

$$= \frac{\text{KLMN}}{\begin{matrix}W & W & W & W\\ | & | & | & |\\ K & L & M & N\end{matrix}} \quad ,$$

where

$$\begin{matrix}|\\ K\end{matrix} = \text{PQR} \quad , \text{ etc..}$$

And thus we have:

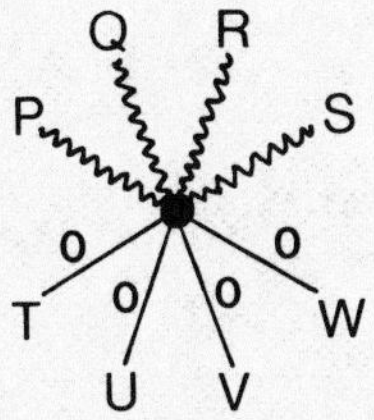

$$= \int_0^\infty\!\!\int_0^\infty\!\!\int_0^\infty d\alpha\, d\beta\, d\gamma$$

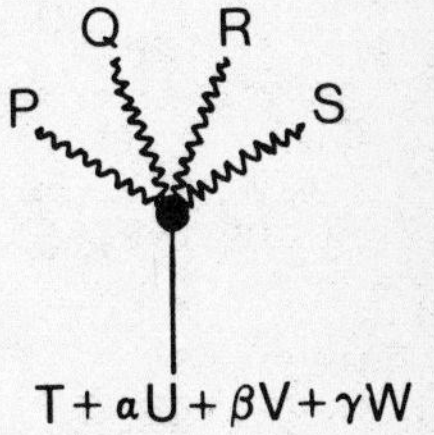

These are actually *compact* integrals. More explicitly this integral is given by:

$$\int_0^\infty \int_0^\infty \int_0^\infty \alpha d\beta d\gamma \quad \frac{\text{KLMN}}{\begin{pmatrix}T+\alpha U+\beta V+\gamma W\\ |\\ K\end{pmatrix}\begin{pmatrix}T+\alpha U+\beta V+\gamma W\\ |\\ L\end{pmatrix}\begin{pmatrix}T+\alpha U+\beta V+\gamma W\\ |\\ M\end{pmatrix}\begin{pmatrix}T+\alpha U+\beta V+\gamma W\\ |\\ N\end{pmatrix}}$$

the result being *symmetric* in T, U, V, W. So now, integrate out all the dual-twistors in Σ leaving

$$\oint \frac{\mathcal{D}\,XZ}{\overline{ABXZ}\;\overline{CDXZ}\;\overline{EFXZ}\;\overline{GHXZ}}\,.$$

Integrate over X using the result above, to retain symmetry, giving

$$\int \mathcal{D}Z \int_0^\infty\int_0^\infty\int_0^\infty \frac{d\alpha\, d\beta\, d\gamma\ \overline{K\,L\,M\,N}}{(\overline{K\,Z\,\boxed{U}}\,)\,(\overline{L\,Z\,\boxed{U}})\,(\overline{M\,Z\,\boxed{U}})(\overline{N\,Z\,\boxed{U}})}$$

where $\boxed{U} = \overline{A\,B} + \alpha\,\overline{C\,D} + \beta\,\overline{E\,F} + \gamma\,\overline{G\,H}$

Suppose that the order of integration can be reversed. The Z integration then just gives

$$\int_0^\infty\int_0^\infty\int_0^\infty \frac{d\alpha\, d\beta\, d\gamma}{\overline{K\ \boxed{U}}\ \overline{L\ \boxed{U}}\ \overline{M\ \boxed{U}}\ \overline{N\ \boxed{U}}}$$

$$=\int_0^\infty\int_0^\infty\int_0^\infty \frac{d\alpha\, d\beta\, d\gamma}{\left(\overline{\boxed{U}\ \ \boxed{U}}\right)^2} = \int_0^\infty d\alpha \int_0^\infty d\beta\ \frac{1}{(1+\alpha+\beta)(a\alpha+b\beta+c\alpha\beta)}$$

This can be evaluated in a manifestly symmetric form as

$$\Delta^{-1}\left\{\begin{aligned}&\mathrm{dog}\left(\frac{\kappa}{\tilde\kappa}\right) - \mathrm{dog}\left(\frac{\tilde\kappa}{\kappa}\right) + \mathrm{dog}\left(\frac{1-\tilde\kappa}{1-\kappa}\right) - \mathrm{dog}\left(\frac{1-\kappa}{1-\tilde\kappa}\right)\\ &\qquad - \mathrm{dog}\left(\frac{(-\kappa)}{(1-\kappa)}\ \frac{(1-\tilde\kappa)}{(-\tilde\kappa)}\right) + \mathrm{dog}\left(\frac{(-\tilde\kappa)}{(1-\tilde\kappa)}\ \frac{(1-\kappa)}{(-\kappa)}\right)\end{aligned}\right\}$$

but also in other forms, e.g.

$$\Delta^{-1}\ \{2\ \mathrm{dog}(\tilde\kappa) - 2\ \mathrm{dog}(\tilde\kappa) + \log\left(\frac{1-\tilde\kappa}{1-\kappa}\right)\log(\kappa\tilde\kappa)\}$$

or

$$\Delta^{-1}\{\mathrm{dog}(\tilde\kappa) - \mathrm{dog}(\kappa) + \mathrm{dog}(1-\kappa) - \mathrm{dog}(1-\tilde\kappa)$$
$$+ \log(1-\tilde\kappa)\log\kappa - \log(1-\kappa)\log\tilde\kappa\}$$

where * $\mathrm{dog}(z) = -\int_0^z \frac{\log(1-t)}{t}\,dt = \sum_1^\infty \frac{z^n}{n^2}$ for $|z| < 1$.

It is this "super-amplitude" which has the crossing property: that is, it can be regarded as the same analytic function in each channel. The amplitudes can be recovered by taking periods, rather as the amplitudes are obtained from the $F(k_i)$ by multiplying F by the delta-function.

---

* The function *dog(z)* is called the *dilogarithm*.

§5.7 COMPTON SCATTERING FOR MASSLESS ELECTRONS *by A.P. Hodges*

We shall find twistor diagrams representing the amplitude for the Compton scattering process. Twistor diagrams are *gauge-invariant*, because they represent the ingoing or outcoming photons in terms of fields ϕ_{AB} and not in terms of a potential A_a. This means that there is no way of representing the single Feynman diagram

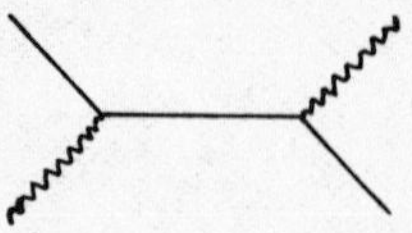

because this is not gauge-invariant. But we can translate the gauge-invariant sum of the two diagrams

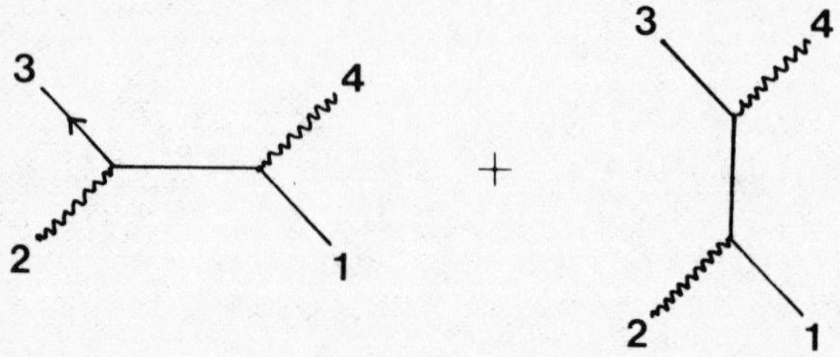

3 4 2 1 + 3 4 2 1

Only this sum is physically significant, anyway.

Particle 1 is an ingoing "electron" with wave function whose Fourier transform is $\tilde{\psi}_A(k_1)\delta^+(k_1{}^2)$. Particle 3 is an outgoing "electron" of the same helicity whose complex conjugate has Fourier transform $\tilde{\psi}_{C'}(k_3)\delta^-(k_3{}^2)$. Particle 2 is an ingoing photon represented by a potential whose Fourier transform is $\tilde{A}_{BB'}(k_2)\delta^+(k_2{}^2)$. This photon is not assumed to be in a helicity state, and the potential is not assumed to be in any particular gauge. Similarly for particle 4, the outgoing photon. Writing down the Feynman rules in 2-spinor form, one obtains for the sum:

$$\int d^4k_1 d^4k_2 d^4k_3 d^4k_4 \delta(k_1+k_2-k_3-k_4)\delta^+(k_1^{\,2})\delta^+(k_2^{\,2})\delta^-(k_3^{\,2})\delta^-(k_4^{\,2})$$

$$\tilde{\psi}_A(k_1)\tilde{A}_{BB'}(k_2)\tilde{\psi}^{C'}(k_3)\tilde{A}^{DD'}(k_4)\left\{\frac{\varepsilon^A_{\ D}\varepsilon^{B'}_{\ C'}(k_1-k_4)^B_{\ D'}}{(k_1-k_4)^2} + \frac{\varepsilon^{AB}\varepsilon_{C'D'}(k_1+k_2)^{B'}_{\ D}}{(k_1+k_2)^2}\right\}$$

Purely algebraic manipulation of the large bracket expression, making use of the relations

$$k_1^{\,2} = k_2^{\,2} = k_3^{\,2} = k_4^{\,2} = 0$$

$$k_{3MC'}\tilde{\psi}^{C'}(k_3) = 0 = k^A_{1N'}\ \tilde{\psi}_A(k_1)$$

produces the following gauge-invariant expression:

$$\int d^4k_1 \ \dots \ \delta^-(k_4^{\,2})\{\tilde{\psi}_A(k_1)\tilde{\phi}_{E'C'}(k_2)\tilde{\psi}^{C'}(k_3)\tilde{\phi}^A_F(k_4)k_3^{FE'}$$

$$-\tilde{\psi}_A(k_1)\tilde{\phi}^A_E(k_2)\tilde{\psi}^{C'}(k_3)\tilde{\phi}^{F'}_{C'}(k_4)k^E_{1F'}\}[(k_1+k_2)^2(k_1-k_4)^2]^{-1},$$

where $\tilde{\phi}_{E'C'}(k_2) = k^B_{2(E'}\tilde{A}_{C')B}(k_2)$, and so forth.

Thus the amplitude splits into two gauge-invariant parts, according to the two helicity parts of the photon. Helicity of the photon is conserved in each part. Suppose now the incoming photon is of definite helicity, of opposite type to the incoming electron (so the second part of the amplitude vanishes). The in- and out-states now translate into twistors as follows:

$$\psi_A(x_1) \longrightarrow f_1(W_\alpha)W\urcorner$$

$$\phi_{D'B'}(x_2) \longrightarrow f_2(Y_\alpha)\partial Y\urcorner\ \partial Y\urcorner$$

$$\psi^{C'}(x_3) \longrightarrow f_3(X^\alpha)X\urcorner$$

$$\phi_{DB}(x_4) \longrightarrow f_4(Z^\alpha)\ \partial Z\lrcorner\ \partial Z\lrcorner \quad ,$$

where f_1 and f_3 are homogeneous of degree -3, and f_2 and f_4 are homogeneous of degree 0. The integral

$$\int d^4k_1 d^4k_2 d^4k_3 d^4k_4 \delta(k_1+k_2-k_3-k_4)\delta^+(k_1{}^2)\delta^+(k_2{}^2)\delta^-(k_3{}^2)\delta^-(k_4{}^2)$$

translates into the expression

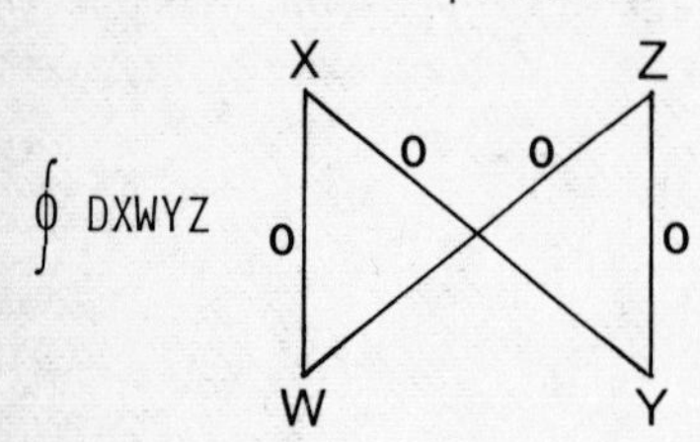

,

and finally we have

$$k_3{}^{FE'} \longrightarrow X\ \partial X$$

$$[(k_1+k_2)^2(k_1-k_4)^2]^{-1} \longrightarrow [XZ\ \partial X \partial Z]^{-1}\ [W\partial Z\ \ Z\partial W]^{-1}.$$

Hence, doing all the contractions correctly, we obtain

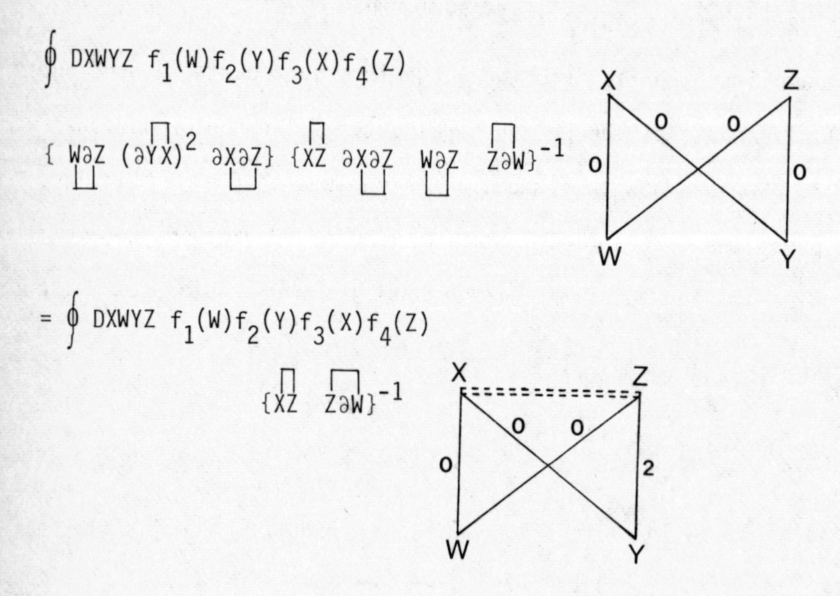

=

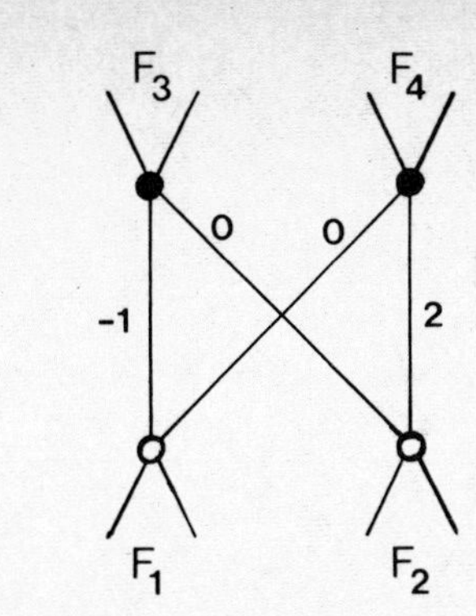

which is finite, gauge-invariant and conformally invariant. Similarly we can deal with the annihilation - creation channel, i.e.

and obtain

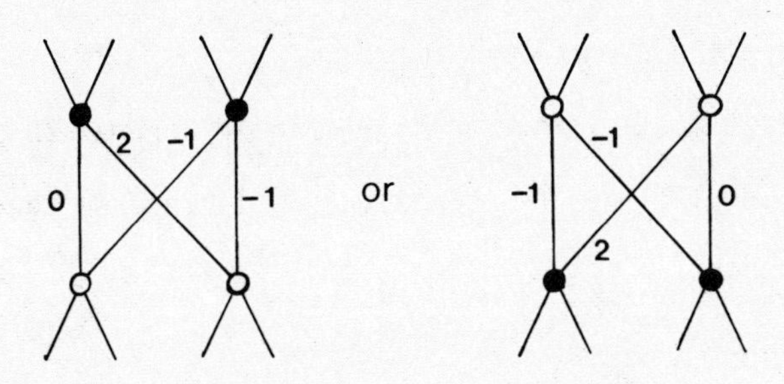

We can if we like put these results into another form by using only functions of homogeneity degrees -3 or -4. Using twistor transforms the first result gives us

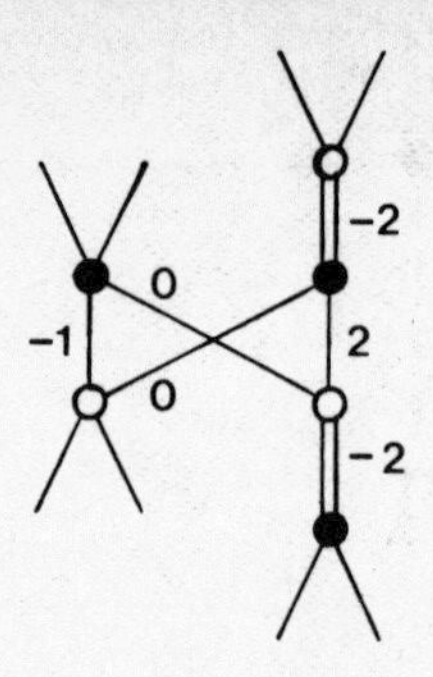

for the opposite-helicity Compton channel, and the second result gives us

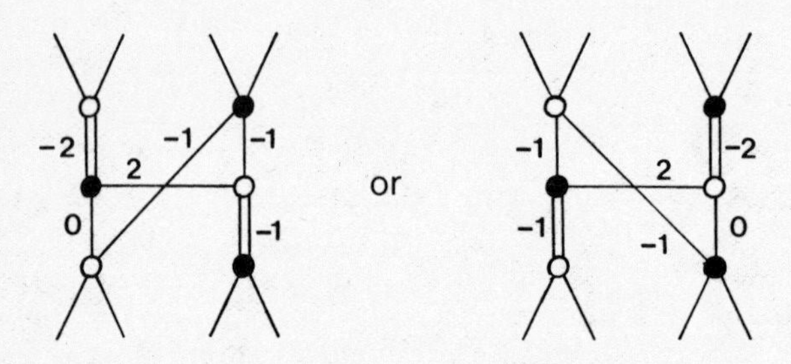

for the annihilation-creation channel.

It will be noticed that many of the lines in these diagrams are boundary-prescriptions and not singular factors. Amazingly, all three diagrams reduce to a single integrand:

$$\oint \frac{f_{-3}(W)g_{-4}(Y)h_{-3}(X)j_{-4}(Z)\ DWYXZUV}{\begin{matrix}W\\ |\\ X\end{matrix}\left(\begin{matrix}V\\ |\\ U\end{matrix}\right)^{3}\begin{matrix}Y\\ |\\ X\end{matrix}}$$

different choices of (boundary) contours supplying the different channel amplitudes. Moreover, this form is the same as that derived in Penrose and MacCallum 1972 from a quite different point of view!

Unfortunately the third channel (same-helicity Compton) does not fit into this neat scheme. It can only be translated into

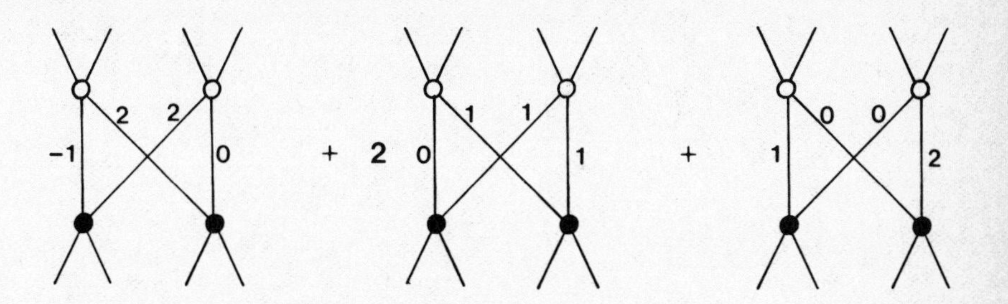

One final remark: Is there really a contour for

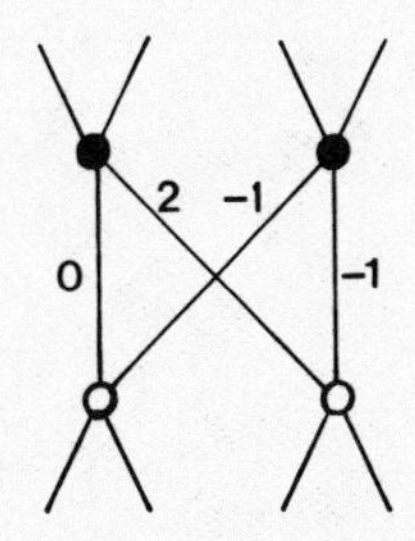

with its two -1 lines meeting at a corner? Yes there is, if we are allowed an extra boundary:

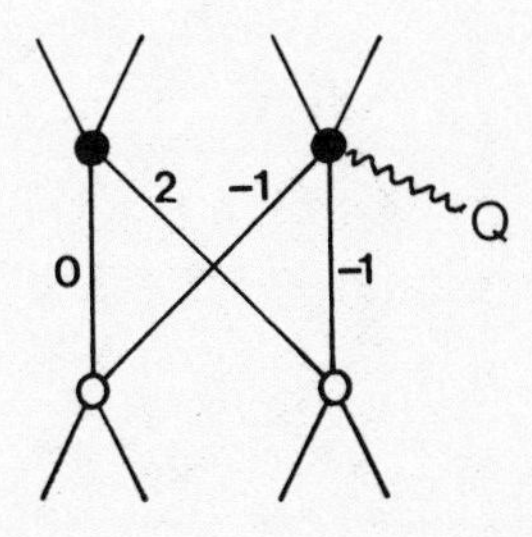

The answer is then independent of Q, and it also agrees with

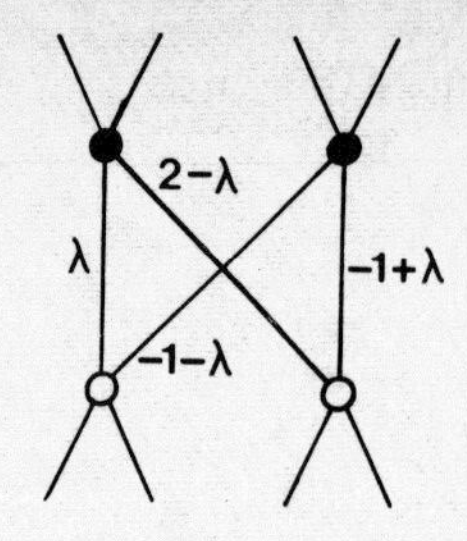

in the limit as $\lambda \to 0$.

REFERENCE

Penrose, R. & MacCallum, M.A.H. 1972 Phys. Reports 6, 241-315.

§5.8 CONTOUR PINCHING AND COHOMOLOGY *by A.P. Hodges*

The point of this note is to exhibit something that can be done with 1-cocycles that could not be done with the old contour integral formalism.

Recall the projective spinor integrals

$$\oint \frac{\pi.d\pi}{(\pi.\alpha)(\pi.\beta)} = \frac{2\pi i}{(\alpha.\beta)} \, , \quad \oint \frac{\lambda.d\lambda \wedge \mu.d\mu}{(\lambda.\mu)^2} = 2\pi i$$

and consider the expression

$$\oint \frac{\pi.d\pi \wedge \lambda.d\lambda \wedge \mu.d\mu \ (\beta.\lambda)}{(\pi.\lambda)(\pi.\beta)(\lambda.\mu)^2} \, . \tag{1}$$

Is there a non-trivial contour? At first sight it looks as though we could integrate over π, and then do the (λ,μ) integration. But this is erroneous. For λ has to pass through all possible values in the course of the (λ,μ) integration, including $\lambda = \beta$. But when $\lambda = \beta$, the π integral contour is *pinched*. There is no sense in which the numerator $(\beta.\lambda)$ can be said to "cancel" this pinching. Numerators cannot affect the existence or non-existence of contours, which depend only on the homology of the integration space with singular regions removed.

Instead, consider the following. Patch π space by $U_1 = \{\pi | \pi.\lambda \neq 0\}$, $U_2 = \{\pi|\pi.\beta \neq 0\}$, $U_3 = \{\pi|\pi.\gamma \neq 0\}$, and define

$$\begin{cases} f_{12} = (\lambda.\beta)(\pi.\lambda)^{-1}(\pi.\beta)^{-1} \text{ on } U_1 \cap U_2 \\ f_{23} = (\beta.\gamma)(\pi.\beta)^{-1}(\pi.\gamma)^{-1} \text{ on } U_2 \cap U_3 \\ f_{31} = (\gamma.\lambda)(\pi.\gamma)^{-1}(\pi.\lambda)^{-1} \text{ on } U_3 \cap U_1. \end{cases}$$

Note that

$$\rho_1 f_{23} + \rho_2 f_{31} + \rho_3 f_{12} = 0,$$

so that we have defined a 1-cocycle; hence the branch-contour (see §2.2) is well-defined (see Figure A) as:

$$\int_{\gamma_{12}} f_{12}\ \pi.d\pi \ + \int_{\gamma_{23}} f_{23}\ \pi.d\pi \ + \int_{\gamma_{31}} f_{31}\ \pi.d\pi \quad .$$

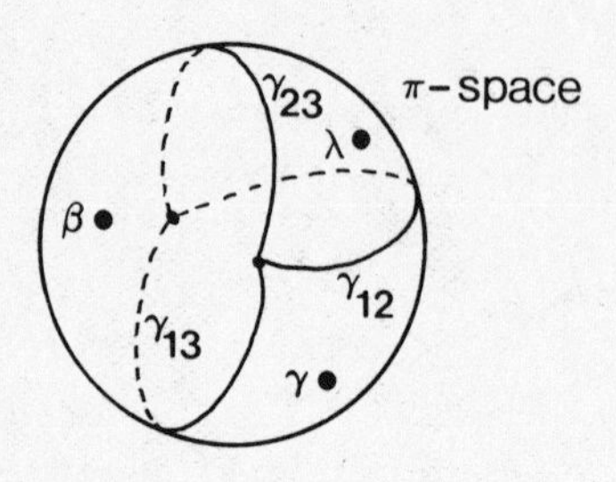

Figure A. *Branch-contour. The section* γ_{23} *has to avoid* β *and* γ *but need not avoid* λ *- similarly for the other two segments.*

Provided $\lambda \neq \beta$, the branch contour can be transformed to

$$\oint f_{12}\ \pi.d\pi \ = \ \oint \frac{\beta.\lambda}{(\pi.\lambda)(\pi.\beta)}\ \pi.d\pi \quad . \tag{2}$$

If $\lambda = \beta$ we cannot make this transformation - but the branch contour still exists perfectly well - there is no "pinching". So we find the cocycle to *extend* the expression (2) in a well-defined way to the case $\lambda = \beta$, and hence to give us a way of obtaining a non-trivial result for the integral (1).

The cocycle can only perform this extension because the numerator factor $\lambda.\beta$ is what it is. If the β in this numerator is considered as a parameter, and is displaced infinitesimally, the cocycle structure disappears. This is in complete contrast to contour integration of functions in the old way, where we can always argue that small variation of parameters must preserve the existence of a contour.

So we have here the example of something manifestly finite (it is still compact integration) which only exists when a certain parameter condition is met. This is exactly what twistor integral theory needs! For instance, it is now logically possible that there can be a twistorial inner product for massive states which exists only if the two masses are equal and is then manifestly finite. This is impossible in the old scheme. Again, another major stumbling-block in twistor integral theory has been the fact that twistor diagrams involving both a massive in-state and a massive out-state could not have contours - contours were always pinched between the singularity at infinity of the in-state and the singularity at infinity of the out-state. But it now appears that when such integrals are in fact finite, there are enough numerator factors (involving the infinity twistor) to allow a reformulation in terms of cocycles in which these numerators "cancel" the pinching. It was in fact a simplified version of this problem that led me to consider the integral (1). For further discussion see §5.9.

§5.9 THE TWISTED CAMEL (or, how to get through the eye of a singularity)

by R. Penrose

I make, here, some remarks concerning the type of circumstance depicted in Figure A, for a function of one or more twistor variables, where two pole singularities pinch together, but at the same time the pinching is compensated by the presence of a zero, i.e. we are concerned with integrating an expression of the form A/BC over some contour, where some relevant common zero set of B and C is also a zero set of A.

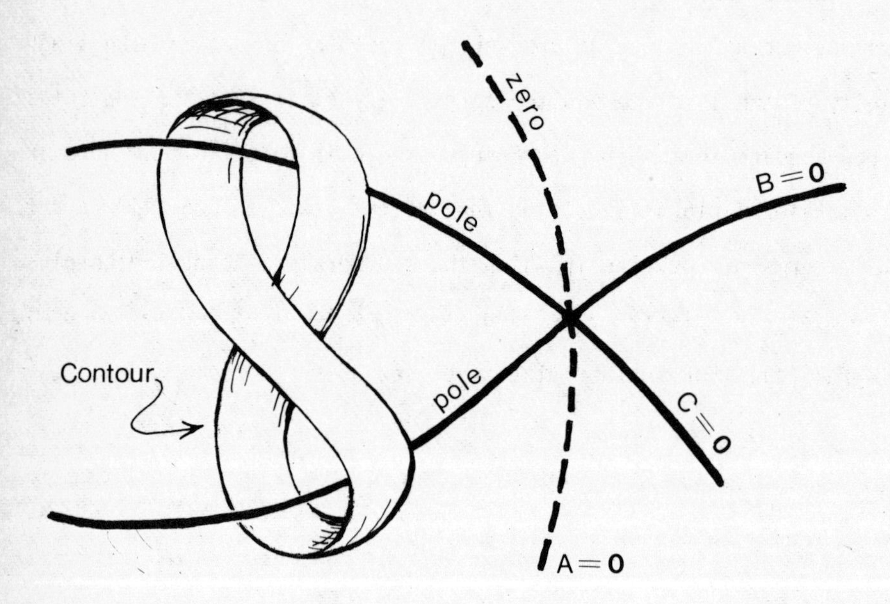

Figure A

The question of how to handle the contour integration when the contour approaches the point of pinching has some interest in a number of contexts. In particular, it has relevance to the integrals that A.P.H. considers in §5.8. In a sense, the contour can (at least in certain such circumstances) be "pushed through" the point of pinching and give a finite result at all stages. I shall indicate *two* ways of so pushing our camel contour through the needle's eye: (i) take the needle away and put it somewhere else; and (ii) blow up the needle!

In method (i), the twistor function is regarded as part of a cochain; we are supposed not to be interested in the actual function itself but in some kind of cohomology group element that it defines. So the function can be replaced by another one in which (one hopes) the pinching now occurs somewhere else. A prototype of this type of behaviour arises when we try to represent certain generalizations of elementary states in terms of twistor functions, such "generalized elementary states" having arisen from a suggestion by M.F. Atiyah as to how best to apply R.S.W.'s Yang-Mills twistor construction (cf. §3.5). [See §5.8 for other uses of method (i) which could well be of considerable significance.] Such a generalized elementary state is obtained from some algebraic curve γ lying in PT^-, the point being that the field is to be singular *only* for complex space-time points which are described by lines in PT meeting γ (see Figure B).

Figure B

I shall consider only one example here, namely when γ is a *twisted cubic* (it had to happen!) and I shall not worry about arranging that γ actually lies in PT^- (which is easy enough to organize by means of a coordinate change). Choose twistor coordinates W, X, Y, Z and let γ be given by $W:X:Y:Z = 1:\theta:\theta^2:\theta^3$ with $\theta \in C\cup\{\infty\}$. Set $Q_1 = X^2 - WY$, $Q_2 = WZ - XY$, $Q_3 = Y^2 - XZ$. Then γ is the common intersection of the three quadrics $Q_i = 0$. Set $U_i = PT - \{Q_i = 0\}$. Then U_1, U_2, U_3 cover $PT - \gamma$. Set $f_{ij} = L_{ij}/Q_iQ_j$ where $L_{12} = W-X$, $L_{23} = Y-Z$, $L_{31} = X-Y$. Then the cocycle condition $f_{12} + f_{23} + f_{31} = 0$ is satisfied. Using f_{ij} as a twistor cocycle function, the field (a neutrino field) may be evaluated using a branched contour integral (see Figure C).

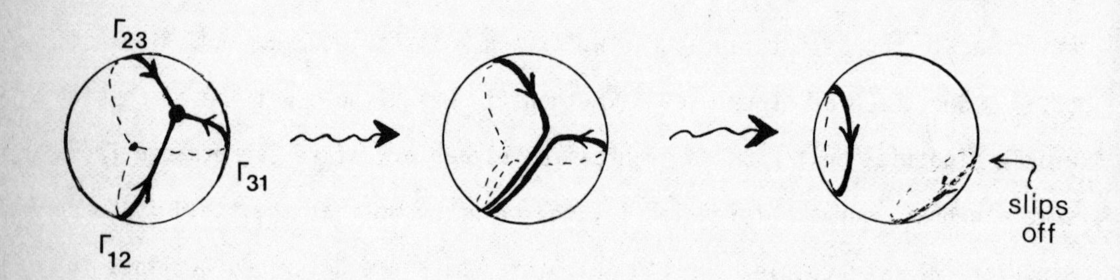

Figure C

But one always finds that the branched contour can be deformed until it is just *one loop* involving just *one* f_{ij}, for a generic space-time point. Pinching occurs only for exceptional space-time points - and then we simply *switch to a different* f_{ij}. Thus, we have an example of (i) above. For further discussion of massless fields based on a twisted cubic curve see §2.16.

To illustrate method (ii) I consider 2 dimensions first, say C^2, where the origin is the point of pinching. Take w,z as coordinates for C^2 and consider the integral

$$\oint_{\Gamma} f(w,z)\,dw\wedge dz ,$$

where

$$f(w,z) = \frac{A(w,z)}{B(w,z)C(w,z)} ,$$

the functions A, B, C having simple zeros on curves a, b, c through (0,0) with distinct tangents there. Consider the small sphere $w\bar{w} + z\bar{z} = \varepsilon^2$ ($\varepsilon > 0$). Holomorphic curves through (0,0) meet this S^3 in Clifford parallels (Hopf S^1's) (essentially) two of which are $b \cap S^3$ and $c \cap S^3$. The contour Γ can be a torus in S^3 avoiding and separating these S^1's, as depicted in Figure D.

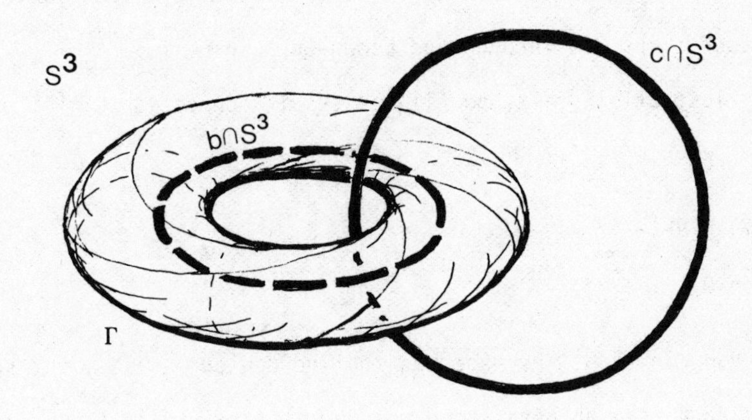

Figure D

If we now let $\varepsilon \to 0$, Γ shrinks down to the origin, which is singular: *pinching*!! So we *blow up* the origin, i.e. replace (0,0) by a whole S^2 (holomorphically!) in which *each* Clifford-Hopf S^1 converges down to a *different* point of S^2. The contour Γ can be fibred (if chosen suitably) by twisted Clifford-Hopf S^1's (cf. Robinson congruence pictures) each of which collapses to a point as $\varepsilon \to 0$, but now b and c are *both avoided* even at $\varepsilon = 0$. So the contour slips off completely in the blown-up space! But we need to check that the form $\phi = f dw_\wedge dz$ is non-singular on the new "origin" S^2 - and here's where A comes in. The blow-up can be achieved explicitly by mapping (w,z) to the point $(w^2 : wz : z^2 : w : z)$ of CP^4. Arrange coordinates so that z = 0 avoids a,b,c,Γ near (0,0), as can easily be done, and choose local coordinates $z = z^2/z$ and $q = w/z$ for the blown-up space. Then

$$\phi = \frac{A(zq,z)z\, dq_\wedge dz}{B(zq,z)C(zq,z)} \quad ,$$

which has a *finite* value on S^2 (which is defined by z = 0).

This shows that every contour integral of the above form must *vanish*, since the contour slips off through the blown-up origin.

A further illustration of method (ii) arises in connection with the spinor integral

$$\oint \frac{\pi.d\pi_\wedge \lambda.d\lambda_\wedge \mu.d\mu(\beta.\lambda)}{(\pi.\lambda)(\pi.\beta)(\lambda.\mu)^2}$$

examined by A.P.H. in §5.8. He observes that no contour exists, but nevertheless the integral can be given good meaning if $(\beta.\lambda)(\pi.\lambda)^{-1}(\pi.\beta)^{-1}$ is interpreted cohomologically. Another way of handling the "contour pinching" that arises here is to use a "blow-up", as suggested above. For this, first fix μ and consider the product space of (π,λ). This is a quadric Q in CP_3, and we have the set-up illustrated in Figure E:

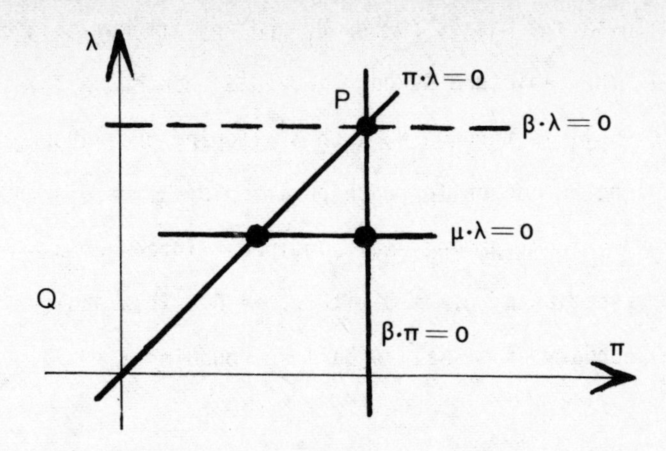

Figure E

The pinching that we wish to remove occurs where the poles $\pi.\lambda = 0$ and $\beta.\pi=0$ come together at the same place P as the zero $\beta.\lambda = 0$. The point P may be blown up into a line by sending *plane sections through* P, of Q, to *lines* in the new plane S. In the process, the two generators of Q through P (namely $\beta.\lambda = 0$ and $\beta.\pi = 0$) get "blown down" to points L and N in S. There remain but two singular lines in S, namely $\pi.\lambda= 0$ and $\mu.\lambda = 0$. (P is now non-singular because of the *numerator* $\beta.\lambda = 0$.) See Figure F.

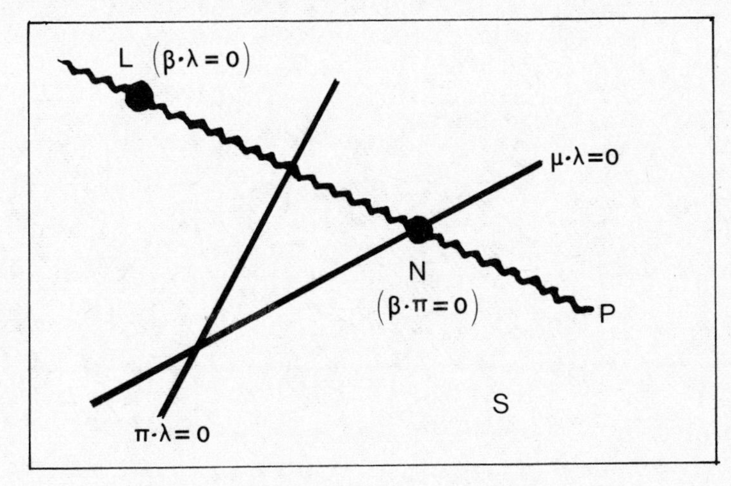

Figure F

Now let μ vary, so the line $\mu.\lambda = 0$ moves in our pictures. In the Q-picture, trouble occurs when this line passes across P, and any attempt at a contour gets pinched. But in the S-picture we can find an S^3 contour. This is Hopf-fibred by S^1's lying on lines through N, each S^1 looping once around N on the line. One such line is chosen for each position of $\mu.\lambda = 0$ - there is a continuous S^2's worth of them, having, say, antipodal intersections with $\pi.\lambda = 0$, to each intersection of $\mu.\lambda = 0$ with $\pi.\lambda = 0$. This whole S^3 may be identified with the boundary of a small 4-ball surrounding N in S.

§5.10 BLOWING UP THE BOX *by S.A. Huggett*

When the massless scalar ϕ^4 vertex is translated into twistors we obtain (cf. Penrose 1975) the twistor diagram called the box (Fig. A). Unfortunately this diagram has the drawback that it is difficult to see why it should

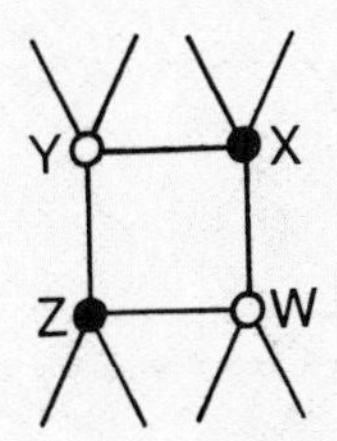

Figure A.

satisfy crossing symmetry (see §5.6). Ideally we would hope for three contours for the box each of which allows the external states to be paired in a different way. However, this does not seem to be the case — if the external lines are integrated out first the remaining integral only has two contours (cf. Sparling 1975) and one of these is hopeless because it does not allow *any* of the states to be paired.

One approach to this problem is to redefine the internal part of the box by blowing up the space in which it is defined. Hopefully we shall then be able to find the missing contours. So what is blowing up and why should it produce contours where there were none before?

The simplest example is drawn in Fig. B. The "Möbius strip" is the blown up version of the plane C^2 - we have replaced a point p in C^2 by a CP^1 in such a way that different directions at p correspond to different points on the CP^1. Let C^2 have coordinates (Z_1,Z_2) centred at p and let CP^1 have coordinates $(\xi:\eta)$. Then the blown up space is

$$\{(Z_1,Z_2;\ \xi:\eta) \in C^2 \times CP^1 :\ Z_1\xi = Z_2\eta\}.$$

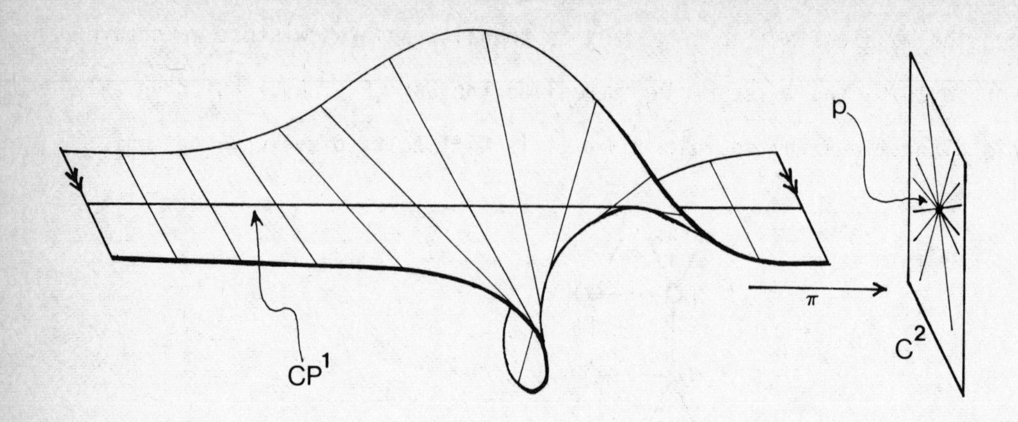

Figure B

It is biholomorphic to C^2 via the projection π except at p. (This particular blow up can also be described in terms of the Hopf fibration of S^3 - see §5.9.) This example also illustrates the creation of a contour: $H_2(C^2;Z)=0$ because C^2 is contractible whereas H_2 (blown up space; Z) = Z because the blown up space is a fibre space with base CP^1 and fibre C.

To generalize we replace C^2 by a variety M and p by a subvariety N of complex codimension m in M. Choose functions $U_1,\ldots,U_m$ whose zeros define N and let $(t_1: \ldots : t_m)$ be homogeneous coordinates for CP^{m-1}. The subvariety M^1 of $M\times CP^{m-1}$ defined by the equations $t_iU_j = t_jU_i$ $i,j = 1,\ldots,m$ is the blown up version of M. In this case N has been replaced by $N \times CP^{m-1}$ so that the different normal directions to N are represented by different points on the CP^{m-1}.

Now for the (non-projective) box. M is the space of (Y,W,X,Z) but there are several different subvarieties N we could blow up. One of these is presented below as an example - we are still not sure exactly what the correct choice for N will be.

Let $N = (Y \propto W) \cap (Y.X = 0) \cap (Y.Z = 0)$. This is of codimension 5 and is defined by the zeros of:

$$U_1 = Y_\alpha Z^\alpha \qquad U_2 = Y_\alpha X^\alpha$$

$$U_3 = Y_{[\alpha}W_{\beta]}A^{[\alpha}B^{\beta]} \quad U_4 = Y_{[\alpha}W_{\beta]}C^{[\alpha}D^{\beta]} \quad U_5 = Y_{[\alpha}W_{\beta]}B^{[\alpha}D^{\beta]}$$

where (ABCD) is a basis. We choose homogeneous coordinates $(t_1:t_2:t_3:t_4:t_5)$ for CP^4 and write down the equations $t_iu_j = t_ju_i$ of which only the following four are independent:

$$u_1t_2 = u_2t_1,\ u_1t_3 = u_3t_1,\ u_1t_4 = u_4t_1,\ u_1t_5 = u_5t_1.$$

Changing to local coordinates for the CP^4 these equations become:

$$u_1\lambda_2 = u_2,\ u_1\lambda_3 = u_3,\ u_1\lambda_4 = u_4,\ u_1\lambda_5 = u_5. \qquad \text{(i)}$$

Now choose functions u_6, u_7, u_8 which make $(u_1,\ldots,u_8)$ a coordinate system for (Y,W) and consider the coordinates $(u_1,\ldots,u_8,X,Z)$ for the whole space M. We blow up our subvariety N by replacing

$$(u_1,\ldots,u_8,X,Z) \text{ by } (u_1,\lambda_2,\lambda_3,\lambda_4,\lambda_5,\ u_6,u_7,u_8,X,Z)$$

using (i). It remains to write the differential form defined by the box in these last coordinates and to examine the homology of the space in which the new form is nonsingular.

REFERENCES

Penrose, R. 1975 In *Quantum Gravity*, Eds. R. Penrose, C.J. Isham & D.W. Sciama. Oxford University Press.

Sparling, G.A.J. 1975 In *Quantum Gravity*, Eds. R. Penrose, C.J. Isham & D.W. Sciama. Oxford University Press.

§5.11 SHEAF HOMOLOGY AND CONTOUR INTEGRALS *by R.O. Jozsa*

Ever since sheaf cohomology was introduced into twistor theory to represent contour integrals in an elegant and rigorous way (see chapter 2 in this volume, and also Jozsa 1976) the interpretation of cohomology groups $H^n(X,S)$ for $n > 1$ and dimension $(X) > 1$ has remained unclear since the cohomological freedom does not manifestly correspond to the kind of freedom in a contour integral. The latter does however look very much like the homological freedom in ordinary singular homology theory (cf. Sparling 1975) at least as far as moving the contour around is concerned. We describe here a sheaf generalization of singular homology which incorporates all the contour integral freedoms (i.e. moving the contour, and adding in functions holomorphic all over one side of the contour) in a manifest way and describe the correspondence between these sheaf homology groups and the sheaf cohomology groups. The main use of this formalism is hopefully to provide a way of identifying some of the more complicated contour integral formulae as higher cohomology groups.

The sheaf homology groups are defined in generality in chapter XI of Swan 1964. The basic construction is as follows. We first define the sheaf C_n of singular n-chains on a space X. For $U \subset X$ open, let $C_n(U)$ be the group of all singular n-chains in X modulo those lying outside U. (i.e. the usual construction of relative singular homology, rel X-U) (See Sparling 1975 for a clear account of all these terms). For $U \subset V$ it is easy to define restriction maps

$$\rho_{VU}: C_n(V) \to C_n(U)$$

making the $C_n(U)$'s into a (complete) presheaf. The direct limit construction

(cf. Jozsa 1976) then gives the sheaf C_n. Thus a non-zero element in the stalk above $x \in X$ corresponds to an n-chain which passes through x, modulo all n-chains not passing through X. The boundary operator of singular homology is well defined between these sheaves and we get the chain complex of sheaves:

$$\xrightarrow{\partial} C_2 \xrightarrow{\partial} C_1 \xrightarrow{\partial} C_0 .$$

Now let S be any sheaf. Form the tensor product $C_n \otimes S$; extend the boundary operator ∂ by the identity on S to $\partial \otimes 1$ and form the global sections $\Gamma(X, C_n \otimes S)$ of $C_n \otimes S$. Thus we get a chain complex:

$$\xrightarrow{\partial\otimes 1} \Gamma(X,C_2 \otimes S) \xrightarrow{\partial\otimes 1} \Gamma(X, C_1 \otimes S \xrightarrow{\partial\otimes 1} \Gamma(X,C_0 \otimes S) .$$

The sheaf homology groups $H_k(X,S)$ are defined to be the homology groups of this complex, i.e. $H_k(X,S)$ = k-cycles/k-boundaries

$$= \ker \partial \otimes 1/\mathrm{Im}\, \partial \otimes 1$$

These groups reduce to the usual singular homology theory (with coefficients in a group) when S is a constant sheaf.

More intuitively: for definiteness take X to be a complex manifold and S a sheaf of germs of holomorphic functions. A global section of C_n is just an n-chain (modulo nothing since $X-U = \phi$) and a global section of $C_n \otimes S$ is an n-chain K_n together with a holomorphic function f defined on a slight "thickening out" of K_n. Note that $C_n \otimes S$ generally has lots of global sections when S may have only very few. The boundary operator $\partial \otimes 1$ forms the usual boundary of K_n and restricts the function f to it. Thus f on K_n is an n-cycle (i.e. has zero boundary) if:

1) K_n has zero boundary (i.e. is a cycle in the usual singular homology sense), or:

2) K_n as a chain has non zero boundary but f on the boundary is zero.

We see that this approach unifies the two ideas of contours without boundary and contours with boundary in the subspace where the integrand vanishes. (These boundary contours were introduced into twistor theory by G. Sparling and play an important role). f on K_n is an n-boundary if there exists g on K_{n+1} such that K_n is spanned by K_{n+1} (i.e. $K_n = \partial K_{n+1}$) and f is the boundary value of g. In this case the integral of f over K_n is zero since f can be holomorphically extended to a whole surface spanning K_n and we see that the sheaf homology freedom corresponds directly to adding in functions holomorphic all over one side of the contour. $H_k(X,\mathcal{S})$ represents all k-dimensional contour integrals in X, unlike standard singular homology theory where the homology has to be recalculated for each new singularity structure of the integrand.

The sheaf homology groups are related to the cohomology groups via Poincaré Duality:

Theorem. If X is a real n-dimensional manifold without boundary then $H_k(X,\mathcal{S})$ is canonically isomorphic to $H_c^{n-k}(X,\mathcal{S})$. (The subscript c denotes cohomology with compact supports).

This gives the basic result that k-dimensional contour integrals in an n-complex dimensional complex manifold should be interpreted as elements of the $2n-k^{th}$ cohomology group and that the cohomological freedom incorporates the contour integral freedom including both closed contours and contours having boundary where the integrand vanishes.

The homological description of the contour integration process can be seen more precisely by considering Ω^k, the sheaf of germs of holomorphic k-forms and the short exact sequence:

$$0 \longrightarrow d\Omega^k \xrightarrow{i} \Omega^{k+1} \xrightarrow{d} d\Omega^{k+1} \longrightarrow 0$$

where d is the holomorphic exterior derivative and i is the injection map. This sequence generates a long homology exact sequence for each value of k:

$$\ldots \xrightarrow{\delta^*} H_i(X,d\Omega^k) \longrightarrow H_i(X,\Omega^{k+1}) \longrightarrow H_i(X,d\Omega^{k+1})$$
$$\xrightarrow{\delta^*} H_{i-1}(X,d\Omega^k) \longrightarrow H_{i-1}(X,\Omega^{k+1}) \longrightarrow H_{i-1}(X,d\Omega^{k+1})$$
$$\xrightarrow{\delta^*} H_{i-2}(X,d\Omega^k) \rightarrow \ldots .$$

For i = k+2, the connecting homomorphism

$$H_{k+2}(X,d\Omega^{k+1}) \xrightarrow{\delta^*} H_{k+1}(X,d\Omega^k)$$

performs the integration of one variable of a (k+2)-form on a (k+2)-dimensional contour to give a (k+1)-form on a (k+1)-dimensional contour, which can be seen explicitly by analysing the construction of δ^* (cf. Spanier 1966, chapter 4, §5, Lemma 3).

As an immediate application of the formalism, consider functions of two twistors $f(Z_1,Z_2)$ which, via two-dimensional contour integrals on $S^2 \times S^2$ submanifolds of PT $\times$ PT can yield spacetime fields corresponding to one-particle massive states or two-particle massless states (for example when $f(Z_1,Z_2) = f_1(Z_1).f_2(Z_2)$). It has been conjectured that in twistor theory two-particle states should be described by H^2's and one-particle states by H^1's (in view of the construction of curved twistor spaces and the interpretation of H^1's as infinitesimal deformations of structures). But both integrals are two-dimensional integrals on the four real-dimensional space $S^2 \times S^2$ and thus naturally represented by $H_2(S^2{\times}S^2,\Omega^2) \simeq H^2(S^2{\times}S^2,\Omega^2)$ i.e.

both are H^2's. The one-particle integrals often use S^2 contours whereas the two-particle integrals use $S^1 \times S^1$ contours. The analysis of the δ^* construction shows that both these integrals are evaluated by the same general procedure and are thus of the same "cohomological type".

REFERENCES

Jozsa, R. 1976 *Applications of Sheaf Cohomology in Twistor Theory.* M.Sc. thesis. University of Oxford.

Spanier, E. 1966 *Algebraic Topology.* McGraw-Hill.

Sparling, G.A.J. 1975 *Homology and Twistor Theory.* In: *Quantum Gravity – An Oxford Symposium*, eds. C.J. Isham, R. Penrose & D.W. Sciama. Oxford University Press.

Swan, R. 1964 *The Theory of Sheaves.* University of Chicago Press.

§5.12 SHEAF COHOMOLOGY AND TWISTOR DIAGRAMS *by M.L. Ginsberg and S.A.Huggett*

Introduction

Now that zero rest mass fields have been identified as elements of sheaf cohomology groups the following question needs to be answered. Given a twistor diagram we can replace all the (external) twistor functions by the cohomology classes they represent. What are the operations between these cohomology classes which correspond to simply integrating the twistor diagram? In this article we shall answer this question for twistor diagrams with no loops, double lines, or boundary lines. These are called tree diagrams.

The Twigs of Twistor Diagrams

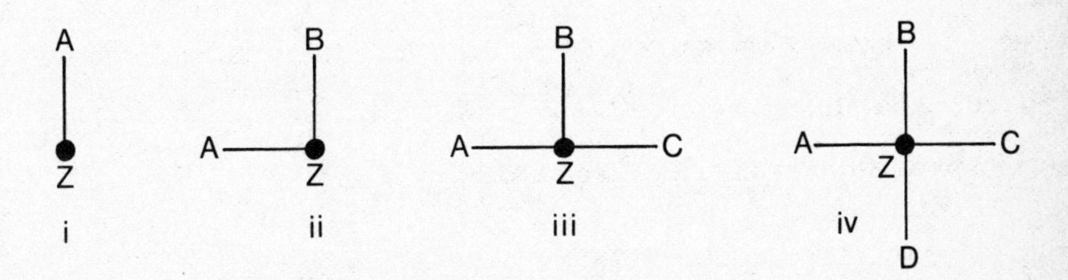

Figure A

The four cases are drawn in Figure A. The first is the function $(A.Z)^{-1}$. This is an element of

$$H^0(PT - \{A.Z = 0\}; \mathcal{O}(-1)) \tag{1}$$

which we shall abbreviate to

$$H^0(PT\text{-}A; \mathcal{O}(-1)). \tag{2}$$

For the second case we cover PT - {A.Z = B.Z = 0} (which we write as PT - AB) by PT - A and PT - B to obtain an element of

$$H^1(PT - AB; \mathcal{O}(-2)). \tag{3}$$

Similarly, the third and fourth cases in Figure A are elements of

$$H^2(PT - ABC; \mathcal{O}(-3)) \tag{4}$$

and

$$H^3(PT - ABCD; \mathcal{O}(-4)) \tag{5}$$

respectively.

The Dot Product

The bigger diagrams in Figure A can be obtained by multiplying smaller ones together. For example the element of $H^1(PT - AB; \mathcal{O}(-2))$ is the product of two elements, one from each of $H^0(PT - A; \mathcal{O}(-1))$ and $H^0(PT - B; \mathcal{O}(-1))$. In order to formalise this we shall need a map

$$H^0(PT - A; \mathcal{O}(-1)) \otimes H^0(PT - B; \mathcal{O}(-1)) \to H^1(PT - AB; \mathcal{O}(-2)) \tag{6}$$

and more generally we shall need a map

$$H^p(V;\mathcal{O}(-m)) \otimes H^q(W,\mathcal{O}(-n)) \to H^{p+q+1}(V \cup W;\mathcal{O}(-m-n)) \tag{7}$$

which we call the dot product. Given covers $\{V_i\}$ for V and $\{W_i\}$ for W we can cover $V \cup W$ with the disjoint union of $\{V_i\}$ and $\{W_i\}$. Using these covers, consider a p-cochain $f_{i_o \dots i_p}$ on V and a q-cochain $g_{i_{p+1} \dots i_{p+q+1}}$ on W. We define a p+q+1 - cochain on $V \cup W$ as follows. Let U be a p+q+2-fold intersection from the cover for $V \cup W$. If σ is a permutation of the indices $i_o \dots i_{p+q+1}$ and if $U = V_{\sigma_{i_o} \dots \sigma_{i_p}} \cap W_{\sigma_{i_{p+1}} \dots \sigma_{i_{p+q+1}}}$ define sign (σ) $f_{i_o \dots i_p} g_{i_{p+1} \dots i_{p+q+1}}$ on U. If on the other hand U is not the intersection of p+1 sets from $\{V_i\}$ with q+1 sets from $\{W_i\}$ choose 0 on U. This map on cochains induces the dot product map on cohomology. Mike Eastwood has shown that the dot product is the same as taking the cup product of the two cohomology classes and then looking at the image under the Mayer-Vietoris coboundary map. In other words the following diagram commutes:

$$\begin{array}{ccc} \Gamma(V_0\cap\ldots\cap V_p;\mathcal{O}(-m))\times\Gamma(W_{p+1}\cap\ldots\cap W_{p+q+1};\mathcal{O}(-n)) & \xrightarrow{\text{multiply}} & \Gamma(V_0\cap\ldots\cap W_{p+q+1};\mathcal{O}(-m-n)) \\ \downarrow \check{\text{C}}\text{ech} & & \downarrow \check{\text{C}}\text{ech} \\ H^p(V;\mathcal{O}(-m))\times H^q(W;\mathcal{O}(-n)) \xrightarrow{\text{cup product}} H^{p+q}(V\cap W;\mathcal{O}(-m-n)) & \xrightarrow{\text{Mayer-Vietoris coboundary}} & H^{p+q+1}(V\cup W;\mathcal{O}(-m-n)) \end{array}$$

Here we have assumed for convenience that $\{V_i\}$ has p+1 sets and $\{W_i\}$ has q+1 sets).

Evaluation of the one-vertex diagram

This diagram is drawn in Figure A(iv) and represents an element of

$$H^3(PT - ABCD;\mathcal{O}(-4)) = H^3(PT;\mathcal{O}(-4)) \tag{8}$$

$$\cong H^0(PT;\mathcal{O}(0))^* \tag{9}$$

$$\cong C \tag{10}$$

from Serre duality.

Evaluation of the two-vertex diagram (Figure B)

This is the first diagram having an internal line and is consequently more complicated. The twigs (external lines) on this diagram have been interpreted as elements of $H^2(PT\text{-}ABC;\mathcal{O}(-3))$ and $H^2(PT^*\text{-}DEF;\mathcal{O}(-3))$. We are looking for a method of combining these elements and obtaining a complex number. This

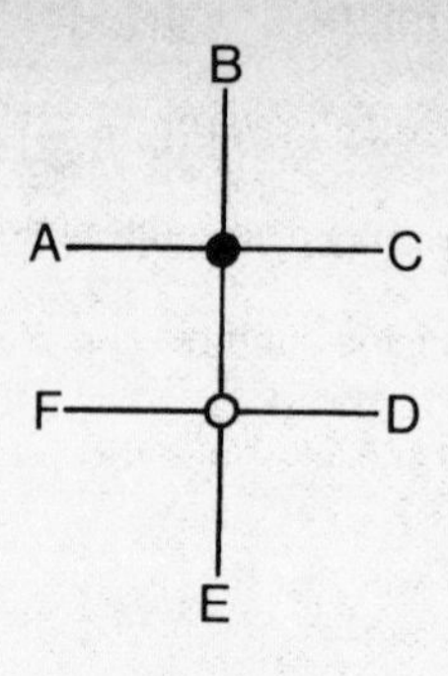

Figure B

method must correspond to integrating the diagram in the ordinary way, using the (unique) contour which separates all the twigs and therefore respects the H^2-ness of the external lines.

We consider the bracket factor, which in this case is the function $(W.Z)^{-1}$, as an element of $H^0(PT \times PT^* - \{W.Z = 0\}; \mathcal{O}(-1,-1))$. We extend the element of $H^2(PT\text{-}ABC; \mathcal{O}(-3))$ to an element of $H^2(\{PT\text{-}ABC\} \times PT^*; \mathcal{O}(-3,0))$ by making it constant in the PT* variables, and we use the dot product followed by a restriction:

$$H^0(PT \times PT^* - \{W.Z=0\};\mathcal{O}(-1,-1)) \otimes H^2(\{PT\text{-}ABC\} \times PT^*; \mathcal{O}(-3,0))$$

$$\rightarrow H^3(PT \times \{PT^* - \{\underset{\sqcup\!\sqcup\!\sqcup}{WABC} = 0\}\}; \mathcal{O}(-4,-1)) \tag{11}$$

$$\cong H^3(PT; \mathcal{O}(-4)) \otimes H^0(PT^* - \{\underset{\sqcup\!\sqcup\!\sqcup}{WABC} = 0\}; \mathcal{O}(-1)) \tag{12}$$

from the Künneth formula

$$\cong C \otimes_C H^0(PT^* - \{\underset{\sqcup\!\sqcup\!\sqcup}{WABC} = 0\}; \mathcal{O}(-1)) \tag{13}$$

from Serre duality

$$\cong H^0(PT^* - \{\underset{\sqcup\!\sqcup\!\sqcup}{WABC} = 0\}; \mathcal{O}(-1)). \tag{14}$$

Now we use the dot product again:

$$H^0(PT^* - \{\underset{\sqcup\sqcup\sqcup}{WABC} = 0\};\ \mathcal{O}(-1)) \otimes H^2(PT^* - DEF;\ \mathcal{O}(-3))$$

$$\to H^3(\{PT^* - \{\underset{\sqcup\sqcup\sqcup}{WABC} = 0\}\} \cup \{PT^* - DEF\};\ \mathcal{O}(-4)) \quad (15)$$

$$\cong H^3(PT^*;\ \mathcal{O}(-4)) \quad (16)$$

if $\overset{ABC}{\underset{DEF}{+\!\!+\!\!+}} \neq 0$

$$\cong C \quad (17)$$

from Serre duality.

The Evaluation of any Tree Diagram

Given any tree diagram we can cut off all its twigs. The remaining diagram is still a tree, and therefore has a vertex (labelled Z say) which is only connected to one other vertex. Having chosen Z we replace all the twigs and obtain Figure C. This is similar to Figure B so far as Z is concerned and

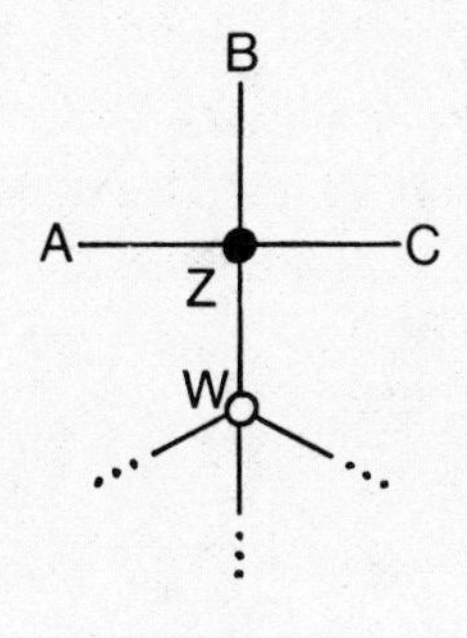

Figure C

we can analyse it in the same way until we get the element of $H^0(PT^* - \{\underset{\sqcup\sqcup\sqcup}{WABC} = 0\};\ \mathcal{O}(-1))$ in (14). This procedure is a twistor transform for tree diagrams - it corresponds to integrating out the Z variable. It has reduced the number of vertices in the tree diagram by one so that we can use an induction argument to finish the evaluation.

Conclusion

A couple of points need to be made. Firstly it is possible to generalise the domains on which our cohomology classes are defined. For example we could replace $H^1(PT\text{-}AB;\mathcal{O}(-2))$ by $H^1(X_2;\mathcal{O}(-2))$ where X_k represents the complement in PT of a k-codimensional subvariety. (Notice that in this notation the general position requirement can be written $X_r \cup X_s = X_{r+s}$.) All the twigs can be generalised in this way and the evaluation procedure will still work. These domains will have to be generalised further, however, when we come to investigate non-tree diagrams. This brings us to the second point, which is that the interpretation of the bracket factor has to be changed a little when we consider non-tree diagrams such as the scalar product and the box.

§5.13 A COHOMOLOGICAL SCALAR PRODUCT CONSTRUCTION *by M.L. Ginsberg*

We first examine the scalar product of two spin-0 fields, for which the relevant twistor diagram (in the case of elementary states) is shown in Figure A.

Figure A Figure B

The value of this diagram is $1/A^{[\alpha}{}_{B}{}^{\beta]}C_{[\alpha}{}^{D}{}_{\beta]}$. The evaluation, as described in Penrose and MacCallum 1972, proceeds in two stages: first the S^1's surrounding the external twigs are integrated around; second, the S^2 integration around the remaining pole in W.Z is performed.

Integration of the external twigs is easy: we simply use the dot product $f\cdot g = \partial^*(f\cup g)$, where ∂^* is the Mayer-Vietoris coboundary, to give us a map

$$H^1(X_2^{AB},\mathcal{O}(-2)) \otimes H^1(X_2^{*CD},\mathcal{O}(-2)) \to H^1(Y_2^{AB},\mathcal{O}(-2,0)) \otimes H^1(Y_2^{CD},\mathcal{O}(0,-2)) \dashrightarrow H^3(Y_4,\mathcal{O}(-2,-2)) \quad [Y = PT \times PT^*]. \tag{1}$$

In order to complete the integration procedure, we see by analogy with the cohomological evaluation of the tree in Figure B that we must interpret $1/(W.Z)^2$ as an $H^2(U;\mathcal{O}(-2,-2))$, where U is some neighbourhood of the complement of Y_4. In fact, since we are dealing with sheaves of germs of functions (and the sections over some closed set A are simply sections over some neighbourhood of A), it will suffice to interpret $1/(W.Z)^2$ as an $H^2(S;\mathcal{O}(-2,-2))$, where $S \equiv Y - Y_4$.

It seems as if we have reduced our original integral to the spinor integral

$$\frac{1}{2\pi i}\oint_{S^2} \frac{\Delta\nu \wedge \Delta\mu}{(\mu.\nu)^2} = 1. \tag{2}$$

In fact, we have not quite done this. Although it is true that S is easily seen to be the product of a CP^1 in X and a CP^1 in X*, the sheaf $\mathcal{O}(-2,-2)$ is the sheaf of germs of homogeneous *twistor* functions, and is not the same as the sheaf $\mathcal{O}_S(-2,-2)$ of germs of homogeneous functions of the spinor variables describing S. Were we to replace the sheaf $\mathcal{O}(-2,-2)$ by $\mathcal{O}(-2,-2)$, we would indeed need do no more than evaluate (2) cohomologically.

The cohomological evaluation of (2) is, in fact, extremely straightforward, for we have an isomorphism

$$H^0(S;\mathcal{O}_S(0,0)) \to C,$$

which is to say, a preferred element of $H^{0*}(S;\mathcal{O}_S(0,0))$. However, by Serre duality,

$$H^{0*}(S;\mathcal{O}_S(0,0)) \cong H^2(S;\mathcal{O}_S(-2,-2)),$$

and we can evaluate (2) using the preferred $H^2(S;\mathcal{O}_S(-2,-2))$ as a cohomological analog to $1/(\mu.\nu)^2$.

To see what to use for our $H^2(S;\mathcal{O}(-2,-2))$, suppose that we have an $H^2(U,\mathcal{O}(-2,-2))$ for some neighbourhood U of S, and also that $L_x \times L_y \subset U \subset PT \times PT^*$ for some x and y. If we have

$$L_x = (ix^{AA'}\mu_{A'}\ \mu_{A'})$$

$$L_y = (\rho_A,\ -iy^{AA'}\rho_A),$$

the integration occurring in the scalar product is

$$\frac{1}{2\pi i}\oint_{S^2} \frac{\Delta\mu \wedge \Delta\rho}{[i\rho_A\mu_{A'}(x^{AA'}-y^{AA'})]^2} . \tag{3}$$

Setting $\nu^{A'} = (x^{AA'}-y^{AA'})\rho_A$, this is easily seen to be

$$\frac{1}{(x-y)^2} . \tag{4}$$

What this means is that we are looking for some $f \in H^2(U;\mathcal{O}(-2,-2))$ for some neighbourhood U of S such that the restriction of f to any quadric of the form $L_x \times L_y$ in U is $1/(x-y)^2$ times the canonical $H^2(L_x \times L_y;\mathcal{O}(-2,-2))$. In other words, the result of contour integrating the restriction of f to $L_x \times L_y$ must be $1/(x-y)^2$.

To see how to do this, suppose $S \subset PT^- \times PT^{*-}$, set $U = PT^- \times PT^{*-}$, and let $f \in H^2(U;\mathcal{O}(-2,-2))$ be arbitrary. Now for $L_x \subset PT^-$ (i.e., x in the past tube) and $L_y \subset PT^{*-}$ (i.e., y in the future tube), let $\phi_f(x,y)$ be the result of contour integrating the restriction of f to $L_x \times L_y$. We now expect from the normal correspondence between elements of cohomology groups and zero rest mass fields that $\phi_f(x,y)$ will satisfy the zero rest mass field equations when considered as a function of x or of y.

To show this, we note that since $H^0(PT^-;\mathcal{O}(-2)) \simeq H^0(PT^{*-};\mathcal{O}(-2)) \simeq 0$, the Künneth formula gives us

$$\begin{aligned} H^2(PT^- \times PT^{*-};\mathcal{O}(-2,-2) &\simeq H^1(PT^-;\mathcal{O}(-2)) \otimes H^1(PT^{*-},\mathcal{O}(-2)) \\ &\simeq Z_0(CM^-) \otimes Z_0(CM^+), \end{aligned}$$

where $Z_0(U)$ is the group of spin-0 zero rest mass fields defined on $U \subset CM$. The (completed) tensor product in the last term allows us to build up an arbitrary zero rest mass field $\phi(x,y)$ as a linear combination of terms of the form $\phi_1(x)\phi_2(y)$, so we actually have

$$H^2(PT^- \times PT^{*-};\mathcal{O}(-2,-2)) \simeq Z_0(CM^- \times CM^+), \tag{5}$$

where the elements of $Z_0(CM^- \times CM^+)$ must satisfy the spin-0 zero rest mass field equations in each variable separately.

Since $1/(x-y)^2$ does satisfy the zero rest mass field equations, if therefore corresponds to a (preferred) $H^2(PT^- \times PT^{*-};\mathcal{O}(-2,-2))$. Using this, we can complete the evaluation of (1) by

$$\begin{aligned} &H^1(X_2^{AB};\mathcal{O}(-2))\otimes H^1(X_2^{*CD};\mathcal{O}(-2))\to H^3(Y_4;\mathcal{O}(-2,-2))\otimes H^2(PT^-\times PT^*;\mathcal{O}(-2,-2)) \\ &\to H^6(Y;\mathcal{O}(-4,-4)) \simeq C. \end{aligned} \tag{6}$$

In fact, we can do more than this. Since an element of $H^1(\overline{PT^+};\mathcal{O}(-2))$ is, by definition, an element of $H^1(U^+;\mathcal{O}(-2))$ for some neighbourhood U^+ of $\overline{PT^+}$, we actually have a map

$$\begin{aligned} &H^1(\overline{PT^+};\mathcal{O}(-2)) \otimes H^1(\overline{PT^{*+}};\mathcal{O}(-2))\to \\ &H^3(\overline{PT^+} \times PT^* \cup PT \times \overline{PT^{*+}};\mathcal{O}(-2,-2)) \otimes H^2(PT^- \times PT^{*-};\mathcal{O}(-2,-2)) \\ &\to H^6(Y;\mathcal{O}(-4,-4)) \simeq C, \end{aligned} \tag{7}$$

which evaluates the scalar product in a considerably more general situation.

Other spins are similar. For example, the key to the evaluation of the digram shown in Figure C below is to find an element of $H^2(PT^- \times PT^*;\mathcal{O}(-3,-3))$ which corresponds to $1/(W.Z)^3$. In the spin-0 case, we used the fact that $H^2(S;\mathcal{O}_S(-2,-2)) \simeq C$ to get a complex number associated with each $L_x \times L_y$ in $PT^- \times PT^{*-}$. Here, $H^2(S;\mathcal{O}_S(-3,-3)) \simeq H^0{*}(S;\mathcal{O}_S(1,1))$ so, for each $L_x \times L_y$ in $PT^- \times PT^{*-}$, we expect to get an element of the dual space of functions of one spinor $\mu_{A'}$ and one spinor ρ_A - in other words, some spinor $\phi_{AA'}$, which we will then interpret as a spinor field $\phi_{AA'}(x,y)$ as before.

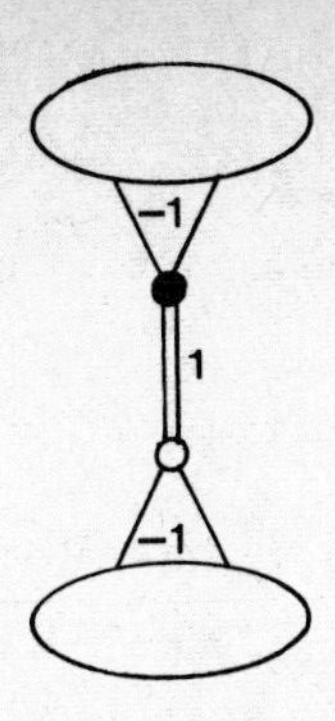

Figure C

Restricting $1(W.Z)^3$ to $L_x \times L_y$ as before to get $1/(i\rho_A\mu_{A'}z^{AA'})^3$ where $z^a = x^a - y^a$, we see that we must now evaluate

$$\frac{1}{2\pi i} \oint_{S^2} \frac{\rho_A\mu_{A'}\ \Delta\mu\wedge\Delta\rho}{(i\rho_A\mu_{A'}z^{AA'})^3} . \tag{8}$$

Substituting $\nu^{A'} = \rho_A z^{AA'}$ and integrating by parts, it is not hard to see that the value of (8) is in fact

$$\phi_{AA'}(x,y) = -\frac{i}{2}\ \frac{(x-y)_{AA'}}{(x-y)^4} . \tag{9}$$

As in the spin-0 case, this will correspond to an element of $H^2(PT^- \times PT^{*-};\ \mathcal{O}(-3,-3))$ precisely if $\phi_{AA'}$ satisfies the zero rest mass field equations

$$\frac{\partial}{\partial x_{AA'}}\ \phi_{BA'} = 0,\quad \frac{\partial}{\partial y_{AA'}}\ \phi_{AB'} = 0, \tag{10}$$

where we have contracted the indices in the above fashion because the isomorphism we are dealing with is

$$H^2(PT^- \times PT^{*-};\mathcal{O}(-3,-3)) \simeq Z_1(CM^-) \otimes Z_{1*}(CM^+)$$

where $Z_{1*}(CM^+)$ is the space of *unprimed* spin-½ zero rest mass fields on CM^+.

As it is clear that the field $\phi_{AA'}(x,y)$ in (9) does indeed satisfy (10), we can evaluate the diagram in Figure C.

In an analogous fashion, $1/(W.Z)^{n+2}$ is easily seen to correspond to the spinor field

$$\phi_{AA'\ldots BB'}(x,y) = \frac{(-i)^{n+1}}{n+1} \frac{\overbrace{(x-y)_{(A(A'} \cdots (x-y)_{B)B')}}^{n \text{ times}}}{(x-y)^{2n+2}}$$

(where the symmetrization is over the primed and unprimed indices separately). Incorporating the constants appearing in the definition of the bracket factor and in the scalar product formulae, we define the *twistor propagator* for spin n/2 to be

$$\phi_{AA'\ldots BB'}(x,y) = \Gamma(n+1) \frac{(x-y)_{(A(A'} \cdots (x-y)_{B)B')}}{(x-y)^{2n+2}}, \tag{11}$$

corresponding to some $\phi_n \in H^2(PT^- \times PT^{*-};\mathcal{O}(-n-2,-n-2))$. For $n \geq 0$, the scalar product is now given by

$$\begin{aligned} &H^1(\overline{PT^+};\mathcal{O}(n-2)) \otimes H^1(\overline{PT^{*+}};\mathcal{O}(n-2)) \xrightarrow{.\otimes\phi_n} \\ &H^3(\overline{PT^+} \times PT^* \cup PT \times \overline{PT^{*+}};\mathcal{O}(n-2,n-2)) \otimes H^2(PT^- \times PT^{*-};\mathcal{O}(-n-2,-n-2)) \\ &\xrightarrow{\cdot} H^6(PT \times PT^*;\mathcal{O}(-4,-4) \simeq C. \end{aligned} \tag{12}$$

For $n < 0$ in (12), the non-cohomological integration involves an integration with boundary. It would seem, therefore, that we need to cohomologically model integration with boundary, but this is not the case. There is a much simpler construction available.

Specifically, to find a map

$$H^1(\overline{PT^+};\mathcal{O}(-3)) \otimes H^1(\overline{PT^{*+}};\mathcal{O}(-3)) \to C,$$

we need some preferred element ϕ_{-1} of $H^2(PT^- \times PT^{*-};\mathcal{O}(-1,-1))$.

For n non-negative, we constructed ϕ_n simply by constructing a zero rest mass field of the appropriate spin, using the twistor function $(W.Z)_{n+2}$ and the contour suggested by the details of the scalar product integration. We can repeat this procedure here. For n = -1, we replace (8) by

$$\frac{1}{2\pi i} \oint_{S^2} \frac{\partial}{\partial\sigma^{A'}} \frac{\partial}{d\omega^{A}} \frac{1}{(W.Z)} \Delta\mu\wedge\Delta\rho \ , \tag{13}$$

where

$$W_\alpha = (\rho_{A'}\ \sigma^{A'}) \quad \text{and} \quad Z^\alpha = (\omega^A, \pi_{A'}).$$

The integral (13) can be rewritten as

$$\frac{1}{2\pi i} \oint_{S^2} \frac{2\rho_A\ \mu_{A'}}{(W.Z)^3} \Delta\mu\wedge\Delta\rho = \frac{1}{2\pi i} \oint_{S^2} \frac{2\rho_A\ \mu_{A'}}{(i\rho_A\mu_{A'}\ z^{AA'})^3} \Delta\mu\wedge\Delta\rho \ ,$$

namely, (8)! (Except for the factor 2, which was absent in (8) because we were considering $1/(W.Z)^3$ instead of $(W.Z)_3$.)

The twistor propagator ϕ_{-1} thus corresponds to the same field $\phi_{AA'}(x,y)$ as does ϕ_{+1}. This in turn corresponds to an element of $H^2(PT^- \times PT^{*-};\mathcal{O}(-3,-3))$ because $\phi_{AA'}(x,y)$ in (11) also satisfies the zero rest mass field equations

$$\frac{\partial}{\partial x_{AA'}} \phi_{AB'} = 0 \qquad \frac{\partial}{\partial y_{AA'}} \phi_{BA'} = 0.$$

For n < 0, we thus see that the scalar product is still given by (12), where ϕ_n and ϕ_{-n} are always chosen to correspond to the same spacetime field. Another way to see that this does indeed give the correct answer is via the following result:

Suppose that $f \in H^1(\overline{PT^+};\mathcal{O}(n-2))$ and $f' \in H^1(\overline{PT^{*-}};\mathcal{O}(-n-2))$ correspond to the same zero rest mass field, and that $g \in H^1(\overline{PT^{*+}};\mathcal{O}(n-2))$ and

$g' \in H^1(\overline{PT^-};\mathcal{O}(-n-2))$ and also $h \in H^2(PT^- \times PT^{*-};\mathcal{O}(-n-2,-n-2))$ and $h' \in H^2(PT^{*+} \times PT^+;\mathcal{O}(n-2,n-2))$ are related similarly. Then $f.g.h = f'.g'.h'$ as an element of $H^6(PT \times PT^*;\mathcal{O}(-4,-4))$.

Employing Serre duality, it is easy to see that this last statement is simply an iteration of:

Let $f \in H^1(\overline{PT^+};\mathcal{O}(n-2))$, $f' \in H^1(\overline{PT^{*-}};\mathcal{O}(-n-2))$, $g \in H^1(PT^-;\mathcal{O}(-n-2))$ and $g' \in H^1(PT^{*+};\mathcal{O}(n-2))$ be related as in the previous paragraph. Then the images in C of f.g and f'.g' under Serre duality are identical.

This, however, is obvious, since both images are simply the scalar product of the two fields involved.

The reason that we have been able to evaluate diagrams of the form shown in Figure D without performing any cohomological boundary integration is that the boundary integration is sidestepped by the construction of the twistor propagator ϕ. Similar techniques cannot be applied to diagrams such as the one shown in Figure E because the singularity regions associated with these diagrams (complements of X_3's, for example) are degenerate.

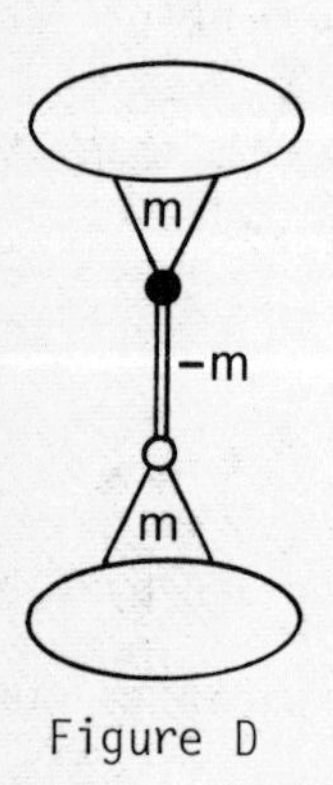

Figure D

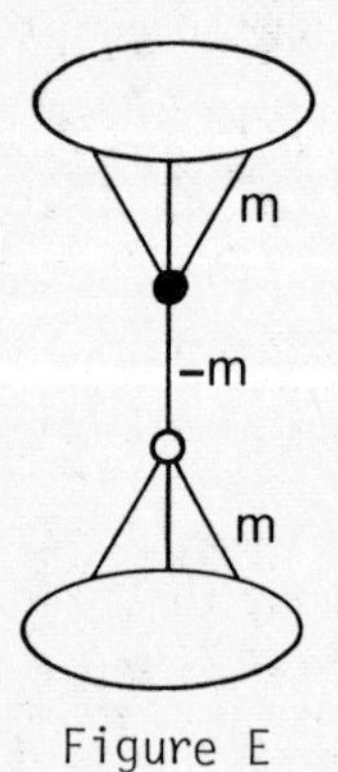

Figure E

REFERENCE

Penrose, R. & MacCallum, M.A.H. 1972. Phys. Reports Vol. 6C, No. 4.

§5.14 COMBINATORIAL QUANTUM THEORY AND QUANTIZED DIRECTIONS *by R. Penrose*

According to conventional quantum theory, angular momentum can take only integral values (measured in units of $1/2\,\hbar$) and the (probabilistic) rules for combining angular moments are of a combinatorial nature. Also, according to quantum theory, a system with zero total angular momentum must be spherically symmetrical. A system must, thus, involve a relatively large angular momentum in order to determine a well-defined direction in space, and we may picture the axis of the angular momentum as giving such a direction. In the limit of large angular momenta, we may therefore expect that the quantum rules for angular momentum will determine the geometry of directions in space. We may imagine these directions to be determined by a number of spinning bodies. The angles between their axes can then be defined in terms of the probabilities that their total angular momenta (i.e. their "spins") will be increased or decreased when, say, an electron is thrown from one body to another. In this way the geometry of directions may be built up, and the problem is then to see whether the geometry so obtained agrees with what we know of the geometry of space and time.

Although there is a standard procedure for the treatment of such a problem (in the non-relativistic case: namely the use of 3-j and 6-j symbols, etc.), it is convenient to make use of an alternative (but equivalent) formalism which can be described very briefly as follows. Consider, first, an n-dimensional Kronecker delta δ_{ab}. Then a set of "isotropic" Cartesian tensors can be built up from this one symbol and scalars by means of the operations of addition, outer multiplication, transposition of indices, and contraction. The only identities satisfied by δ_{ab} which hold independently of n are $\delta_{ab} = \delta_{ba}$ and $\delta_{ab}\,\delta_{bc} = \delta_{ac}$ (summation convention assumed). Also, we have $\delta_{aa} = n$ (> 0) and a certain identity, which depends on n, holds. Now consider an abstract structure which behaves formally exactly like the above set of tensors except that the formal equation $\delta_{aa} = -2$ is imposed, together with the one identity: $\delta_{ab}\,\delta_{cd} + \delta_{ac}\,\delta_{db} + \delta_{ad}\,\delta_{bc} = 0$. The "scalars" of this abstract structure will be taken to be the *rational* numbers. The tensor-like quantities which can be constructed in this way I call *binors*. It can then be shown that a necessary and sufficient condition for a binor $A_{ab\ldots d}$ to vanish is that its *norm* $\|A \ldots\| = A_{ab..d}A_{ab..d}$ (a rational number) should vanish.

To obtain the physical interpretation of the binors, I envisage the following type of situation. Imagine the universe to be represented as a

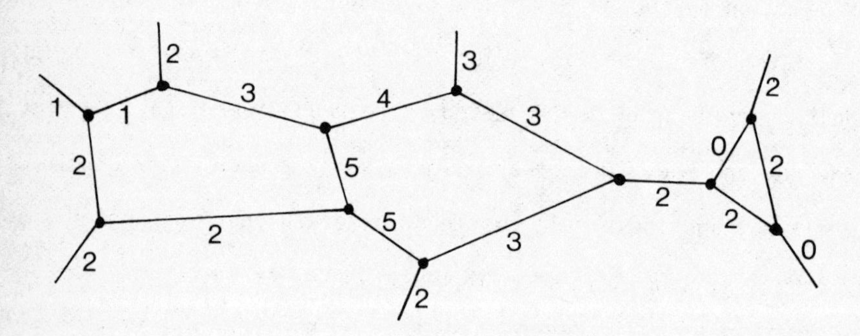

Figure A

network of segments (see Fig. A) where, for simplicity, it will be assumed that each internal segment connects two vertices and each vertex joins just three segment ends. Associated with each segment is a non-negative integer. Each segment is to be thought of as representing the world line of a particle, nucleus, atom, etc. Or, generally, some structure which may be momentarily considered as stationary and isolated from the rest of the universe. The network thus gives a kind of combinatorial space-time picture of the universe. The integer associated with each segment is twice the total intrinsic angular momentum quantum number of the particle (or structure). Hence, for a pi-meson or ground state helium atom this integer is zero; for an electron or proton it is one; for a deuteron it is two; and so on. In order for the proposed calculus to give an accurate description of a part of the universe, it must be supposed that it is possible to neglect effects due to relative velocities of the particles (or structures). In this sense the theory is a non-relativistic one.

Corresponding to each possible such (open) network is associated a binor which is a contracted product of binors described as follows. Each segment numbered 0 is represented by the scalar 1. Each segment numbered 1 is represented by a δ_{ab}. Segments marked 2 by $\delta_{ac,bd} = 1/2!\,(\delta_{ab}\,\delta_{cd} - \delta_{ad}\,\delta_{cb})$, marked 3 by

$$\delta_{abc,\,def} = 1/3! \begin{vmatrix} \delta_{ad} & \delta_{ae} & \delta_{af} \\ \delta_{bd} & \delta_{be} & \delta_{bf} \\ \delta_{cd} & \delta_{ce} & \delta_{cf} \end{vmatrix}$$

and so on. The first group of indices of $\delta_{ab..d,ef..h}$ is to be associated with one end of the segment and the second group with the other end. (This is symmetrical since $\delta_{ab} = \delta_{ba}$ implies $\delta_{ab..d,ef..h} = \delta_{ef..h,ab..d}$.) At each vertex the three relevant groups of indices involved must all be paired off and contracted so that none of these indices remains uncontracted, and no two belonging to the same $\delta_{...,...}$ are contracted together. This implies that the sum of the three integers involved is even, and is at least twice the greatest of them. Then, the result of these contractions will be unique up to sign. The free indices of the resultant binor are then just those corresponding to the free ends of the network.

Consider, now, a situation in which a portion of the universe is represented by a known network X (see Fig. B) having, among its free ends, one numbered m and one numbered n. Suppose the particles or structures represented by these two ends combine together to form a new structure. (See Fig. C). We wish to know, for any allowable p, what is the probability

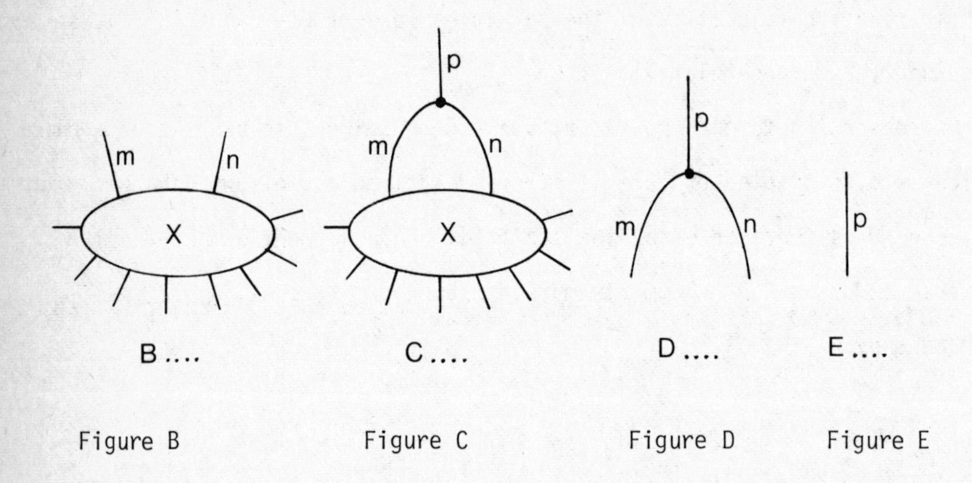

Figure B Figure C Figure D Figure E

that the angular momentum number of this new structure be p. Fig. D is the network representing the combining of these two structures to form the third, but ignoring the rest of the universe; Fig. E is the "network" for the final structure alone. Let B..., C..., D..., E..., be the binors representing the networks of Figs. B, C, D and E, respectively. Then,

$$\textit{required probability} = \frac{\| C\ldots \|}{\| B\ldots \|} \times \frac{\| E\ldots \|}{\| D\ldots \|}$$

(in fact $\|E\ldots\| = (-1)^p(p + 1)$ and $\|D\ldots\| = (-1)^{(m+n+p)/2}(\frac{m+n-p}{2})!$ $(\frac{n+p-m}{2})!\ (\frac{p+m-n}{2})!\ (\frac{m+n+p+2}{2})!/m!n!p!)$. As a particular case of this we can deduce the result that the binor corresponding to a network vanishes if and only if the situation is "forbidden" according to the rules of quantum theory.

Consider now the situation of Fig. F involving two bodies with *large* angular momenta M and N. We might define the *angle* between the axes of the

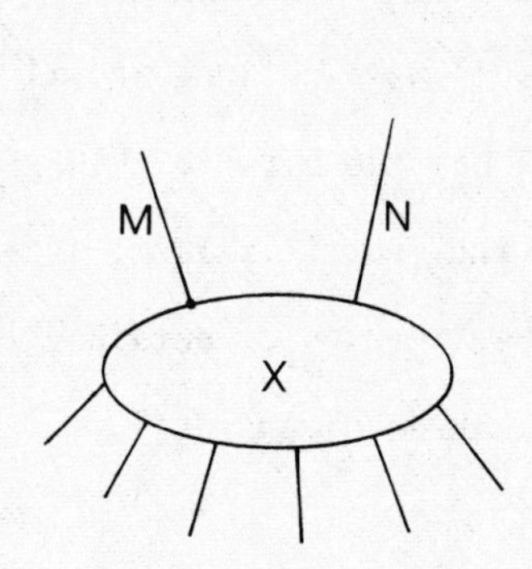

Figure F

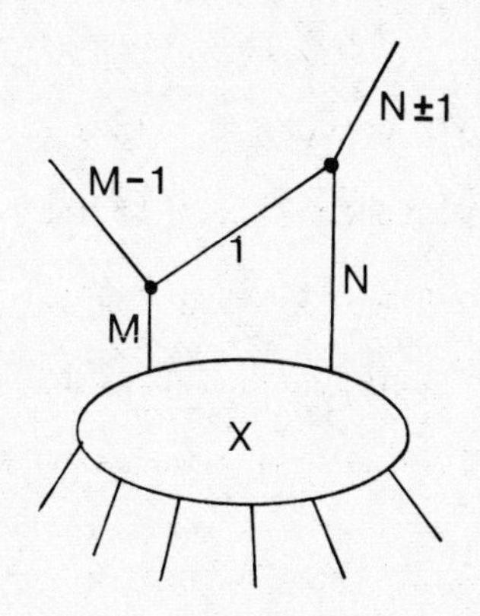

Figure G

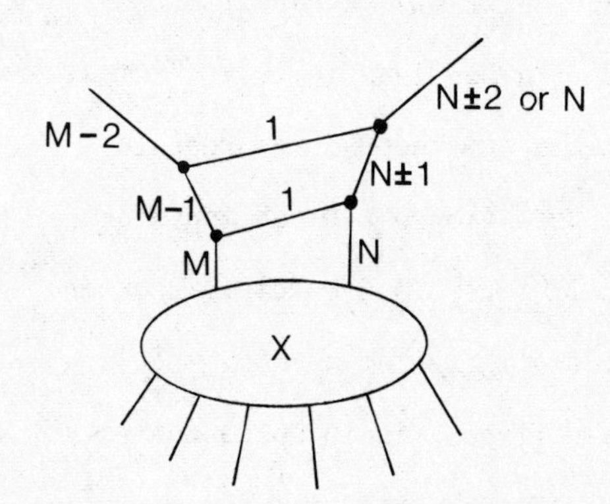

Figure H

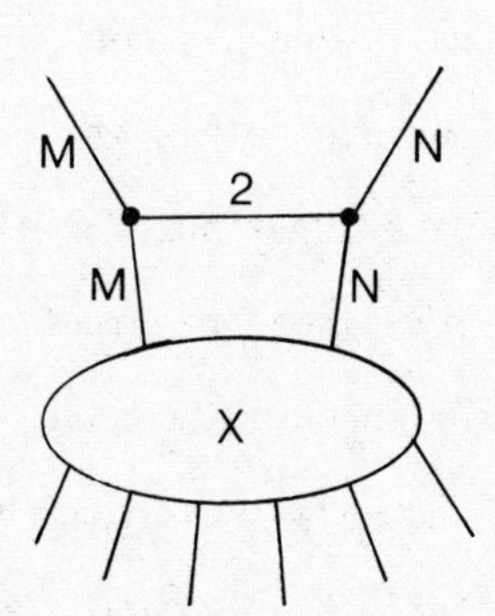

Figure I

bodies in terms of the relative probability (defined in terms of an ensemble of systems with the same Fig. F network) of occurrence of N + 1 and N - 1, respectively, in the "experiment" given by Fig. G. However, part of this probability may be due to *ignorance* of the relationship between the bodies. (This may manifest itself in the absence of sufficient connecting links in the known network X.) To eliminate the possibility that part of the probability be due to ignorance we envisage a repetition of the "experiment" as given in Fig. H. If the probability given in the second experiment is essentially unaffected by the result of the first experiment then we may say that the angle θ between the axes of the bodies is well-defined and is determined by this probability. It then turns out that the binor of Fig. I is essentially 1/2 cos θ times the binor of Fig. F and from this fact, and certain binor identities, it is possible to show that the angles obtained in this way satisfy the same laws as do angles in a three-dimensional Euclidean space.

This is very satisfactory and is perhaps a little surprising in at least two respects. In the first instance, since no complex numbers were used in the binor calculus it is somewhat remarkable that a full array of directions in three-dimensional space has been built up, rather than in, say, just a two-dimensional subspace, since in ordinary quantum theory, in order to build up all the wave functions for all the possible spin directions for an electron, *complex* linear combinations must be used. Secondly, according to standard quantum theory, the wave function of a system of high angular momentum will *not* normally determine a well-defined axis in space contrary to what has apparently been assumed here.

The answer to both these points seems to lie in the fact that the "directions" that emerge in the theory described here are things which are defined by the systems in relation to one another and they will not generally agree with the directions in a *previously* given (and unnecessary!) background space. The space that is obtained here is to be thought of (indeed *must* be thought of) as being the one *determined* by the systems themselves.

It is to be hoped that some modification to the above scheme might enable the effects due to relative velocities of systems to be taken into account so that perhaps a four-dimensional space-time might be constructed. (Time is absent from the above theory even to the extent that the time ordering of events is irrelevant!) Two additional features would have to be involved. The first is that the relativistic addition of angular momenta includes the possibility of multiple pair creation and many of the additional complications implied by relativistic field theory. The second is that relative velocity implies the possibility of a mixing of spin with orbital angular momentum so that the idea of "distance" between the world lines or particles is involved. Particularly because of this second feature, the fully relativistic theory would seem to be of a different order of difficulty from the one treated above.

REFERENCES

Penrose, R. 1971a. In *Combinatorial Mathematics and its Applications*, ed. D.J.A. Welsh. Academic Press.

Penrose, R. 1971b. In *Quantum Theory and Beyond*, ed. T. Bastin. Cambridge University Press.

§5.15 THE CHROMATIC EVALUATION OF STRAND NETWORKS *by J.P. Moussouris*

This note describes Penrose's "chromatic" method for evaluating strand networks, which is mentioned in his papers on spin networks [Penrose 1971a,b & 1972], but has hitherto remained unpublished. We consider the example shown in Figure A; in §5.16 it will be shown that the evaluation of this network in fact amounts to a calculation of the familiar Clebsch-Gordan coefficients.

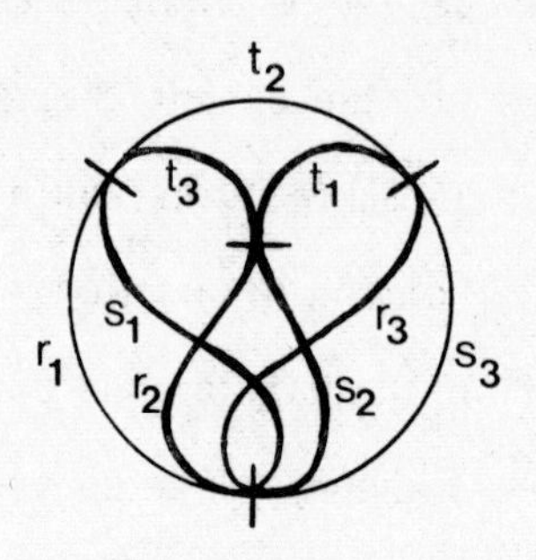

Figure A

Recall that each edge labelled by an integer e represents a bundle of e strands, and each bar b signifies the alternating sum of all cross-connections (subtracting for odd parity). Since there are no open edges, the network becomes a polynomial P in closed loops when all the bars are expanded. The *value* of the network is defined as P(-2), divided by $\Pi e!$, where the product is taken over all edges.

A polynomial is completely determined by its behavior at integers $N \geq 0$. Moreover when closed loops are given positive values N, P(N) can be interpreted as a scalar in an ordinary N-dimensional Cartesian space: each bar in the strand network is regarded as a generalized delta

$$\delta^{j_1 j_2 \cdots j_r}_{i_1 i_2 \cdots i_r} \text{ (} = \pm 1 \text{ for } i_1 \cdots i_r \text{ an } {\text{even} \atop \text{odd}} \text{ permutation of } j_1 \cdots j_r, 0 \text{ otherwise)}$$

and each edge as a bundle of indices to be contracted. Since

$$\delta^{j_1 j_2 \ldots j_r}_{i_1 i_2 \ldots i_r} = r!\, \delta^{[j_1}_{i_1} \delta^{j_2}_{i_2} \ldots \delta^{j_r]}_{i_r} \text{ and } \delta_i^{\ j} \delta_j^{\ k} = \delta_i^{\ k}$$

the contractions reduce to a polynomial in the closed loop $\delta_i^{\ i} = N$, precisely $P(N)$ as defined above. The chromatic method relates $P(N)$ for $N \geq 0$ to the number of N-colourings of a set of graphs.

Expanding in components, we regard the index values from 1 to N as a set of N distinct colours. Then each non-zero term in the contraction of generalized δ's corresponds to an N-colouring of the strands such that for each strand entering a bar, there is exactly one strand of the same colour leaving the opposite side of the bar. Hence we may follow a strand through consecutive bars until it closes into a *cycle* - that is, a sequence $e_1 b_1 e_2 b_2 \ldots e_m b_m e_{m+1} = e_1$ of edges and bars such that e_i and e_{i+1} enter opposite sides of b_i and no two bars are the same. Cycles that have a bar in common (*incident* cycles) must have distinct colours. For example, the network in Fig. A has only six kinds of cycle (shown in Fig. B), all of which are incident (since the bottom bar is common).

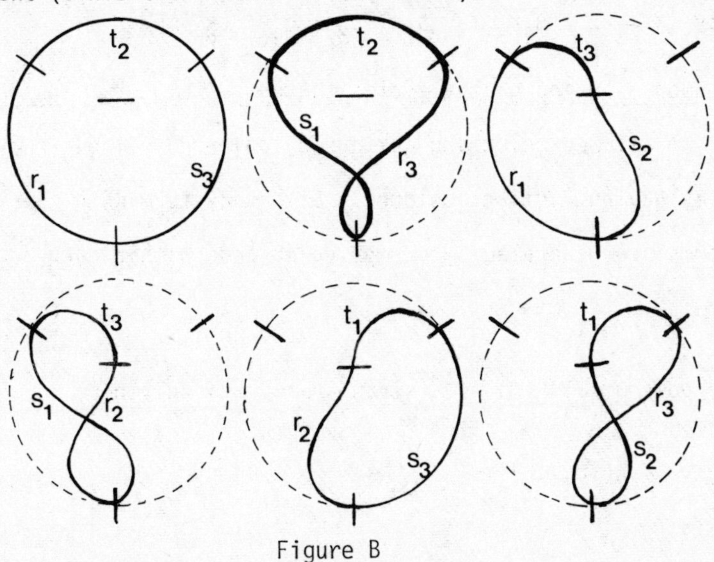

Figure B

Now in each ±1 term in the δ-contraction we can count the number of cycles of each type, say c_i of the ith cycle. We notice that:

(1) All terms with cycle numbers c_i have the same sign. This sign is the parity of the number of crossings that occur at bars. But the total number of crossings between different cycles is even: they always appear and disappear in *pairs* as we move cycles apart. For simplicity, invoke the convention that the individual strands within each edge bundle are non-intersecting. Then the total parity of bar-crossings must equal the parity of "spurious" intersections of different cycles which occur away from the bars (e.g. in Fig. B cycles 2 and 3 have a spurious intersection where edge r_3 crosses s_2). But the number of spurious intersections is simply

$$\sum_{e_i \cap e_j} e_i e_j - \sum_{\text{self-intersecting cycles}} c_k^2 ,$$

where the first term involves products of strand numbers of spuriously intersecting edges, and the second eliminates self-intersecting cycles. Hence the overall parity of the term is completely determined by $\{c_n\}$. [In our example, the parity is $s_1 r_2 + s_1 r_3 + r_3 s_2 + c_2 + c_4 + c_6$ (mod 2).]

(2) For each allowed colouring of the cycles the number of terms in the sum is precisely $\Pi e_j!$ An allowed colouring assigns c_i colours to the ith cycles, with incident cycles having distinct colours. Given any term with this colouring, $\Pi e_j!$ terms are generated by permuting strands within each edge. We have proven:

Theorem: The loop polynomial P(N) of a strand network is given by

$$\frac{P(N)}{\Pi e_j!} = \sigma \sum_{\{c_i\}} \epsilon \, K(\{c_i\},N)$$

where $\sigma = (-1)^{\sum_{e_i \cap e_j} e_i . e_j}$, $\epsilon = (-1)^{\sum C_k \text{ self-intersecting}}$,

$K(\{c_i\},N)$ is the number of allowed N-colourings of the $\{c_i\}$ cycles, and the sum is over all non-negative cycle numbers $\{c_i\}$ such that for each edge

$$e_i = \sum_{\text{cycles through } e_i} C_j .$$

From this formula we obtain the value of the strand network by simply setting $N = -2$. An immediate corollary is that this value is an integer, since any polynomial integer-valued at positive integers is integer-valued at negative integers as well.(Express $P(N)$ in the form $\sum_{k=0}^{m} b_k\binom{N}{k}$, where $b_0=P(0),\ldots,b_k = P(k) - \sum_{j=0}^{k-1} b_j\binom{k}{j}\ldots$ are all integers, as are $\binom{-N}{k} = \frac{(-N)(-N-1)\ldots(-N-k+1)}{k!} = (-1)^k\binom{N+k-1}{k}$, so $P(-N)$ is integral also.)

In Fig. A there are 9 edges and 4 bars (across which sums of strand numbers must be equal), leaving 5 independent parameters. Hence we expect that if one cycle number is fixed the other 5 will be determined. Indeed if we set $c_1 = z$, then since $t_2 = c_1 + c_2$ we have $c_2 = t_2 - z$, and similarly for the other cycles $c_3 = r_1 - z$, $c_4 = t_3 - r_1 + z$ $c_5 = s_3 - z$, $c_6 = t_1 - s_3 + z$. Hence in this case the chromatic sum is simply over the range of 2 which gives non-negative values for all c_i. Recall

$$t = (-1)^{c_2+c_4+c_6} = (-1)^{t_1+t_2+t_3-r_1-s_3+z} \text{ and } \sigma = (-1)^{s_1r_2+s_1r_3+r_3s_2} .$$

Finally, since all cycles are incident, $K(\{c_1\},N)$ is simply the number of ways of distributing N objects into six bins:

$$\frac{N!/(N-J)!}{c_1!c_2!c_3!c_4!c_5!c_6!}$$

Here $J = \Sigma c_i = t_1 + t_2 + t_3$ is independent of z, so the numerator $N!/(N-J)!$

factors out of the chromatic sum. Setting $N = -2$, we simply get $(-2)(-3)\ldots(-J-1) = (-1)^J(J+1)!$. Putting all this together we obtain from the value of our strand network

$$(-1)^q(J+1)! \sum_z \frac{(-1)^2}{z!(t_2-z)!(r_1-z)!(t_3-r_1+z)!(s_3-z)!(t_1-s_3+z)}$$

where $q \equiv s_1r_2 + s_1r_3 + r_2s_2 + r_1 + s_3 \pmod 2$.

Although this method can be applied to any spin network, it becomes more complicated to evaluate $K(\{c_i\}, N)$ when not all cycles are incident. Define a graph G_c whose points are the cycle types of the network, with lines joining incident cycles. Then $K(\{c_i\},N)$ is the number of n-colourings of G_c which assigns precisely c_i colours to the ith point, with distinct colours for adjacent points. Equivalently we may define a graph $G\{c_i\}$ obtained from G_c by expanding the ith point into a complete subgraph of c_i points and connecting all points in adjacent subgraphs. Then

$$K(\{c_i\};N) = \frac{C(G\{c_i\},N)}{\Pi c_i!} \quad ,$$

where $C(G\{c_i\},N)$ is the "chromatic polynomial" (cf. Biggs 1974) giving the number of ordinary N-colourings of $G\{c_i\}$.

REFERENCES

Biggs, N. 1974 *Algebraic Graph Theory*, Cambridge University Press.

Penrose, R. 1971a. In *Combinatorial Mathematics and its Applications*, ed. D.J.A. Welsh. Academic Press

Penrose, R. 1971b. In *Quantum Theory and Beyond*, ed. T. Bastin. Cambridge University Press.

Penrose, R. 1972 In *Magic Without Magic*, ed. J. Klauder, Freeman.

§5.16 VECTOR COUPLING COEFFICIENTS AND SPIN NETWORKS *by J.P. Moussouris*

Spin networks provide a graphical method for dealing with Young symmetrizers for the group SU(2) — that is, with operations of combining and splitting angular momenta in non-relativistic quantum mechanics.

Recall that if $|\frac{1}{2},\frac{1}{2}\rangle = u^A$ and $|\frac{1}{2},-\frac{1}{2}\rangle = d^A$ are the "up" and "down" states of a spin ½ system (cf. §4.5), then basis states $|jm\rangle$ for an arbitrary spin j system can be constructed from *symmetrized* products:

$$|jm\rangle = \underbrace{u^{(A}u^{B}\ldots u^{L}}_{r}\ \underbrace{d^{M}d^{N}\ldots d^{P)}}_{s} ,$$

where we define r and s by $j = (r+s)/2$ and $m = (r-s)/2$. The corresponding diagrammatic representation is shown in Figure A.

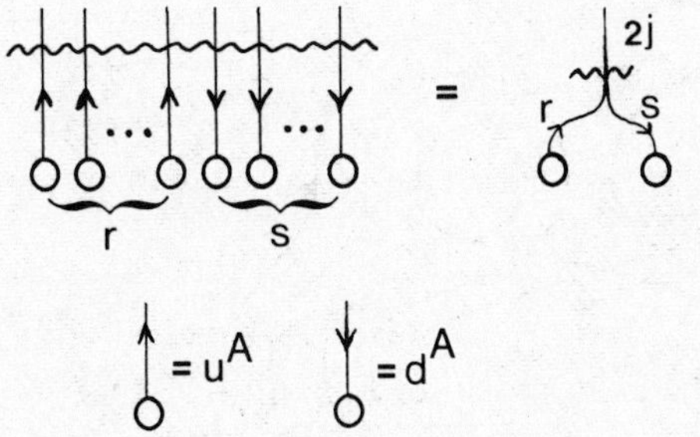

Figure A

In order to compute the vector coupling coefficient or Wigner coefficient $\begin{Bmatrix} j_1 & j_2 & j_3 \\ m_1 & m_2 & m_3 \end{Bmatrix}$ giving the amplitude for three spins to combine to zero, we simply construct the product state $|j_1m_1\rangle\ |j_2m_2\rangle\ |j_3m_3\rangle$, as shown in Figure B and then apply antisymmetrizers (i.e. spinor epsilons) to reduce it to a scalar.

Note that in Figure B we have the following formulae holding:

$$j_i = (r_i + s_i)/2 \quad , \quad m_i = (r_i - s_i)/2$$

$$J = \sum_i j_i = \sum_i r_i = \sum_i s_i,\ 0 = \Sigma m_i .$$

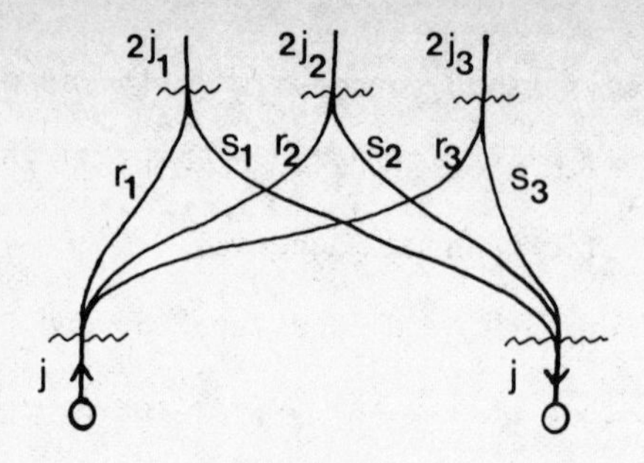

Figure B

If we consider, for example, the vector coupling coefficient

$$\begin{Bmatrix} 3/2 & 2 & 3/2 \\ 1/2 & 0 & -1/2 \end{Bmatrix}$$

then the corresponding diagram is as shown in Figure C:

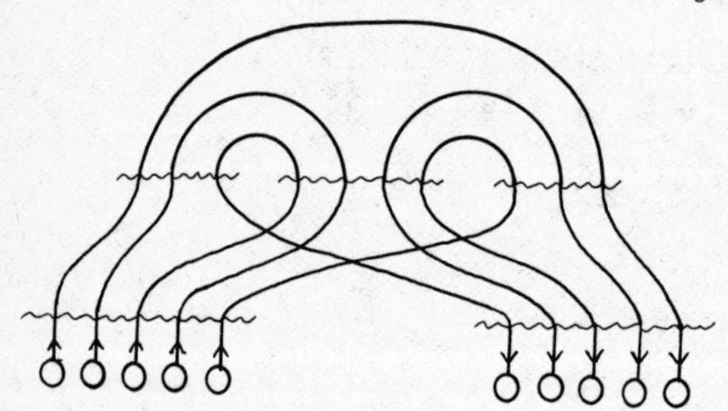

Figure C

We now note that for the general vector coupling coefficient expression we must have the relationship shown in Figure D, since every ε_{AB} term that has an index at the "up" end on the left must have the other at the "down" end (otherwise it would be symmetrized).

Note also that $t_1 = j_2 + j_3 - j_1$, etc., determined by j's. Hence the coupling computation reduces to finding λ, and introducing proper normalization.

First we must examine our graphical method more closely. One motivation for this notation is that it captures identities such as $\delta_A^B \, \varepsilon_{BC} = \varepsilon_{AC}$ in the form indicated in Figure E.

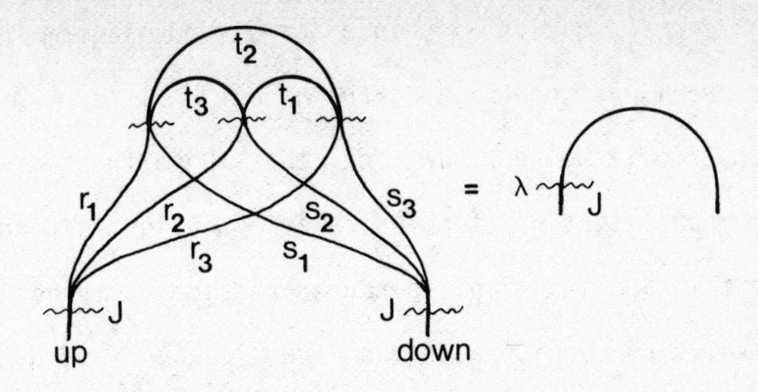

Figure D

Figure E

However, certain sign problems arise if we insist that graphical expressions retain their identity under more general deformations. In particular, observe the expressions in Figure F corresponding to the formulae $\delta_A^D \, \varepsilon_{CD} \, \delta_B^C = -\varepsilon_{AB}$ and $\varepsilon_{AD} \, \varepsilon_{BC} \, \varepsilon^{CD} = -\varepsilon_{AB}$, respectively.

Figure F

But we can eliminate these problems by simply adopting the convention of introducing an extra negative sign into our strand diagram for each appearance of (A) a *crossing*, and (B) a *contravariant* ε^{AB}.

It is not difficult to show, in fact, that these two sign conventions are sufficient to ensure that *all* topologically equivalent strand diagrams correspond to the same invariant spinor expression (provided crossings are simple, vertical tangents isolated, and that free ends occur at the extreme top and bottom of the diagram).

An immediate consequence of (B) is that an elementary closed loop has the value -2. Hence we may interpret our result as an isomorphism between invariant spinor algebra and the algebra of a "-2 dimensional" Cartesian system of *binors*. Because of (A), the spinor symmetrizers become *anti*-symmetrizers in binor algebra, while the antisymmetric ϵ_{AB} becomes the elementary symmetric binor. In binor algebra all the sign problems of spin combinations are isolated in the negative trace of this binorial "Kronecker delta".

Returning finally to the vector coupling network, we determine λ by contracting the free ends of the diagram and replacing symmetrizers with binor antisymmetrizers (bars), as shown in Figure G:

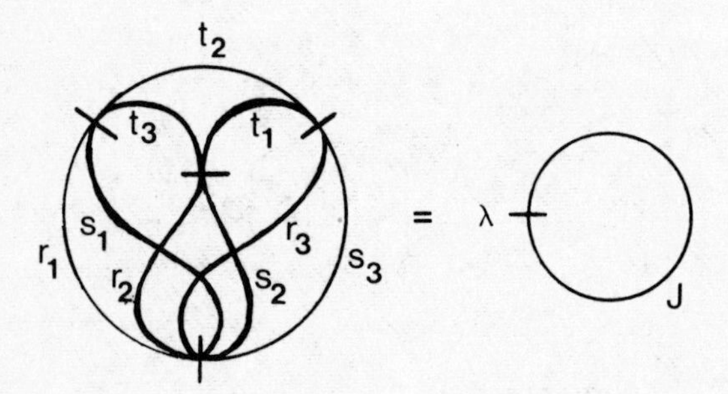

Figure G

The network on the left in Figure G is precisely the one evaluated by the chromatic method in §5.15. The result gives Racah's famous formula for the vector coupling coefficients, modulo a factor which can be computed by normalizing the incoming states and requiring unitarity.

These methods (which were invented by R.P.) can clearly be applied to computation of higher-order Clebsch-Gordan coefficients as well.

§5.17 INTEGRALS FOR STRAND NETWORKS *by S.A. Huggett*

1. *SU(2) diagrams from Spin Networks*

We can evaluate the norm of a closed spin network by calculating the value of the corresponding SU(2) invariant diagram. To obtain this diagram the typical spin network vertex shown in Fig. A is replaced by the diagram in Fig. B where the lines represent $\varepsilon^{A}{}_{B}$, ε_{AB} and ε^{AB} so that the original spin network becomes a contracted symmetrized product of spinor epsilons and deltas. These diagrams are explained in §5.16 by J.P. Moussouris. In fact, rather than calculate their values directly we make use of certain reduction formulae (analogous to those described in Penrose 1970 for binors). The diagrams can be simplified a little if we redraw Fig. B as Fig. C.

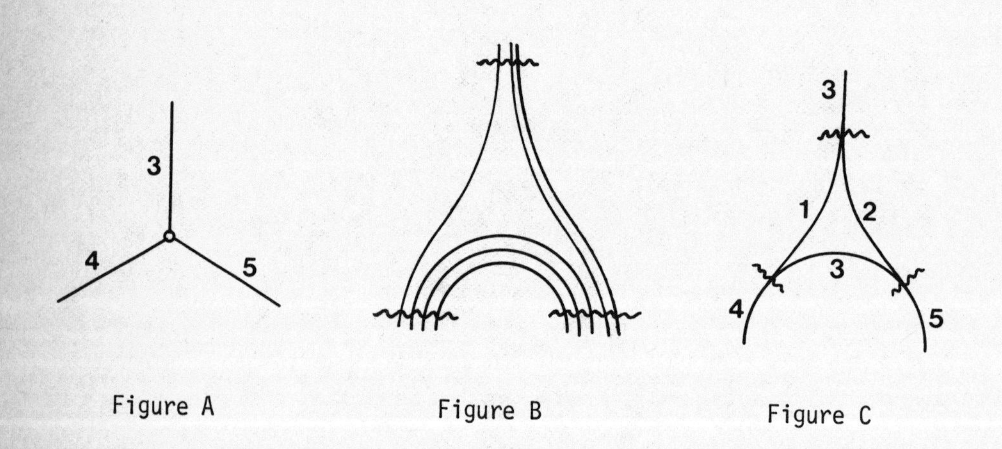

Figure A   Figure B   Figure C

2. *Integrals for SU(2) diagrams*

G.A.J. Sparling discovered another way of evaluating an SU(2) diagram. We associate a spinor contour integral with each symmetrizer. For example the diagram in Fig. D has spinors attached to it as in Fig. E so that the associated integral is

$$\frac{1}{2\pi i}\oint\frac{(\alpha.\zeta)^{a}(\beta.\zeta)^{b}(\eta.\gamma)^{c}(\eta.\delta)^{d}(a+b+1)!\zeta.d\zeta_{\wedge}\eta.d\eta}{a!\quad b!\quad c!\quad d!\qquad(\eta.\zeta)^{a+b+2}}$$

The indices of the spinors $\alpha,\ldots,\zeta$ have been omitted and the contour is an S^2 in the ζ,η space. (ζ and η are to be thought of as being one either side of the symmetrizer in Fig. E.) Suppose Fig. F represents more of the diagram in Fig. E. Now the spinors α and κ are also integrated out (again over an S^2 contour) and in a closed diagram all the spinors are integrated out.

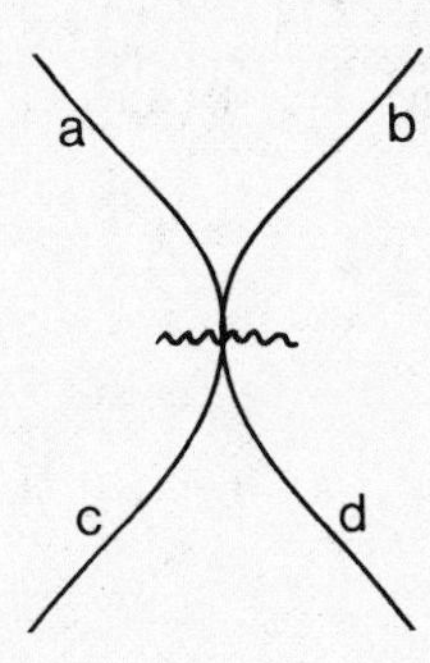

Figure D

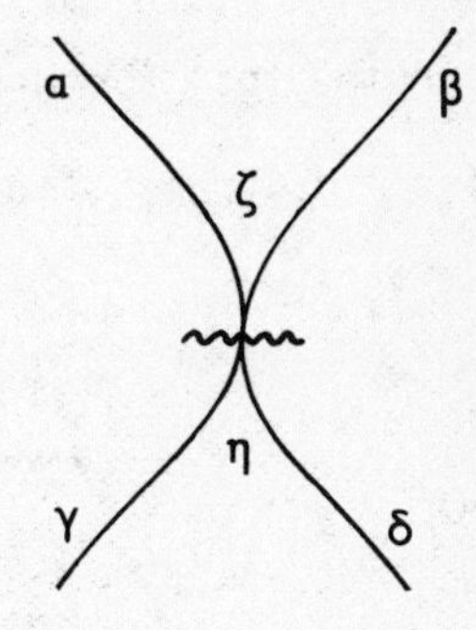

Figure E

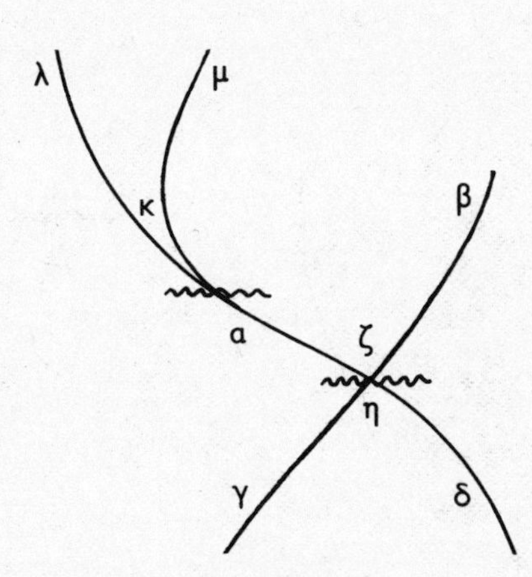

Figure F

The integrals obey certain identities some of which are listed in Fig. G. Numbers (i) to (iv) are obvious and (vi) follows from (v) and (i). Number (v) is proved using an integration by parts argument which is messy but not hard. The point is that these identities are precisely the reduction formulae referred to in 1 for SU(2) diagrams. Therefore the value of an SU(2) diagram can be calculated by doing the associated integral.

As a useful example (to which we shall return in 3) consider the diagram in Fig. H. The associated integral can be represented as in Fig. I where vertices are spinor variables being integrated and straight (wavy) lines are denominator (numerator) factors in the integrand. (This is more or less standard twistor diagram notation.)

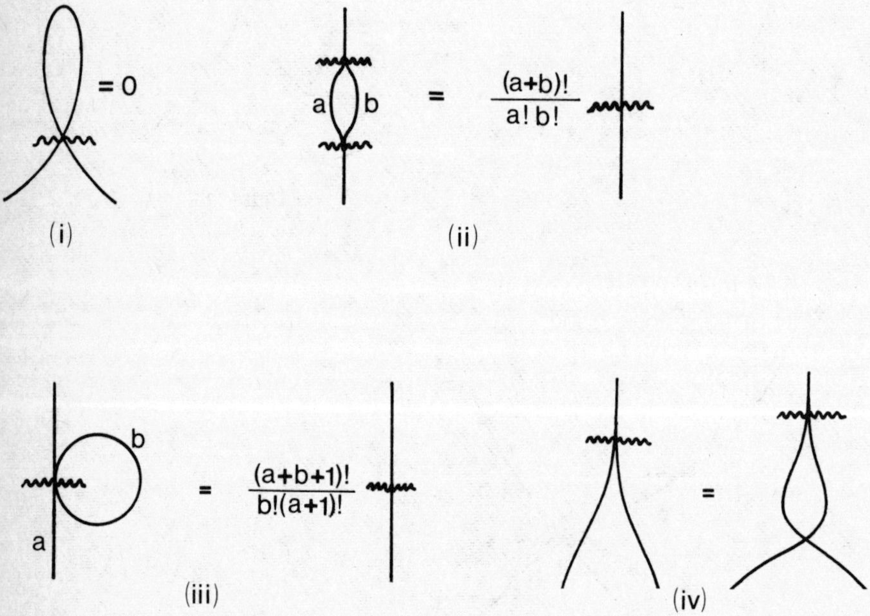

Figure G

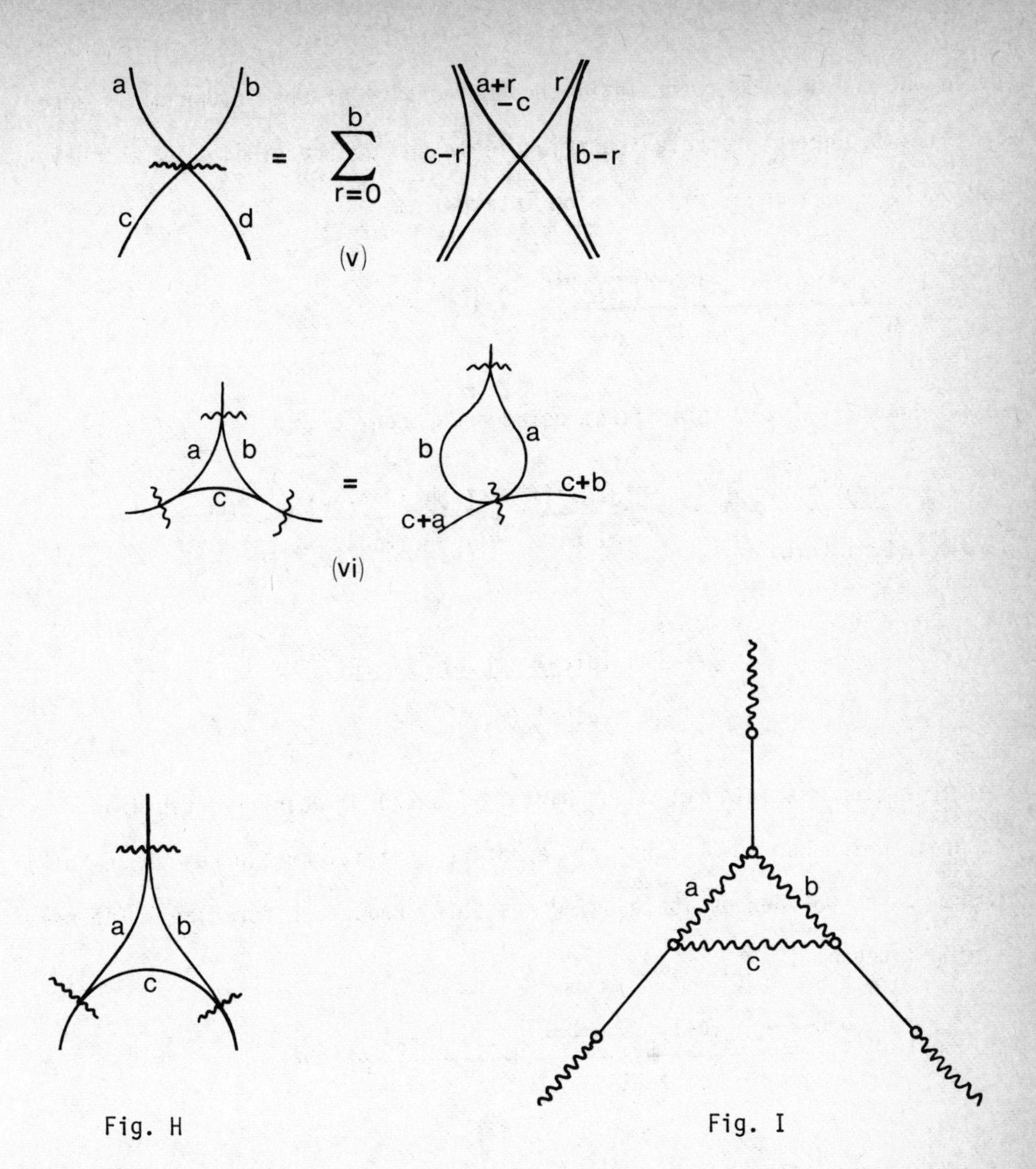

Fig. H

Fig. I

3. Integrals for SU(1,1) diagrams

As part of the general program of extending the spin network results to other groups we can try to find the analogous reduction formulae between strand networks representing SU(1,1) coupling coefficients. J.P. Moussouris has pointed out that to obtain some of these coupling coefficients it is enough to simply exchange numerator and denominator factors in our integrals (and to adjust the new numerator until the homogeneity of the integrand is zero again). We shall integrate the triangle in the centre of Fig. J, which is the SU(1,1) version of Fig. I. The integral is

$$\oint \frac{(\xi.\beta)^{a+c-2}(\eta.\gamma)^{a+b-2}(\zeta.\alpha)^{b+c-2}\xi.d\xi_\wedge\eta.d\eta_\wedge\zeta.d\zeta}{(\xi.\eta)^a(\eta.\zeta)^b(\zeta.\xi)^c} .$$

We can do the ξ integration first using an S^1 contour:

$$\oint \frac{(\xi.\beta)^{a+c-2}\ \xi.d\xi}{(\xi.\eta)^a(\zeta.\xi)^c} = \frac{(a+c-2)!}{a!\ c!} \oint \frac{(\zeta.\beta)^{a-1}(\eta.\beta)^{c-1}\ \xi.d\xi}{(\eta.\zeta)^{a+c-2}(\xi.\zeta)(\xi.\eta)}$$

$$= 2\pi i\ \frac{(a+c-2)!(\zeta.\beta)^{a-1}(\eta.\beta)^{c-1}}{a!\ c!\ (\eta.\zeta)^{a+c-1}}$$

The first step was integration by parts. Instead of doing the remaining integral (which has an S^2 contour) we notice that it corresponds to an SU(2) diagram which we then evaluate using the SU(2) reduction formulae. The remaining integral is

$$\oint \frac{(\zeta.\alpha)^{b+c-2}\ (\zeta.\beta)^{a-1}(\eta.\gamma)^{a+b-2}(\eta.\beta)^{c-1}\eta.d\eta_\wedge\zeta.d\zeta}{(\eta.\zeta)^{a+b+c-1}}$$

which has the SU(2) diagram in Fig. K. Then Fig. G number (v) converts Fig. K to Fig. L, in which there is a loop on the spinor β with the number a-1-r on it. We use Fig. G number (i) to show that r = a-1 is the only non-zero term in the sum in Fig. L, and obtain Fig. M, which has the value:

$$(\alpha.\beta)^{c-1}\ (\beta.\gamma)^{a-1}\ (\gamma.\alpha)^{b-1} .$$

It is not clear, however, whether this example is sufficient to generate all the integrals arising in the SU(1,1) case.

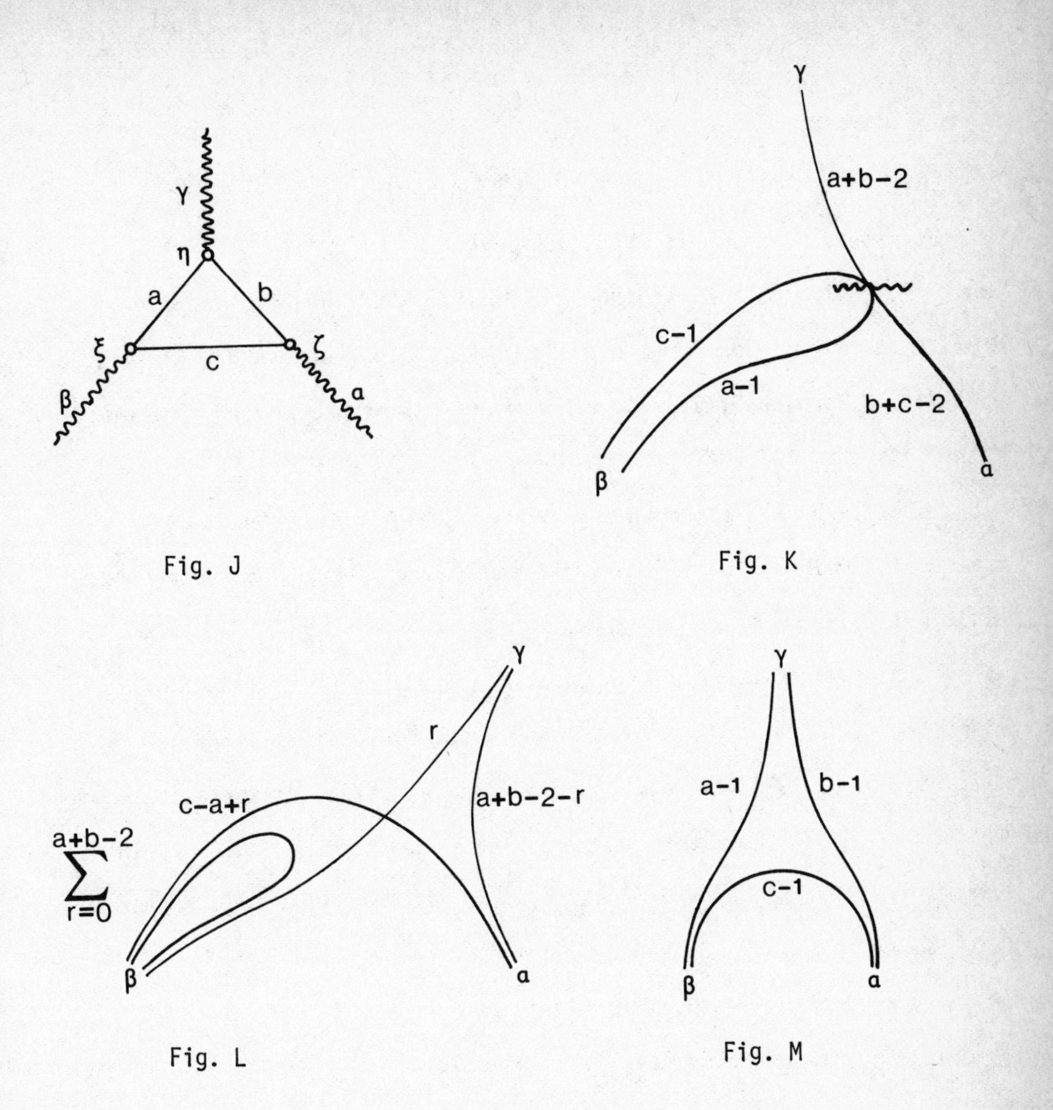

REFERENCE

Penrose, R. 1970. In *Quantum Theory and Beyond*, ed. E.T. Bastin. Cambridge University Press.

Bibliography

Atiyah, M.F., Hitchin, N.J., Drinfeld, V.G. & Manin, Yu.I. 1978 *Construction of instantons.* Phys. Lett. A65, 185-187.

Atiyah, M.F., Hitchin, N.J. & Singer, I.M. 1978 *Self-duality in four-dimensional Riemannian geometry.* Proc. Roy. Soc. Lond. A362, 425-461.

Atiyah, M.F. & Ward, R.S. 1977 *Instantons and algebraic geometry.* Comm. Math. Phys. 55, 117-124.

Bramson, B.D. 1975 *The alinement of frames of reference at null infinity for asymptotically flat Einstein-Maxwell manifolds.* Proc. Roy. Soc. Lond. A341, 451-461.

Bramson, B.D. 1975 *Relativistic angular momentum for asymptotically flat Einstein-Maxwell manifolds.* Proc. Roy. Soc. London. A341, 463-490.

Burnett-Stuart, G. 1978 *Deformed twistor spaces.* M.Sc. dissertation, Oxford.

Burns, D. 1979 *Some background and examples in deformation theory.* In: *Complex manifold techniques in theoretical physics,* eds. D.E. Lerner & P.D. Sommers, pp.135-153. London: Pitman.

Carey, A.L. & Hannabuss, K.C. 1978 *Twistors and geometric quantization theory.* Reports on Math. Phys. 13, 199-231.

Christ, N.H. 1979 *Self-dual Yang-Mills solutions.* In: *Complex manifold techniques in theoretical physics,* eds. D.E. Lerner & P.D. Sommers, pp.45-54. London: Pitman.

Christ, N.H., Weinberg, E.J. & Stanton, N.K. 1978 *General self-dual Yang-Mills solutions.* Phys. Rev. D 18, 2013-2025.

Clarke, C.J.S. 1976 *On the four-valuedness of twistors.* Comm. Math. Phys. 47, 229-231.

Corrigan, E.F., Fairlie, D.B., Templeton, S. & Goddard, P. 1978 *A Green function for the general self-dual gauge field.* Nucl. Phys. B140, 31-44.

Corrigan, E.F., Fairlie, D.B., Yates, R.G. & Goddard, P. 1978 *Bäcklund transformations and the construction of the Atiyah-Ward ansätze for self-dual SU(2) gauge fields.* Phys. Lett. B72, 354-356.

Corrigan, E.F., Fairlie, D.B., Yates, R.G. & Goddard, P. 1978 *The construction of self-dual solutions to SU(2) gauge theory.* Comm. Math. Phys. 58, 223-240.

Crampin, M. & Pirani, F.A.E. 1971 *Twistors, symplectic structure and Lagrange's identity.* In: *Relativity and gravitation,* eds. Ch.G. Kuper & A. Peres, pp. 105-110. London: Gordon & Breach.

Curtis, G.E. 1975 *Twistor theory and the collision of plane-fronted impulsive gravitational waves.* Oxford: D. Phil. thesis.

Curtis, G.E. 1978 *Twistors and multipole moments.* Proc. Roy. Soc. Lond. A359, 133-149.

Curtis, G.E. 1978 *Twistors and linearized Einstein theory on plane-fronted impulsive wave backgrounds.* Gen. Rel. Grav. 9, 987-997.

Curtis, G.E. 1978 *Ultrarelativistic black-hole encounters.* Gen. Rel. Grav. 9, 999-1008.

Curtis, W.D., Lerner, D.E. & Miller, F.R. 1978 *Complex pp waves and the non-linear graviton construction.* J. Math. Phys. 19, 2024-2027.

Curtis, W.D., Lerner, D.E. & Miller, F.R. 1978 *Some remarks on the nonlinear graviton.* Gen. Rel. Grav., to appear.

Dighton, K. 1974 *An introduction to the theory of local twistors.* Int. J. Theor. Phys. 11, 31-43.

Drinfeld, V.G. & Manin, Yu. I. 1978 *A description of instantons.* Comm. Math. Phys. 63, 177-192.

Ferber, A. 1978 *Supertwistors and conformal supersymmetry.* Nucl. Phys. B132, 55-64.

Flaherty, E.J. 1976 *Hermitian and Kählerian geometry in relativity.* Springer lecture notes in physics, no. 46. Berlin: Springer-Verlag.

Flaherty, E.J. 1978 *The nonlinear graviton in interaction with a photon.* Gen. Rel. Grav. 9, 961-978.

Hansen, R.O. & Newman, E.T. 1975 *A complex Minkowski space approach to twistors.* Gen. Rel. Grav. 6, 361-385.

Hansen, R.O., Newman, E.T., Penrose, R. & Tod, K.P. 1978 *The metric and curvature properties of H-space.* Proc. Roy. Soc. Lond. A363, 445-468.

Harris, A.S. 1975 *Topology in twistor theory.* Oxford: M.Sc. thesis.

Hartshorne, R. 1978 *Stable vector bundles and instantons.* Comm. Math. Phys. 59, 1-15.

Hayashi, M.J. 1978 *General relativity as gauge field theory in curved twistor space.* Phys. Rev. D18, 3523-3528.

Hitchin, N.J. 1979 *Polygons and gravitons.* Math. Proc. Camb. Phil. Soc. 85, 465-476.

Hodges, A.P. 1975 *The description of mass in the theory of twistors.* London: Ph.D. thesis.

Huggett, S.A. 1976 *Twistor diagrams.* Oxford: M.Sc. thesis.

Hughston, L.P. 1976 *A particle classification scheme based on the theory of twistors.* Oxford: D. Phil. thesis.

Hughston, L.P. 1979 *Some new contour integral formulae.* In: *Complex manifold techniques in theoretical physics*, eds. D.E. Lerner & P.D. Sommers, pp. 115-125. London: Pitman.

Hughston, L.P. 1979 *Twistors and particles.* Springer lecture notes in physics, no. 97. Berlin: Springer-Verlag.

Hughston, L.P., Penrose, R., Sommers, P. & Walker, M. 1972 *On a quadratic first integral for the charged particle orbits in the charged Kerr solution.* Comm. Math. Phys. 27, 303-308.

Hughston, L.P. & Sheppard, M. 1979 *On the magnetic moments of hadrons.* Reports on Math. Phys., to appear.

Hughston, L.P. & Sommers, P. 1973 *Spacetimes with Killing tensors.* Comm. Math. Phys. 32, 147-152.

Hughston, L.P. & Sommers, P. 1973 *The symmetries of Kerr black holes.* Comm. Math. Phys. 33, 129-133.

Isenberg, J. & Yasskin, P.B. 1979 *Twistor description of non-self-dual Yang-Mills fields.* In: *Complex manifold techniques in theoretical physics,* eds. D.E. Lerner & P.D. Sommers, pp. 180-206. London: Pitman.

Isenberg, J., Yasskin, P.B. & Green, P.S. 1978 *Non-self-dual gauge fields.* Phys. Lett B78, 462-464.

Jozsa, R. 1976 *Applications of sheaf cohomology in twistor theory.* Oxford: M.Sc. thesis.

Klotz, F.S. 1974 *Twistors and the conformal group.* J. Math. Phys. 15, 2242-2247.

Ko, M., Ludvigsen, M., Newman, E.T. & Tod, K.P. 1979 *The theory of H-space.* Phys. Reports, to appear.

Ko, M., Newman, E.T. & Penrose, R. 1977 *The Kähler structure of asymptotic twistor space.* J. Math. Phys. 18, 58-64.

Kopczyński, W. & Woronowicz, L.S. 1971 *A geometrical approach to the twistor programme.* Reports on Math. Phys. 2, 35-51.

Lerner, D.E. 1977 *Twistors and induced representations of SU(2,2).* J. Math. Phys. 18, 1812-1817.

Madore, J., Richard, J.L. & Stora, R. 1979 *An introduction to the twistor programme.* Phys. Reports 49, 113-130.

Moore, R.R. 1978 *Fibre bundles and the geometry of twistor space.* Oxford: M.Sc. thesis.

Newman, E.T. 1976 *Heaven and its properties.* Gen. Rel. Grav. 7, 107-111.

Newman, E.T. 1979 *Deformed twistor space and H-space.* In: *Complex manifold techniques in theoretical physics,* eds. D.E. Lerner & P.D. Sommers, pp. 154-165. London: Pitman.

Newman, E.T., Porter, J.R. & Tod, K.P. 1978 *Twistor surfaces and right-left spaces.* Gen. Rel. Grav. 9, 1129-1142.

Newman, E.T. & Winicour, J. 1974 *A curiosity concerning angular momentum.* J. Math. Phys. 15, 1113-1115.

Patton, C.M. 1979 *Zero rest mass fields and the Bargmann complex structure.* In: *Complex manifold techniques in theoretical physics,* eds. D.E. Lerner & P.D. Sommers, pp. 126-134. London: Pitman.

Penrose, R. 1965 *Zero-rest-mass fields including gravitation: asymptotic behaviour.* Proc. Roy. Soc. Lond. A284, 159-203.

Penrose, R. 1967 *Twistor algebra.* J. Math. Phys. 8, 345-366.

Penrose, R. 1968 *Twistor quantization and curved space-time.* Int. J. Theor. Phys. 1, 61-99.

Penrose, R. 1969 *Contour integrals for zero-rest-mass fields.* J. Math. Phys. 10, 38-39.

Penrose, R. 1971 *Angular momentum: an approach to combinatorial space-time.* In: *Quantum theory and beyond,* ed. T. Bastin, pp.151-180. Cambridge: University Press.

Penrose, R. 1972 *On the nature of quantum geometry.* In: *Magic without magic: J.A. Wheeler,* ed. J.R. Klauder, pp. 333-354. San Francisco: Freeman.

Penrose, R. 1974 *Relativistic symmetry groups.* In: *Group theory in non-linear problems,* ed. A.O. Barut, pp. 1-58. Dordrecht: D. Reidel.

Penrose, R. 1975 *Twistor theory, its aims and achievements.* In: *Quantum gravity, eds.* C.J. Isham, R. Penrose & D.W. Sciama, pp. 268-407. Oxford: University Press.

Penrose, R. 1975 *Twistors and particles — an outline.* In: *Quantum theory and the structures of time and space,* eds. L. Castell, M. Drieschner & C.F. von Weizsäcker, pp. 129-145. Munich: Carl Hanser Verlag.

Penrose, R. 1976 *Nonlinear gravitons and curved twistor theory.* Gen. Rel. Grav. 7, 31-52.

Penrose, R. 1976 *The nonlinear graviton.* Gen. Rel. Grav. 7, 171-176.

Penrose, R. 1977 *The twistor programme.* Reports on Math. Phys. 12, 65-76.

Penrose, R. 1979 *On the twistor descriptions of massless fields.* In: *Complex manifold techniques in theoretical physics,* eds. D.E. Lerner & P.D. Sommers, pp. 55-91. London: Pitman.

Penrose, R. & MacCallum, M.A.H. 1972 *Twistor theory: an approach to the quantization of fields and space-time.* Phys. Reports 6, 241-315.

Penrose, R. & Sparling, G.A.J. 1979 *Twistors and particles.* Rev. Mod. Phys., to appear.

Penrose, R., Sparling, G.A.J. & Tsou S.T. 1978 *Extended Regge trajectories.* J. Phys. A11, L231-235.

Penrose, R. & Ward, R.S. 1979 *Twistors for flat and curved space-time.* To be published in an Einstein centennial volume, eds. P.G. Bergmann, J.N. Goldberg & A.P. Held.

Perjés, Z. 1972 *An application of twistor theory in weak interactions.* In: *Neutrino* 1972, eds. A. Frenkel & G. Marx, Vol. II, pp. 183-184.

Perjés, Z. 1975 *Twistor variables of relativistic mechanics.* Phys. Rev. D11, 2031-2041.

Perjés, Z. 1977 *Perspectives of Penrose theory in particle physics.* Reports on Math. Phys. 12, 193-211.

Popovich, A.S. 1978 *Twistor particle theory.* Oxford: M.Sc. thesis.

Qadir, A. 1978 *Penrose graphs.* Phys. Reports 39C, 131-167.

Rawnsley, J.H. 1979 *On the Atiyah-Hitchin-Drinfeld-Manin vanishing theorem for cohomology groups of instanton bundles.* Math. Ann. 241, 43-56.

Ryman, A. 1975 *Twistor theory: a topological study of some twistor diagrams.* Oxford: D. Phil. thesis.

Sommers, P.D. 1973 *On Killing tensors and constants of motion.* J. Math. Phys. 14, 787-790.

Sommers, P.D. 1976 *Properties of shear-free congruences of null geodesics.* Proc. Roy. Soc. Lond. A349, 309-318.

Sparling, G.A.J. 1974 *Ph. D. Thesis.* Birkbeck College, London.

Sparling, G.A.J. 1975 *Homology and twistor theory.* In: *Quantum gravity*, eds. G.J. Isham, R. Penrose & D.W. Sciama, pp.408-499. Oxford: University Press.

Tod, K.P. 1975 *Massive particles with spin in general relativity and twistor theory.* Oxford: D. Phil. thesis.

Tod, K.P. 1977 *Some symplectic forms arising in twistor theory.* Reports on Math. Phys. 11, 339-346.

Tod, K.P. 1979 *Remarks on asymptotically flat* H-*spaces.* In: *Complex manifold techniques in theoretical physics*, eds. D.E. Lerner & P.D. Sommers, pp. 166-179. London: Pitman.

Tod, K.P. & Perjés, Z. 1976 *Two examples of massive scattering using twistor Hamiltonians.* Gen. Rel. Grav. 7, 903-913.

Tod, K.P. & Ward, R.S. 1979 *Self-dual metrics with self-dual Killing vectors.* Proc. Roy. Soc. Lond. A, to appear.

Walker, M. & Penrose, R. 1970 *On quadratic first integrals of the geodesic equations for type {2 2} spacetimes.* Comm. Math. Phys. 18, 265-274.

Ward, R.S. 1977 *Curved twistor spaces.* Oxford: D. Phil. thesis.

Ward, R.S. 1977 *On self-dual gauge fields.* Phys. Lett. A61, 81-82.

Ward, R.S. 1978 *A class of self-dual solutions of Einstein's equations.* Proc. Roy. Soc. Lond. A 363, 289-295.

Ward, R.S. 1979 *The self-dual Yang-Mills and Einstein equations.* In: *Complex manifold techniques in theoretical physics*, eds. D.E. Lerner & P.D. Sommers, pp. 12-34. London: Pitman.

Weber, W.J. 1978 *Some applications of sheaf cohomology in twistor theory.* Oxford: M.Sc. thesis.

Wells, R.O., jr. 1979 *Cohomology and the Penrose transform.* In: *Complex manifold techniques in theoretical physics*, eds. D.E. Lerner & P.D. Sommers, pp. 92-114. London: Pitman.

Wells, R.O., jr. 1979 *Complex manifolds and mathematical physics.* Bull. Amer. Math. Soc., to appear.

Witten, E. 1978 *An interpretation of classical Yang-Mills theory.* Phys. Lett. B77, 394-398.

Witten, E. 1979 *Some comments on the recent twistor space constructions.* In: *Complex manifold techniques in theoretical physics*, eds. D.E. Lerner & P.D. Sommers, pp. 207-218. London: Pitman.

Woodhouse, N.M.J. 1976 *Twistor theory and geometric quantization.* In: *Group theoretical methods in physics*, eds. A. Janner, T. Janssen & M. Boon, pp. 149-163. Springer lecture notes in physics, no. 50, Berlin: Springer-Verlag.

Index